Mandana Banedj-Schafii

System transferability of public hospital facility management between Germany and Iran

**Karlsruher Reihe
Bauwirtschaft, Immobilien und Facility Management
Band 4**

Karlsruher Institut für Technologie (KIT),
Institut für Technologie und Management im Baubetrieb

Hrsg. Prof. Dr.-Ing. Dipl.-Wi.-Ing. Kunibert Lennerts

Eine Übersicht über alle bisher in dieser Schriftenreihe erschienenen Bände
finden Sie am Ende des Buchs.

System transferability of public hospital facility management between Germany and Iran

by
Mandana Banedj-Schafii

Dissertation, genehmigt von der Fakultät für
Bauingenieur-, Geo- und Umweltwissenschaften
der Universität Fridericiana zu Karlsruhe (TH)
Tag der mündlichen Prüfung: 15. Mai 2009
Referenten: Prof. Dr.-Ing. Dipl.-Wi.-Ing. Kunibert Lennerts,
              Prof. Dr. med. Farid Abolhassani

Impressum

Karlsruher Institut für Technologie (KIT)
KIT Scientific Publishing
Straße am Forum 2
D-76131 Karlsruhe
www.uvka.de

KIT – Universität des Landes Baden-Württemberg und nationales
Forschungszentrum in der Helmholtz-Gemeinschaft

KIT Scientific Publishing 2010
Print on Demand

ISSN: 1867-5867
ISBN: 978-3-86644-395-2

"Without health everything is nothingness"

Arthur Schopenhauer

ACKNOWLEDGMENT

I have pursued this thesis as a personal challenge related to my home countries. The research developed into an adventure that drove me to desperation so that I would have been tempted to leave the work incomplete if I had not enjoyed the kind support of the persons who are named in the following acknowledgement.

First I would like to thank my supervisor *Professor Dr.-Ing. Dipl.-Wi.-Ing. Kunibert Lennerts*. He gave me the opportunity to undertake the research project that led to the work which is presented in this thesis. I owe him experiences which go far beyond the FM area and include guidance in the art of negotiation, in international scientific and professional interaction and in pursuit of my interests through suitable presentations. This interaction with Professor Lennerts has formed me and its consequences will accompany me beyond my PhD thesis. Likewise I am grateful for the fast support of my Iranian co- advisor *Professor Dr. Abolhassani*. He guided me in difficult circumstances and motivated me to find solutions for complicated situations.

Without *Dr. Zare* and the managers of the hospitals Shariati, Tebie Kudakan and Vali Asr the research projects would not have been possible. My sincere thanks go particularly to the active participants of the Workshops, especially *Mr. Moghaddam, Mr. Farrokhdust, Mr. Khooi, Mr. Haghighifar, Mr. Kuseler* and *Mr. Baba* who explained the process flows with full application. Of course the diligent students of the University of Teheran may not be forgotten. Sincere thanks are given to them all. Numerous conversations and expert talks with many persons in the hospitals, ministries and universities have influenced this thesis significantly. Even though they are not named here and may not be aware of it, these persons gave important impulses for this work. Also I would like to thank *Prof. Dr. R. G. Heinze,* chairman of the Institute of Work and Economic Sociology at the Ruhr-University Bochum, *Prof. Dr. rer. pol. Dipl.-Soz. G. Wachtler,* Institute of Sociology at the Bergische-University Wuppertal and *Prof. Dr. H. Lietzmann,* Institute of Political Science at the Bergische-University Wuppertal. On the basis of their broad experience they gave helpful suggestions for the development of the transferability model.

A big thank-you of course goes to my family, my brothers and sister and my dear parents *Ute* and *Shahrokh Banedj-Schafii*; particularly to my father who supported me in Germany, as well as in Iran with translations and administrative work and aided me in all of my enterprises in every possible way. And of course I must mention also my dear husband *Bijan Namazi,* for his patient, full appreciation and for his indefatigable encouragement.

TABLE OF CONTENT

1   Introducation

2   Facility Management (FM)

3   Object of study: Hospitals

4   The research project OPIK

5   Evaluation and relevancy of the project results

6   Concept development and solution finding

INDEX OF TABLES

LIST OF ABBREVIATIONS

AHK	*Außenhandelskammer*
AIG	*Arbeitsgemeinschaft Instandhaltung Gebäudetechnik*
AiP	*Arzt im Praktikum*
BAT	*Bundesangestelltentarif*
BGV	*Berufsgenossenschaftliche Vorschriften*
BMBF	*Bundesministerium für Bildung und Forschung*
BMG	*Bundesministerium für Gesundheit*
CAFM	*Computer Aided Facility Management*
CEN/TC	*Committee Europeen de Normalization / Technical Committee*
DDR	*Deutsche Demokratische Republik*
DIN	*Deutsche Industrie Norm*
DKG	*Deutsche Krankenhaus Gesellschaft*
DRG	*Diagnostic Related Groups*
ECO	*Economic Cooperation Organisation*
EN	*European Norm*
EU	*European Union*
FKT	*Fachvereinigung für Krankenhaustechnik*
FM	*Facility Management*
GB	*Great Britain*
GDP	*Gross domestic product*
GEFMA	*German Facility Management Association*
GKV	*Gesetzliche Krankenversicherung*
GOÄ	*Gebührenordnung für Ärzte*
HBG	*Health Benefit Group*
HCFA	*Health Care Financing Administration*

HNF	Hauptnutzfläche
HRG	Healthcare Resource Group
HW	Hauswirtschaft
IFMA	International Facility Management Association
IHK	Industrie und Handelskammer
IMF	International Monetary Fund
IRI	Islamic Republic of Iran
ISO	International Organization for Standardization
IT	Information Technology
KBV	Kassenärztliche Bundesvereinigung
KHG	Gesetz zur wirtschaftlichen Sicherung der Krankenhäuser
KrW-/ AbfG	Kreislaufwirtschafts- und Abfallgesetz
LDF	Leistungsorientierte Diagnosefallgruppe
MCI	Management Consulting Institute
MOHME	Ministry of Health and Medical Education
MPBetreibV	Medizinprodukte- Betreiberverordnung
MPG	Medizinproduktegesetz
MPV	Medizinprodukteverordnung
MRT	Ministry of Road and Transportation
MTArb	Manteltarifvertrag der Arbeiter
NATO	North Atlantic Treaty Organization
NEN	Netherlands Norm
NFMA	National Facility Management Association
NGO	Non Governmental Organizations
OECD	Organisation for Economic Co-operation and Development
OP	Operation Room

OPEC	Organization of the Petroleum Exporting Countries
OPIK	Optimierung und Analyse der Prozesse im Krankenhaus
PLD	Pflegedienstleistung
PHC	Primary Health Care
PPP	Public Private Partnership
QM	Quality Management
RKI	Robert Koch Institut
RöV	Röntgenverordnung
SCI	Statistic Centre of Iran
SGB	Sozialgesetzbuch
STK	Sicherheitstechnische Kontrolle
Str.SchV	Strahlenschutzverordnung
TB	Technisches Büro
UN	United Nations
UNICEF	United Nations International Children's Emergency Fund
VDE	Verband der Elektrotechnik Elektronik Informationstechnik
VDI	Verein Deutscher Ingenieure
VDS	Vertrauen durch Sicherheit
WHO	World Health Organisation

SUMMARY (English)

The subject of this PhD-thesis is the transferability of Facility Management (FM) by using the example of hospitals in Germany and Iran. The intention is to determine the status and significance of facility management in the project countries, to analyse similarities, semblances and differences in order to develop a system transferability model. Resulting from this comparison a transfer of knowledge, information and know-how should occur which effects an examination and optimization of the own system, methods and structures.

The research is done on the basis of the research project OPIK (Optimisation and Analysis of the Processes in Hospitals) which has been successfully running since 2001 with cooperation partners from science, hospitals, professional associations and industry by the professorship of Facility Management at the University of Karlsruhe (TH).

Three processes that were analyzed in the OPIK-Germany project (maintenance of medical equipment, maintenance and repair of technical facilities and laundry management) were chosen exemplarily and compared with their pendants in the OPIK-Iran project that started 2006 in Tehran.

Beneath the analysis of the processes, including process steps and responsibilities, characteristic variables (cost and quality factors) and the interfaces, an extensive data acquisition was performed and faced to the German results.

With the perceptions of the process analysis and the comparisons of the hospital management, the health system and country-specific conditions nine main parameters were recognized that influenced the transferability of facility management. These are management, economy, politics, culture, judicative, education, public and private institutions, infrastructure and geography. By using defined models and specific analysis, methods were developed to rank these main parameters.

These methods were verified by using the example of the medical equipment department. Based on those results proposals were worked out for the implementation of facility management in the Iranian hospitals.

SUMMARY (German/ Deutsch)

Die Systemübertragbarkeit von Facility Management (FM) am Beispiel von Krankenhäusern in Deutschland und Iran ist Gegenstand dieser Arbeit. Ziel ist es, den Stand und den Stellenwert von FM in den Projektländern zu untersuchen, Gemeinsamkeiten, Ähnlichkeiten oder Unterschiede zu analysieren um daraus ein Systemübertragbarkeitsmodel zu entwickeln. Durch diese Gegenüberstellung soll ein Wissen-, Informations- und Know-how-Transfer erfolgen, der zur Auseinandersetzung und Optimierung der eigenen Systeme, Methoden und Strukturen anregt.

Die Untersuchung erfolgt anhand des Forschungsprojekts OPIK (Optimierung und Analyse der Prozesse im Krankenhaus), welches seit 2001 erfolgreich mit Kooperationspartnern aus Wissenschaft, Krankenhäusern, Fachvereinigungen und Industrie an der Professur für Facility Management der Universität Karlsruhe (TH) durchgeführt wird.

Drei im OPIK-Deutschland Projekt durchgeführte Prozesse (Wartung der Medizintechnik, Instandhaltung und Wartung technischer Anlagen und Wäscheversorgung) wurden beispielhaft gewählt und mit ihren Pendants im Projekt OPIK-Iran, welches 2006 im Teheran begann, verglichen. Neben den Analysen der Prozesse, samt Prozessschritten, Verantwortlichkeiten, Prozessfaktoren (Kosten und Qualität) und Schnittstellen erfolgte eine umfassende Datenerhebung, die den deutschen Ergebnissen gegenübergestellt wurde.

Aus den Erkenntnissen der Prozessanalysen und Vergleichen zwischen dem Krankenhausmanagement, den Gesundheitssystemen und den landesspezifischen Gegebenheiten wurden neun Hauptparameter ersichtlich, die Einfluss auf die Übertragbarkeit von Facility Management aufweisen. Diese sind: Wirtschaftlichkeit, Management, Politik, Kultur, Gesetzgebung, Bildung, öffentliche und private Institutionen, Infrastruktur und Geographie. Mit Hilfe von Modellen und Analysen wurden Methoden entwickelt, um diesen Hauptparametern Rangfolgen zuzuordnen.

Abschließend wurde diese Methode am Bespiel der Abteilung Medizintechnik durchgeführt, und ein Ausblick für die Übertragung von Facility Management in iranische Krankenhäuser aufgeführt.

1 Introduction

1.1 Statement of the problem

Starting to establish a Facility Management department in the BHRC[1] (Building and Housing Research Center) in Teheran/ Iran in 2003 - after having studied civil engineering and lived in Germany for twenty years - the idea emerged: "Is the technology or a management system learned and used in Germany able to be transferred to Iran one to one? Why does a technology work or a system run in one country but is failing or halting in another one? Which are the parameters or indicators that cause this difference? Does this phenomenon concern all countries in the global market that is coming closer and closer together? "

In the age of globalisation knowledge of the worldwide state of science and technology is the basic condition for transfer, cooperation, and development. On the one hand, knowledge of the structures and market potentials that are available abroad can be used to enhance local methods and to adapt local applications to the existing needs. On the other hand, this knowledge facilitates the detection of international opportunities to fill and conquer new markets.

Globalisation offers the advanced countries that are export oriented not only great opportunities to protect their high rank in the product market, but also the chances to conquer the international service market as well. Thus especially German companies are very active in the FM markets in Eastern Europe and the Middle and Far East. But which conditions have to be considered abroad to enter these markets successful?

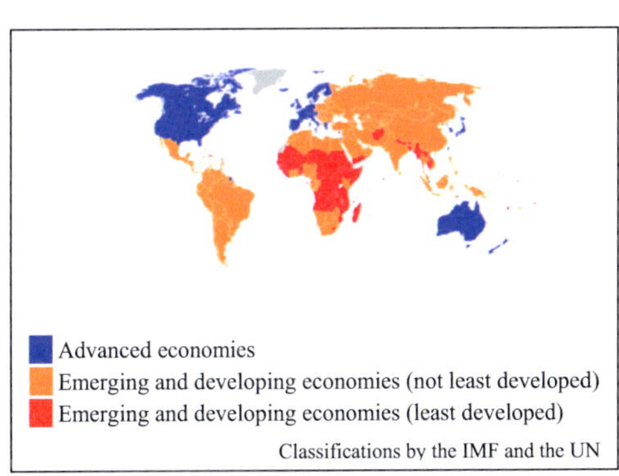

Figure 1-1: Classification of the world economies [IMF 2008]

---

[1] The **B**uilding and **H**ousing **R**esearch **C**enter, affiliated to the Ministry of Housing and Urban Development, is a national center for research and assessment of products and systems in the building and housing field.

Attention must also be drawn to the view and the position of the developing countries and the emerging markets[2]. Since decades these countries acquire basically technical products, like technical installations, machines and also biomedical technology from the industrial countries. The transfer of high-class management - for example the ISO certification - not only for products, but also for management methods, know-how and services is likewise becoming more and more important. At the same time the trend towards privatisation - especially outsourcing[3] - can be seen (in Iran especially during the period of President Khatamie (1997-2005)).

This situation analysis also applies to facility management (FM), a young but strongly growing field within the building and service industry. Facility management integrates technical, commercial and infrastructural processes – all the "secondary processes" running in an organisation - and offers them as an "all – round service package".

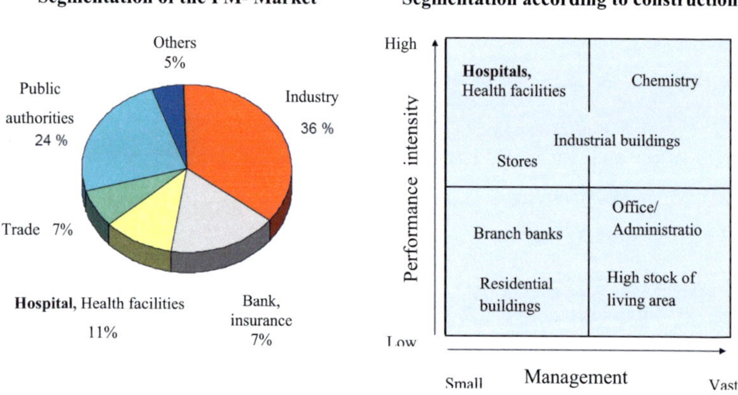

Figure 1-2: Segmentation of FM and construction types [HELBING2000]

---

[2] "There is no established convention for the designation of "developed" and "developing" countries or areas in the United Nations system. In common practice, Japan in Asia, Canada and the United States in northern America, Australia and New Zealand in Oceania and Europe are considered "developed" regions or areas. In international trade statistics, the Southern African Customs Union is also treated as developed region and Israel as a developed country; countries emerging from the former Yugoslavia are treated as developing countries; and countries of Eastern Europe and the former USSR countries in Europe are not included under either developed or developing regions."[UN2005]

[3] **Outsourcing** involves the transfer of the management and/or day-to-day execution of an entire business function to an external service provider [OVERBY2007]

As can be seen in Figure 1-2 hospitals - with a proportion of 11% - belong to the "facilities" which have the highest performance intensity. These are objects that have to bear high management costs and require infrastructures that are strongly dependent on technology. Because these facilities directly influence human life, high safety standards must be maintained. In order to be able to resist the immense cost pressure and to guarantee economic continuity, concentration on the core business is necessary. The core business of hospitals consists of cure from illness and medical care. The outsourcing of the "secondary areas" and the gradual establishment of FM departments in hospitals offer considerable saving potentials, especially through optimisation of the processes that provide cure and medical care.

The need for FM systems exist independent of the specific circumstances in a country, so that the question must be addressed how FM can be introduced and integrated into the existing structures. Can FM models or FM systems be transplanted from their region? To what extent can the available FM systems be taken over? Which methods are usable as they are, and which methods must be adapted or changed?

1.2   Purpose and approach

In order to be able to offer results and attempts at solutions, it is very important to limit the scope of the investigation to suitable parts of the wide range of the topic and to define the purpose of the investigation exactly at the outset.

This PhD-thesis aims to invent and develop methods for the evaluation of the transferability of facility management systems in hospitals, particularly methods that make it possible to find conceptual solutions for the analysed processes. Besides, the following questions should be answered:

- Which common criteria, differences or resemblances can be detected?
- How can the basic conditions be compared and regularities marked?
- Which influence do the different parameters have? Which factors are vital?
- How are the relations and interactions of the parameters?
- To which extent can the results be compared objectively and critically to extend the investigation beyond the statistical analyses and reports?
- What are the main characteristics of an approach that leads to a concept or a method for the evaluation of system transferability?

- Which experiences can be used for the founding of FM departments in hospitals?
- How can feasible prognoses or trend developments be predicted?

In addition to an international comparison, a detailed investigation is performed with the Federal Republic of Germany as an example of an advanced economy and the Islamic Republic of Iran as an example of an emerging or developing economy; two countries that move in different economic, political and cultural frames. The facility management systems in the hospitals of both of these countries are examined within the scope of the project "OPIK-Optimisation and Analysis of the Processes in Hospitals ". In order to gain insights and find solutions, information and data were collected, evaluated and compared in the Iranian and German project hospitals. The international comparisons proved to be complicated because many data are collected and evaluated statistically by different institutions in different ways. Hence, the data and their interpretation must be systematised. In this connection, it must be stressed that it was difficult to obtain information about this special area not only in Germany. The collection of information in Iran may be regarded as a "special challenge" of this thesis, which is structured as follows:

In agreement with the medical methods, an anamnesis and diagnosis (analysis of the present situation) is performed first and afterwards a therapy and drugs are recommended (solution suggestion).

After the introduction of the thesis, the initiation, development and current situation of facility management in hospitals of both countries are discussed in chapter two.

Chapter three first introduces the existing international health systems. It then focuses on the health systems of the project countries. The systems which are in use are described and compared. A closer investigation of the hospitals which are considered in the thesis research is another main focus area of this chapter.

In order not to be restricted to the basic conditions of the processes in the hospitals as research parameters, but to be able also to consider concrete comparative data for the analysis, the research project OPIK that is being conducted in Germany since 2001 was started in Iran in 2006. This activity is discussed in Chapter four where the following processes are analysed in detail:

- Maintenance and repair of technical facilities
- Maintenance of medical equipment
- Laundry management

The processes - as well as respective conditions and specific features - are evaluated in chapter five where graphic instruments are used as main tools for the intercultural comparison. This thesis is devoted primarily to the comparative analysis of FM in hospitals. This analysis should overview the different economic, social, cultural, political, sociological and organizational basic conditions which directly or indirectly influence the introduction of FM and the successful application of FM solutions. The comparative approach in this thesis requires the development of methods and instruments that make it possible to recognize development trends in areas which influence FM in hospitals strongly, to specify significant differences as well as similarities, and to identify as well as evaluate system transferability.

To prove the practicability of the developed methods and tools, a case study is performed for the example of medical equipment in chapter six, that closes with a conclusion.

This thesis is based on experiences gained in different trainings and projects like HSG[4] in 2004 and the preparation and realisation of the World Bank project Trainee Course "Facility Management and Healthcare Management"[5] with 250 high-ranking employees of the Iranian Ministry of Health and Medical Education (MOHME), as well as hospices in German hospitals and companies (like Städtisches Klinikum Karlsruhe[6], Bardusch[7]). Particularly the attendance of the political and business delegation of the Ministry of Economics in Baden-Württemberg to Iran fall into this domain.

---

[4] 2008 HSG joint Zander and is now belonging to the largest FM companies in Germany, named HSG Zander. A three month training was successfully completed in the former HSG Company by the author in 2004. This training containing practical work in hospitals initiated workaday situations and problems that facility managers are faced with, lead in the topic of Facility Management in hospitals.

[5] In 2004 and 2005 a post graduated trainee course was realized at the University of Karlsruhe (TH). Beside the content of the different lectures held in these courses, the acceptance, the handling and transfer of the participants helped to understand transferability of new ideas and systems.

[6] Training in the technical and medical equipment department of the Städtisches Klinikum in Karlsruhe (the municipal hospital of Karlsruhe) delivered inside into the everyday work from the in-house viewpoint.

[7] The internship with Bardusch, a textile rental service company, showed the automated mass laundry (washing 420 t daily) and pointed the optimized processes out in such a huge amount in comparison to small in-house laundries in hospitals.

2 Facility Management (FM)

The purpose of this chapter is to give an overview of the origin and development of FM worldwide. Special attention is paid to the state of FM in the two countries – Germany and Iran - exemplifying advanced economies and emerging and developing economies. The focus is directed towards FM in hospitals.

2.1 Development and definition of Facility Management

The emergence of FM can be traced back to intensified competition in all parts of the world and the enormous cost pressure on companies. In order to ensure their existence, they are obliged to focus on their core business. They are forced either to outsource the burdensome secondary areas or to develop new entrepreneurial management tools and strategies. Besides the migration of areas of activity from the production to the service sector, the significance of real estate (that has assumed rank five in the list of enterprise resources after work, capital, technology and information [LYNE1995]) makes the introduction and use of FM indispensable.

The beginning of Facility Management is dated to 1952 when "after the war in Korea and the rise of the cold war the airline PAN AM as service provider took charge of the driving, managing and maintenance of the site of the US-Air-Forces in Cape Canaveral, Florida, which is still pursued by the present successor." (Johnson Controls, since 1996) [LOCHMANN1998, p. 44 and the following]

The "Avalanche of FM" was, however, triggered by the conference "Facilities impact on productivity" of the world's biggest office furniture manufacturer Hermann Miller. This event in the year 1978, referred to in many literature references as a „starting signal of FM", pointed for the first time to the connection between the "Facilities" of the working environment and the productiveness of the employees. [HELLERFORTH2006, p.11]

Shortly after this event the FM-Institute was established in 1979 at Ann Arbour, Michigan followed in 1980 by the foundation of the NFMA (National Facility Management Association). At the time of the accession of Canada in 1982 it was renamed to "International Facility Management Association" (IFMA). [COTTS1999, p.3] With headquarters in

Houston, Texas, agencies in more than 60 countries of the world and more than 18.500 members [IFMA2007] it is regarded as the major of the FM federations.

In Europe the concept of Facility Management was taken up in the middle of the 1980s. The first basic research in the German-speaking world was conducted by the study group CSG of the University of Vienna. The research was continued by the Management Consulting Institute (MCI) in Vienna, which regarded Facility Management in a more far-reaching context than the North-Americans. Their approach using concepts which are oriented towards the life cycle of facilities gave a more comprehensive meaning to FM. Beginning in England, where the Federation of the Facility Managers was founded in 1986, and expanding over the Netherlands, national facility management associations were established in numerous European countries, like for example the German Federation for Facility Management inc. (GEFMA) in 1989. The objectives of these associations are the advancement of facility management activities, as well as the development of uniform standards and guidelines for training and for continuing education opportunities in this area. Furthermore the "IFMA Deutschland e.V." was founded in 1996 as the German Chapter of the IFMA Houston. [LOCHMANN1998, p. 45 and the following]   With the adoption of mainly American concepts by big enterprises and management consulting companies (restructuring, lean management, reengineering, total quality management ...) and the introduction of the topic of the added value processes first in the private sector, but later also in state and municipal organisations (PPP- Public Private Partnership) FM has been established on the market since the end of the 1980ies. The definition of "FM" lags behind this development. The reason for this lag is the different definitions of the countries or associations.

„Facility Management was defined before 2003 by the IFMA as follows: *"Facility Management is the practice of coordinating the physical workplace with the people and work of the organization; it integrates the principles of business administration, architecture, and the behavioral and engineering sciences."* Since 2003, the definition by the IFMA is: *"Facility Management is a profession that encompasses multiple disciplines to ensure functionality of the work environment by integrating people, place, process and technology."* In order to demonstrate the different approaches, the definition of GEFMA is shown as an additional example: *"Facility management is an entrepreneurial process which aims at an improved flexibility of utilisation, productivity of labour and profitability of capital through the integration of planning, control and management of buildings, sites and facilities, giving consideration to the work place and the work environment. "Facilities" are integrated as strategical resources into the overall process of the enterprise".*

In order to bring light in the "definition forest" and to obtain a uniform definition, at least for the European area, the CEN/TC 348 (technical committee) under the secretariat of the NEN (Nederland Norm) approved the following definition which was implemented in spring 2007. Facility management is defined in EN15221-1:2006 as follows: *"The integration of processes within an organisation to maintain and develop the agreed services which support and improve the effectiveness of its primary activities."* [CEN2006]. This definition is based on the FM model in Figure 2-1.

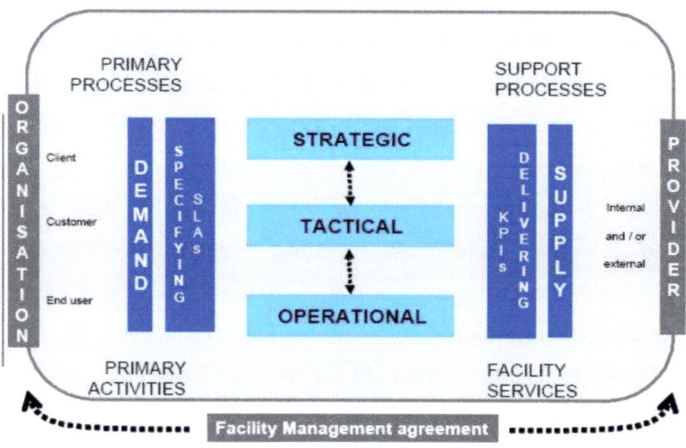

Figure 2-1: FM-model according to EN 15221/1

Consequently, FM is understood as a support of the core business and a potential organisation model for companies. The definition is kept very open and corresponds to a "top down" approach. FM as entity contains a "...strategic, tactical and operative level which makes it possible to adapt the extent of the services to the customer requirements. This does not mean, however, that FM includes all support services". [CHRISTEN2006]

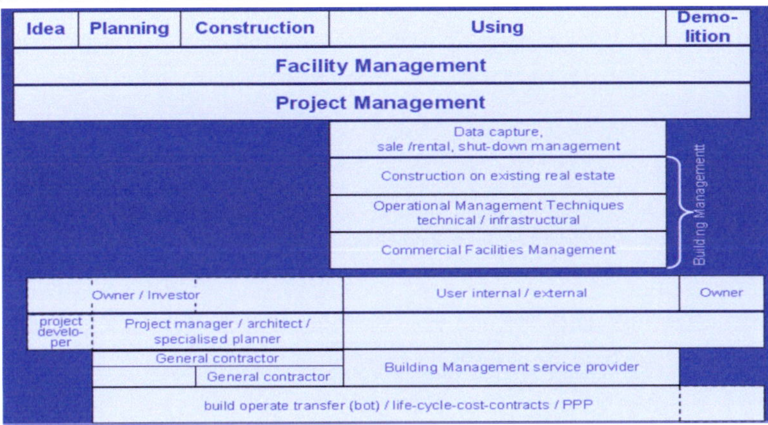

Figure 2-2: FM-overview [LENNERTS2006]

Figure 2-2 summarises the activities which are parts of supply management and structure management.

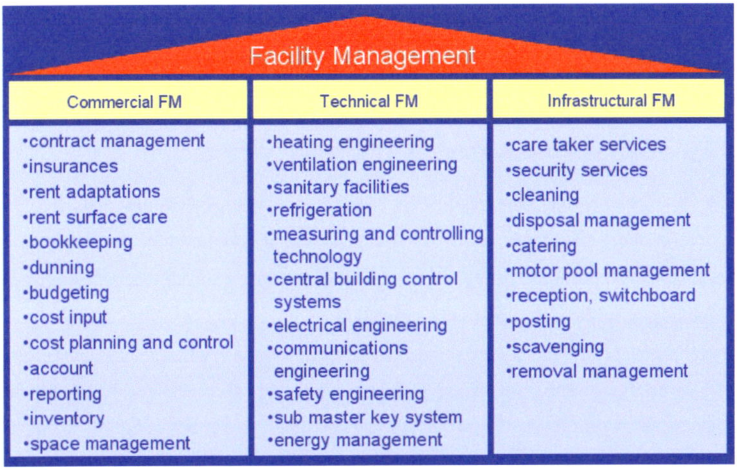

Figure2-3: Classical definition of facility management in the operating phase [Lennerts2006]

The work group maintenance and building technology AIG (Arbeitsgemeinschaft Instandhaltung Gebäudetechnik) defines operative FM as building management and distinguishes between, commercial, technical and infrastructural building management. [AIG1996]

As shown in Figure 2-2 and Figure 2-4, FM covers the entire life cycle of an object. In the ideal case, the FM team accompanies the management of the secondary processes from the first idea, the planning and the implementation of the object over the entire useful life span up to the demolition. This farsighted, long-term and integral planning covering the entire life cycle is, besides a dynamic, demand-oriented organisation (lean team), the foundation for cost-effective and client-oriented use of real estate.

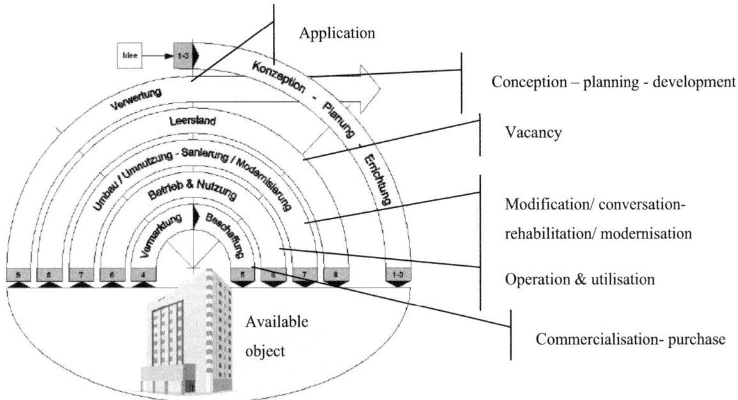

Figure 2-4: Life cycle of facility management [GEFMA 2004]

Hence these facilities can be optimised with respect to benefit (economic and for the enterprise), cost, ecology and conservation of value, relieving at the same time the operational core areas and minimising the use of resources. Cost reduction is achieved through improved processes, higher quality and rising customer satisfactions.

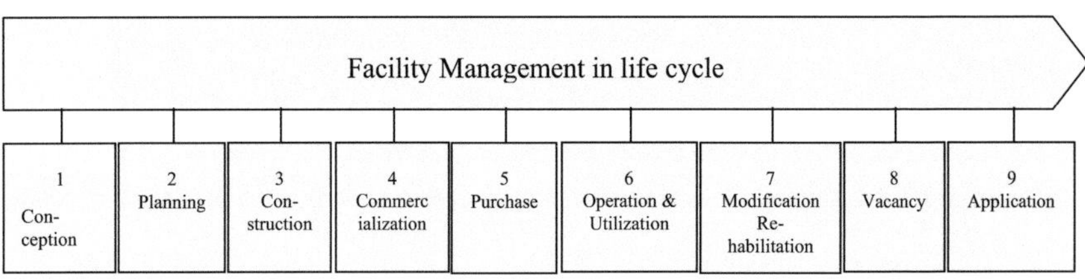

Figure 2-5: Representation of the life breakdown of facility management [GEFMA2004]

2.2 Facility Management in Germany

2.2.1 The State of Facility Management in Germany

The start into the "FM age" took place in Germany around the year 1995 [LOCHMANN1998, S.49ff] through the establishment and restructuring of many FM enterprises (e.g., RGM ` 94, Sasse Facility Management GmbH ` 95, DIB society for location support ` 95, DYWIDAG - Service GmbH ` 95).

This trend was enhanced by the formation of specialized subsidiaries of big companies (Lufthansa Building Management GmbH ` 95, ABB building services engineering AG ` 97, Thyssen Facility Management GmbH ` 97), or the privatisation of many state facilities (DeTe Immobilien und Service GmbH ` 96, German railways real estate company mbH ` 96 and Deutsche Bahn Anlagen und Hausservice GmbH` 96).

The activities of the FM associations, GEFMA and IFMA, as well as the formation of numerous educational and training institutions for FM, have established FM in Germany and have made it well-known.

In 2004 the facility management sector achieved a turnover of more than 50 billion Euros - with rising trend. The marketing research institute of Lünendonk forecasts an annual sales growth of more than eight percent up to the year 2009 [LÜNENDONK2004]. With this potential facility management will continue to belong to the growing branches of industry in the future.

2.2.2 Facility Management in German Hospitals

The alteration of the legal basic conditions (primarily caused by health reforms and legally required energy saving), as well as the rising expenses and the resulting cost pressure in the health service, make the hospitals face the challenge to implement new management concepts and processes in order to fulfil their extremely ambitious service task more economically and at the same time guarantee the best possible quality. Figure 2-6 shows that according to statistics of the German Hospital Society (DKG 2005/2006) 34,6% of the entire expenditures in a hospital are allotted to the area of the general FM services, 2.2% to the energy needs and media requirements. Furthermore there are 24% of material costs. 41.3% of the expenditures are personnel costs. The possibility to assign the costs of the real estate and its operation to their causes is a difficult but very decisive issue in the context of cost optimisation through FM.

With rising maintenance costs, rapid innovation and an increasing complexity of the facilities, which demand a holistic approach facility management gets more and more into the focus of hospital management. It is recognised that "the possibilities of... management in German hospitals … are not always exhausted " [GUDAT2005, S.29] and there is a backlog demand for the standardization of processes and of applied products, as well as for standardization of contract negotiation and contract formulation.

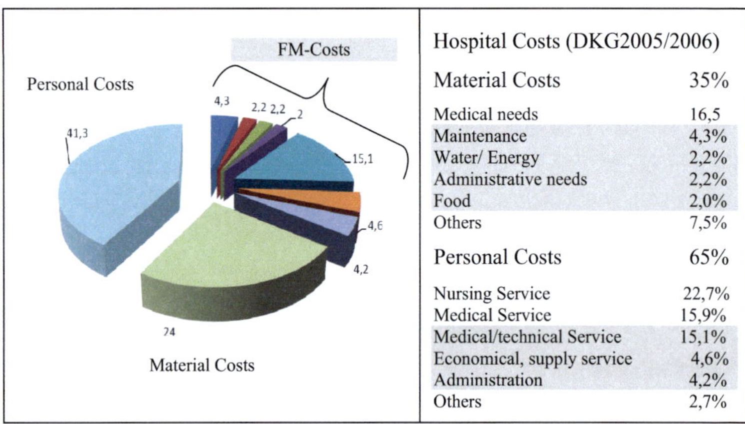

Hospital Costs (DKG2005/2006)	
Material Costs	**35%**
Medical needs	16,5
Maintenance	4,3%
Water/ Energy	2,2%
Administrative needs	2,2%
Food	2,0%
Others	7,5%
Personal Costs	**65%**
Nursing Service	22,7%
Medical Service	15,9%
Medical/technical Service	15,1%
Economical, supply service	4,6%
Administration	4,2%
Others	2,7%

Figure 2-6: Hospital expenses in Germany [DKG2005]

Service processes in health constitutions have mostly grown historically. Nevertheless, an increasing number of facility management departments have been established especially since the end of the 1990ies in hospitals which deal with the secondary processes that includes all processes except "care and curing" - which can be defined as core processes of a hospital.

Some examples for secondary processes are catering, hygiene and cleaning, personnel service, logistics services and technical building management. In some hospitals departments like radiology are also outsourced. The success of clinical care can depend directly or indirectly on the efficient execution of these activities. To protect sterility and hygiene in sensitive areas, to keep complicated medical technology available, and to guarantee perfect material logistics, adaptable organisation and thought-out processes are required.

The saving potentials can therefore be achieved; on the one hand, by optimisation of the individual processes (horizontal level) - what will be done in this PhD-thesis by using the OPIK-Project - on the other hand by systematic linking of the individual processes to integral flows (vertically). In this context the processes must be linked not only on the horizontal level, but also vertically (see Figure 2-7), that means that also all the secondary processes/ FM-interfaces should be analysed and optimized as one whole FM-process. This process must tie to the core process without redundancy and is a different approach to the "clinical or critical pathways" that passes the patient route dependent to the respective DRG basis through the duration of the hospital stay.

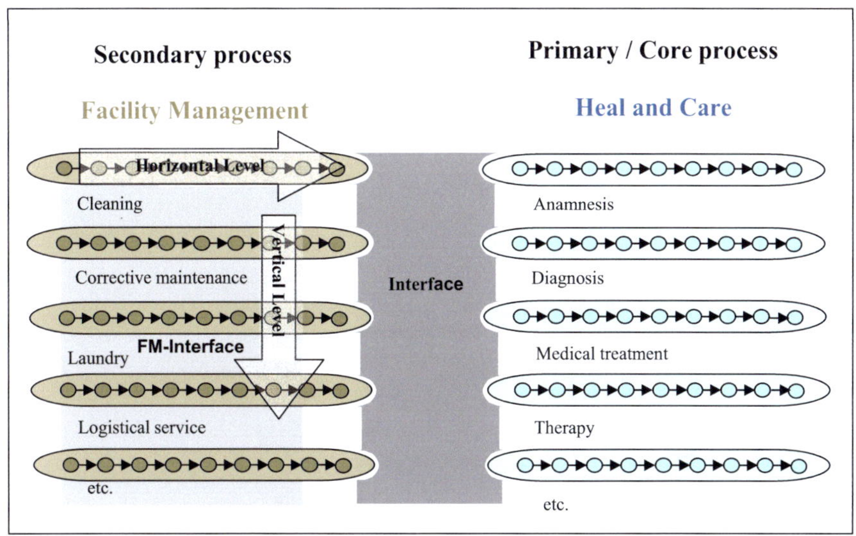

Figure 2-7: Processes in the hospital [Author2007]

The residence times of the patients therefore acquire a new meaning, which permits transparency in the utilisation of space through knowledge about the period of use of individual spaces. Spaces which are not fully occupied can thus be used and rented "warm" to third parties. Concepts for the usage of buildings are becoming part of the FM tasks in a hospital and require "number input".

Technical solutions are aimed at and weighted with respect to the requirements of "affordable expenditures", because not everything which is technically possible does have an effect on the effort required to earn the flat rate of the case. Facility management is therefore used with the help of benchmarks to optimize the available infrastructures and assists the hospital to remain "affordable". To sum up, it can be said that the costs are to be held low and that the constantly rising demands for quality and service standby times are to be met. These wide spectrums of organizational-technical features, linked with high responsibility, distinguish clinical facility management from conventional FM in industry and administration. It requires innovative management concepts and intelligent organizational solutions considering the costs over the entire life cycle of the real estate.

The "FM department" can be operated by different user models. Three different models are mainly applied in practice. These are service societies based on partnership (Model 1), individual assignments to external special companies (Model 2) and internal service societies (Model 3). The choice of the model is dependent on the specific structural conditions and needs of the particular hospital. The model is chosen so that it provides the best advantages for the clinic, the patient and the insurance company.

Model 1: service societies based on partnership (Share 51:49)	Advantages	• Knowledge is introduced by external company • Cost reduction • No sales tax
	Disadvantages	• Take-over of employees • Quality of rationalisation
Model 2: individual assignment to external special companies	Advantages	• No personnel obligation • high rationalisation and saving potential • easily exchangeable
	Disadvantages	• Personnel takeover • Coordination and control effort • Loss of know-how through migration • Effort for quality control • Cost issues (added value tax)[8]
Model 3: internal service societies	Advantages	• Know-how remains • Trained structures • Experienced personal • Complete cost transparency • Negotiable prices

---

[8] Hospitals in Germany do not have to pay added value tax

		• No added value tax
	Disadvantages	• Big inertia of old structures and teams • Continuance of old customs • Change of the rate and reimbursement structure → *difficulty in cost reduction*

Table 2-1: Model of FM-departments [FROSCH2001, S.20ff]

Model 1 (see table 2-1) offers a good base for quality improvement, as well as service developments, while social interests are fulfilled at the same time. A big, financially strong partner with experience and financial means for investments, increase of capital and reengineering programs is, however, indispensable for the application of this model. A long-term connection and the persistence of the partner should be guaranteed as well. According to Frosch, Model 3 asserts itself more and more in the German hospital scenery.

The organisation of FM in hospitals is organised due to the individual situation, structure and policy. To see how an organisation chart could look like, the structure of the university clinic of Frankfurt is shown in Figure 2-8.

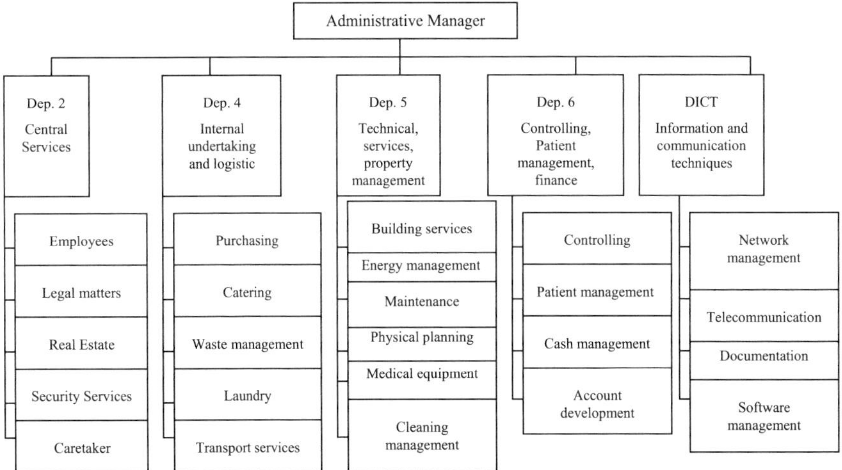

Figure 2-8: Organisation chart for FM at the University Hospital of Frankfurt [GOETHE UNIERSITÄT FRANKFURT 2007]

2.3   Facility Management in Iran

2.3.1.1 State of Facility Management in Iran

The concept of Facility Management was translated with مدیریت امکانات و تجهیزات (management of the possibilities and facilities)[9]. At present, no official translations, guidelines or even norms exist.

Iran is a country in which the government exercises a very great influence not only in the public but also in the private sector. This is also valid for hospitals (see chapter 3.2). The national budget finances itself primarily with the income from oil and gas. Environmental protection does not yet play a big role and the philosophy of life cycle thinking value added chains or optimisation of resources practically did not exist in the past. This changed in the course of the privatisation efforts during the term of office of President Khatamie (1997-2005). Big outsourcing attempts and a growing interest in the process optimisation which was caused primarily by the introduction of quality management (mostly ISO certifications) asserted themselves in private as well as in the governmental organizations.

FM for the purposes of the existing definition: "All supporting services co-ordinated by one hand" nevertheless, must be substituted with "All supporting services from several hands", i.e. service providers have established themselves since the middle of the 1990's also in Iran (services: 43%, industry: 36%, agriculture: 21%) [AHK2006]. Their number is large, but they mostly offer only single services.

The first attempts at the implementation of FM began in 2003 in the BHRC-Building and Housing Research centre of the Ministry of Housing and Urban Development, with the "FM division" which was part of the department for construction informatics.

One of the projects of this department was the development of CAFM software which was to be performed in cooperation with the University of Kaiserslautern and the University of Berlin and be supplied by the "FM division" with the necessary know-how. Unfortunately the project was not completed and the entire department was dissolved in 2005. In the course of this project the first FM conference took place 2003 in Teheran.

---

[9] Translation was performed in BHRC, Building and Housing Research Center

2.3.2 Facility Management in Iranian Hospitals

As mentioned in the introduction, contact with the Ministry of Health and Medical Education could be made in the context of this thesis. This ministry took up the idea of the process optimisation in hospitals partially and implemented it in 2003 in a pilot project called "structural correction" اصلاح ساختار for 40 hospitals in the country. Afterwards the subject FM could be taught in a World Bank project for 250 high-ranking employees of the health ministry in a postgraduate training courses called "restructuring in hospital management" in 2004 and 2005.

Facility management according to definition EN15221-1 does not yet exist in Iran. Nevertheless, there exists a department (معاونت پشتیبانی), the "support secretary" which is available in the hospitals, as well as in the university and owns the management of the service providers.

3　Object of study: Hospitals

This chapter gives an overview of the scope of the research project and of existing conditions which have influenced it. At first, typical health systems are introduced to permit an insight into the abundance of the possible versions. Besides, the German and the Iranian health system are treated in detail. Because past developments (historical background) have a big influence on the existing systems, the history of hospitals in each project country is described in a sub-chapter of its own, followed by the present hospital management in the project country.

3.1　Health Care System

The health system (health service) of a country includes "all persons, organisations, facilities, regulations and processes, which have the task of support and preservation of the health or the prevention and treatment of diseases and injuries" [BROCKHAUS2006]. These are means or methods that are chosen by countries or governments to guarantee medical care for their inhabitants. Many countries guarantee the medical care in their constitution. Every country has found its own methods of solution based on its historical, political and cultural development, and particularly on its economical resources. However, all methods of solution have in common that they are subject to constant innovation and optimisation. Hence, health care addresses a very dynamic market which is in constant movement and subject to continuous modification.

Some systems are introduced as examples to provide an impression of the diversity. Mainly models of advanced economies are introduced, because in the newly industrializing as well as the less developed economies mostly variations of these models or less complicated systems are encountered. According to Dr. A.O. Core [KERN2003] health system models are essentially of the following types:

M1 -Beveridge-Model

M2-Bismarck-Model

M3-Market-Model

M1-The Beveridge Model

The Beveridge model of a national health service is characterized by predominant state financing and often state production / allocation of services (e.g., Iran, Denmark, Great Britain, Ireland, Spain, Portugal, Finland, Sweden, Norway, Italy, and Australia). Central and East European countries mostly operate a state health system of the "Semashko type" with a complete state steering system covering financing and organisation.

Characteristics of this model are: [cf. STAPF-FINE2003]

- Only public hospital planning and investment financing
- Most of the hospitals capital funded by public sponsorship
- Regional sub-categorization of the health service
- Financing mainly from state tax
- Capital costs are assumed by regional health agencies
- Reimbursement of the stationary care funded by the state household[10]

The model can be split further as follows:

(a) Countries with regional health service: (Iran, Great Britain, Spain, Italy, Finland, Denmark, Norway, and Sweden)

- Regions, districts, provinces (health service zones) are responsible for the hospital planning as well as the organisation of the health care and the admission of facilities.
- Budgets are negotiated generally with health authorities.[11]
- In the case of central states (Italy) regions implement only the given health plans; in the countries with municipal management / local government (Scandinavian countries) the state gives only frame recommendations
- Investment costs are funded by the central state, or local authority districts (public funds, taxes)
- Appropriation of funds with the help of capitation
- HCFA -DRG-System, FinDRGs, Nord-DRGs

---

[10] HRG (Healthcare Resource Groups) [the classification which refers about procedures and diagnoses [defined] HBG (Health Benefits Groups) [not related to the hospital cases but to the need of health care of the population of a region]; both systems can be tied together in a matrix. (GB); HCFA (Health Care Financing administration)-DRGs (Ireland, Portugal); AN-DRGs (Australian)

[11] (Problems can appear by underfunding of investments, deficit of the health service (migration of patients) and waiting lists)

(b) Countries with prevailing tax financing and health insurance (Greece)

- Central state bears complete responsibility for planning and licensing of the hospitals, mandatory central government guidelines for location, structure and medicine technology

- Payments to hospitals are based on budgets for personnel costs and capital costs

- Adjustment of capital according to inflation rate

- Hospitals are reimbursed by social security funds on the basis of identical daily care and nursing rates

M2-The Bismarck Model

The Bismarck model is used mostly in countries with "financed by contribution" health insurance. It describes a social security system which is financed to a great extent by compulsory contributions dependent on the income of employees and/or employers[12]. The health services which are under state supervision are mostly provided privately. [SCHÖNIG2001,p.202] (e.g., Germany, France, Austria, Switzerland, Japan, Luxembourg, Belgium, the Netherlands).

The financing is performed primarily through the compulsory insurance (accident, health and complementary insurances)

- Mostly free choice of the doctor who provides primarily ambulant care (Austria, Luxembourg)

- The central state is in charge of hospital planning, hospital construction and hospital operation (Belgium, France, Luxembourg, and Netherlands). Permissions of the health ministry and ministry of social security are necessary; task of the state or the canton (Austria Switzerland)

- Dual financing system (Belgium, Luxembourg), monistic system (Netherlands, France)

- DRG Forms, LDF (achievement-oriented diagnosis case groups) [Austria], except Belgium that has no efforts.

---

[12] The differentiation of the financing according to employee and employer contributions is little goal oriented in the economical sense, because employees must also gain the employer's share by their achievement.

M3-The Market Model

The market model is a private insurance model with basic social care which is performed by private financing and private supply of health services. Governmental interventions or controls happen relatively seldom (United States of America). [BARKMANN2004, p.13]

- The most deregulated health care
- Group insurance through employer, private assurance, social security (Medicare) for elderly persons, basic health care for destitute population
- 15% without insurance [STAPF-FINE2003, S.57]
- The federal state is in charge of hospital planning and control of requirements
- Limited choice of the doctors and restriction of services
- Around 3/4 of the hospital investments are financed by dept on the capital market, less than 10% by charity and public authorities [STAPF-FINE2003, S.14]
- Hospital maintains contractual relationship with insurers
- Central state does not intervene in the hospital sector; planning standards
- Monistic financing
- HCFA (Health Care Financing administration) DRG (1983, research project of the University of Yale).

3.1.1 Health Care System of the Federal Republic of Germany

Health care plays a significant role in German labor politics. Approximately every ninth employee (about 4.5 mil. [13]) works in the health care branch. According to the Federal Office of Statistics, the expenses for health care in 2006 have amounted to a total of about 245 bn. Euros. The expenses per person have been about 2.900 Euros. [STATISTISCHES BUNDESAMT2006]

The developments in the health service depend – as mentioned in the introduction - on many external parameters. Especially the demography of a country plays a central role in this connection. In Germany (as well as in many other advanced economies) a decline of the population can be recognized. From 2005 to 2020 a decline in population of 1 million people is expected. [NORTH / LB in 2002, S.10]

At the same time, the age distribution is subjected to alteration. A growing number of the older people with higher life expectancy affects the available social security system. The probability of need for intensive care rises. This is already taking place: the duration of hospital visits by people older than 80 years rose by 2.2 days during the 1990's.

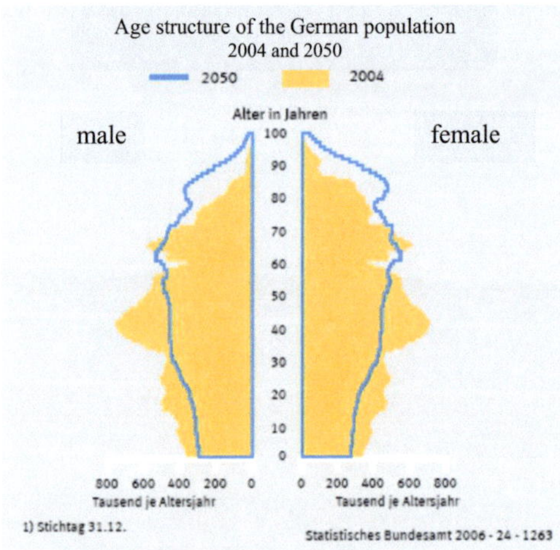

A growing spectrum of disease patterns – due to the improved facilities of diagnostics and therapy (phenomenon of multi-morbidity especially for older patients) presents new challenges for the health branch.

Ecological parameters like the income of private households as a function of economic growth influence the development as well as medical-technical progress.

Figure 3-1: Age structure of the population [STATISTISCHES BUNDESAMT2006]

---

[13] According to the Statistisches Bundesamt (2006)

The market has also developed strongly for „individual health services" (fitness, wellness, Anti-Aging, esthetic surgery, cosmetics, drugs, alternative treatments and esoteric methods). These innovations in medicine technology (new and improved treatments and products) allow new process developments. In the area of ambulant operation rooms, for example, the miniaturization of the OP instruments, the improvement of the technology and the price reduction lead to shorter stationary stays in the hospitals. With an annual turnover of 12 bn. Euros (2002) [the NORD / LB in 2002, S.12] the German companies for medicine technology stand at the head of the developments. This position is also due to the investments in research and development. Germany invests 10% of the annual turnover in this area. [FEDERAL MINISTRY OF HEALTH2007]

Furthermore the application of digital information systems influences the health service. Improved communication and documentation allows the optimisation of the working processes and the creation of transparent cost structures; it is therefore a very helpful tool for an efficient hospital management.

3.1.1.1 Administration

The whole responsibility for the health service lies in the hand of the central government and the federal states. They are the carriers of the health policy and have to guarantee the medical care of the citizens. They transfer state duties (e.g. the performance and financing of medical services) to other institutions: On the one hand, to institutions of public law (health insurance schemes, association of physicians, insurance associations, etc.) and on the other hand to private suppliers (medical practises, pharmacists, pharmaceutical companies, hospitals, etc.).

Health care is separated to a great extent into ambulant and stationary care. The care is provided primarily by freelance doctors and pharmacists as well as private big corporations in the pharmaceutical or medical technology industry. The government is involved through health centres, municipal hospitals and university medical centres.

3.1.1.2 Financing

Contrary to other countries, Germany does not have a public health system (financed by taxes; state authorities provide the services - as for example in Great Britain) and no market-economically oriented health system (with the self-pay patients or privately insured persons having access to private services - as for example in the USA), but a social security system, which is financed primarily by contributions from compulsory and private health insurances. In this system; the services are mainly provided by institutions of public law and by private suppliers.

The financing of the health service is achieved to 68.2% through the compulsory health insurance, to 10% by taxes, to 10.8% by additional payment and to 6.6% by private health insurances. [STAPF-FINE2003, S.9]

Health Insurance Schemes

Germany is the country of origin of social security (introduced in 1883). 90% of the population is insured by compulsory health insurance[14]. The contributions are paid in equal parts by employers and employees. They are geared to the respective income. About 9% of the population is insured by private insurers.

The premiums depend on the contracted service, the general state of health, the gender and the entry age of the insured person. Only about 0.1 to 0.3% of the population are without health insurance protection. [BMG2005]

In addition to the insurance services, self participation or additional payments of patients demonstrate a growing interest in the financing of the health system.

The compulsory Health Insurance Schemes are made up of seven different health insurance fund species, namely the "local Health Insurance Schemes, company Health Insurance Schemes, the guild Health Insurance Schemes, the agricultural Health Insurance Schemes, the Sea Insurance Fund, Social Miners Insurance and the Substitute Fund " [SOCIAL CODE SGB §4]. In Germany there are all together more than 500 different Health Insurance Schemes. The compulsory Health Insurance Schemes define their contribution rate according to the economic imperatives [SOCIAL CODE SGB §220 following]. Hence, these rates are different (for the old federal states in 1996 on an average 13.4%, for the new federal states 13.3%). Because of the different contribution rates complete freedom of choice between all compulsory insurances was implemented with the "health structure law". [BMG2005]

---

[14] (Gesetzliche Krankenversicherung)

Private Insurance

The health insurance of the "not compulsory insured persons" is covered to a great extent by private health insurances; however, not everybody can choose a private insurer. The legislator allows only to the following groups of persons to change to private health insurance:

- *Students* (after expiry of the family health insurance).

- *Self-employed persons, free-lancing persons, enterprisers* (who show the biggest interest in private insurance).

- *Employees and workers with an income above the compulsory insurance limit* (this group includes all persons who achieve a regular annual income of at least 48.150 EUR. These persons take a big interest of private health insurance. A big advantage for these persons is that the employer pays half of the insurance contribution.

- *Government employees* (employees in the civil service with social security claim can insure privately).

- *Physicians, dentists and residents* (this group of persons is treated as a rule like employees or self-employed).

Beside these full insurances the private health insurance also offers complementary optional services for the hospital (more comfortable accommodation and/or free doctor's choice).

In the private health insurance the refund principle is applied. There is no contractual relationship between the private health insurance and the doctor, but only one contract between the doctor and the patient. The patient receives the bill from the doctor and submits this to his insurance company. The private health insurers take over the full costs of the treatment. However, the doctors are obliged to stay within the fee framework established by the Federal Government[15]. [KRANKENKASSEN2008]

---

[15] (Gebührenordnung für Ärzte)

3.1.1.3 Medical education

The science ministries are responsible for the medical and dental-medical education and the university clinics.

Medical Study

The study of medicine consists of at least six years with two years of preclinical and four years of clinical study.

After completion of a six-year medicine study the doctor performs 18-months of practical work as a resident[16]. After that he can apply for the "licence to practise medicine". This licence to practise medicine entitles him to independent professional activity as a physician at hospitals as well as in free establishments (however, not as a panel doctor). After the regional (state) authority has granted the licence to practise medicine the doctor can apply for "advanced training" controlled by the medical association. This training ends with the award of a specialist's certificate (general medicine counts as a special field of its own). The end of such advanced training has recently become a condition for licensing as a panel doctor. The advanced education takes between four and six years depending on the profession. [BUNDESAGENTUR FÜR ARBEIT 2008]

3.1.2   Health system of the Islamic Republic of Iran

In a country that is almost as large as Great Britain, France, Germany and Sweden together, and where the cultural and economical differences between the rural and urban population are huge, medical care has proved to be a difficult issue. Shortly after the second world war mobile doctor-teams moved into the country to fight against epidemic malaria and tuberculosis. [SHADPOUR1994, P.11 ff.]

The Behdar (healer) training project was started as early as 1940. It provided a four year training for missions in rural regions. The project failed, however, because the Behdar went to the universities after their training, became doctors and settled in the big cities. A further attempt was initiated in 1964 with the HEALTH CORPS ACT. The concept of this project was that medical graduates and people who passed the A-level should perform their military service by working for the health ministry. After their training they were sent into the country

---

[16] (AiP- Arzt in Praxis)

and worked as "Health Worker". The problem was, however, that frequently the young people could not adapt to the rural life. The project was made more difficult by the high fluctuation of the staff (who left after they had completed their military service). [AMINI1983]

After the revolution in 1979 „the right of health care" was guaranteed to every citizen in the constitution.

The rapid increase in population (1975-2000: industrial nations: 20%; Iran 92%) [MOHME2005] and the massive migration from the countryside forced the government to act despite countless difficulties (Iran was in war with Iraq (first Golf war 1980-1988) and under sanction).

Under these conditions the „West Azerbaijan Project", an idea from 1972, was reactivated by the WHO, Tehran University, the Healthcare Ministry and "Azerbaijan" province and implemented as Primary Healthcare Network in 1985. The PHC-network has been especially useful for the people in the country side.

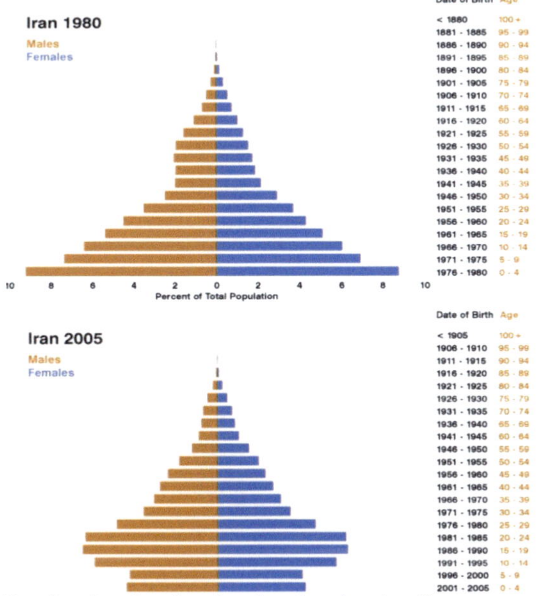

The demographical structure and development of Iran can be gathered from the fig. 3-2. The baby boom that started after the Islamic Revolution in 1979 has dropped since the end of the nineties. This trend, the agricultural recession and the rural exodus tend to change Iran from a country with big to a country with small families.

Figure 3-2: Age Structure 1980-2005 [STATISTICAL CENTER OF IRAN2006]

The most frequent causes of death are heart disease (21%) followed by traffic accidents (11%) [NAGHAWI2005].

Large problems with drug addiction exist among others due to the neighborhood to Afghanistan. Iran is one of the countries with the largest number of heroin and opium addicts. [UNDEP2000]

AIDS plays a big role besides Hepatitis C. Although 4846 HIV and 678 Aids infected persons in 2003 are not very large numbers, the rapid growth of the number of infected persons is alarming. [UNDP2004].

3.1.2.1 Administration

The PHC network is active on the national, provincial and district level. The country is divided into 30 provinces, 280 districts, 676 cities and 66000 villages. [MOHME 2006]

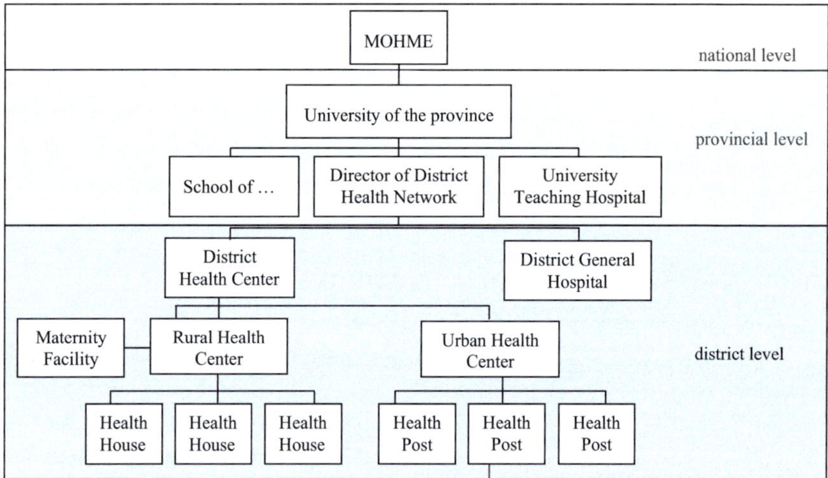

Figure 3-3: Organigram of the health network of Iran [based on ABOLHASSANI2006]

The national level

The healthcare ministry is under the authority of the minister who has the supervision of nine departments and several offices. The ministry prescribes the guidelines, planning and leadership of the regional healthcare organizations and the medical universities.

The provincial level

At the moment there are 39 medical universities and 4 medical schools in 23 provinces, 30 healthcare centres (provincial level), 327 district healthcare centres, 2316 urban, 1383 municipal healthcare post, 2424 rural healthcare centres and 17000 healthcare houses. [MOHME 2006]

In most of the provinces the chancellor of the medical university is at the same time the executive director of the regional healthcare organization.

The district level

The executive units of the district level are the health houses, rural and urban health centers, Behvarz (healer) training centre, District health centers and district hospitals, which are all part of the health network.

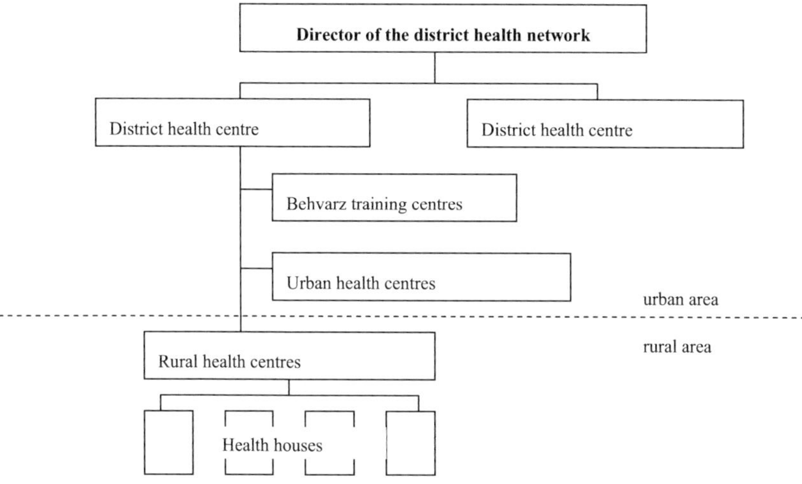

Figure 3-4: Organization Chart of the Health Network [ABOLHASSANI2006]

The district health center

The district health center is an autonomous administrative unit which is responsible for the planning, control and support of the rural and urban health centers, the Behvarz training centers and the health houses. Its tasks include technical and logistical support of the health centers, special health service which is beyond the resources of the health centers, epidemic related studies of endemic diseases, planning, control and support of healthcare projects, international quarantine measurement at the borders of the country, monitoring of healthcare standards and control and analysis of data from healthcare centers.

The district hospital

The district hospital is a general hospital with at least 7 departments (general medicine, obstetrics and gynecology, internal medicine, pediatrics, anesthesia, radiology and a clinical lab). It mainly takes over the cases which have been referred from the health centers (out- and in-patient treatments), admission and treatment of emergency patients.

The rural health centers (Khaneye Behdashte Markazi)

The health centers take care of approximately 9000 people. The team consists of a doctor who heads the centre and staff for family planning, disease control, environment health, mouth hygiene, labs, care and administration. The rural health center, the second level of service provision, is a village-based facility that supervises one to five health houses or health posts in the same or neighboring villages, not farther away than 40 km.

The main function of the health centers is the support of the health houses and their control, taking over of cases which are beyond the capability of the health houses, as well as the contact to the higher levels of the healthcare system. Besides that, they perform fundamental lab examinations, train the staff, control and investigate foods. They also conduct hygienic examinations in schools and at work places and prepare statistical evaluations and regular reports.

The maternity facility

For financial reasons maternity facilities were originally not established in the health centers. After a childbirth mortality of 90 out of 100,000 children was detected in 1988 (based on a statistical survey), the necessity of these facilities were realized, and they were established in 1990 by the health ministry. [ABOLHASSANI 2006]

800 maternity facilities were integrated into the health centers (which take care of at least 20.000 people) in a 5 year program. Exceptions were made for centers for 12000 people that

were difficult to access due to climatic circumstances or poverty. Additionally, 120 training centers were founded which trained the staff in a 6 month course. This staff is concerned with family planning.

The maternity facilities are provided with 3 midwives who work in 8 shifts round the clock, 3 cleaners and an ambulance driver who is on stand-by duty round the clock. Complicated cases are transferred to the next hospital.

The urban health centers

The urban health centers take care of about 12,000 people. They are structured–with regard to staff- like the rural health centers, but employ a radiologist in addition. The task of the health centers consist of the treatment of the urban people, if necessary referral to hospitals, training in the healthcare area, family health service, disease control, clinical diagnostics, statistical surveys, documentation and reports. It must be emphasized that the private sector provides a large part of the urban population. Therefore the public facilities have a minor influence in those areas.

A large problem in the cities is the migration from the land. With the help of UNICEF the "Community healthcare posts" were founded, which should guaranty the primary healthcare. [SHADPOUR1994, S.27 ff.]

Health Houses (Khaneye Behdasht)

The health house is the smallest unit in the PHC-network and can provide for up to 1500 people. As most of the villages have less than 1500 habitants the main health houses are responsible for several satellite villages. The health houses are provided with a female and a male Behvarz (healer) who come from the village. In this way the Behvarz know the entire community and can assess the healthcare situation very well. The main task of the health houses is the primary medical care of the community, which consists of a yearly census of the population, public information, education and training, Family health (policies relating to the family (precaution, aftercare), care of the children under 5 years, care of the schoolchildren, house-calls), disease control (cause research, blood smears, simple treatments), activities for environmental hygiene (control of the foodstuff production for the community, controls at the schools and at work places, water control (especially drinking-water), collection and documentation of health information, preparation of reports.

3.1.2.2 Financing

Health services are financed through the public budget (which is adopted by parliament after prior consultation with the government – yearly budget plan for the ministry). The ministry forwards the budget to the medical universities which are responsible for the financing of their region.

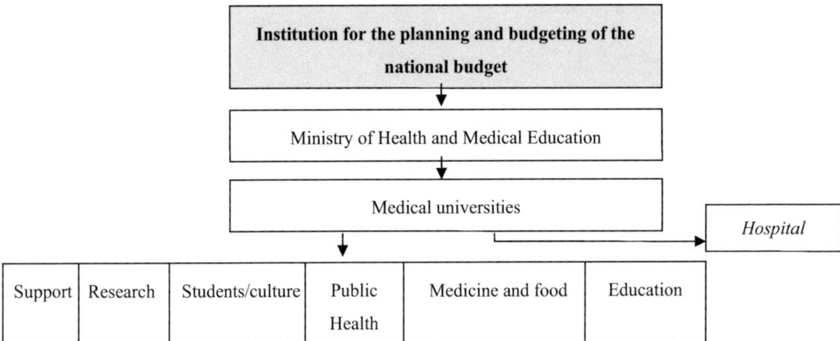

Figure 3-5: Finance path of hospitals [AUTHOR2007]

Other financial sources for the hospitals are the insurances; treatment costs of private individuals (out of pocket expenses), and contributions of the religious and international organizations and charity.

Health Insurance

The employer in Iran is required to insure the employees. This mainly occurs through the "Tamine Ejtemai" (health- and additional insurance, single and group insurances).

The insurances became established at the beginning of the 20th century by foreign insurance companies. The first Iranian joint-stock company of insurance was registered in 1935.

In 1986 the law of "general health insurance" was declared whose purpose was health insurance for all people in the country. In order to implement this goal solid financing and correct education were needed.

Different insurance forms can be distinguished:

- Governmental insurances (Tamine Ejtemai, Military Insurances, Komitee Emdad Emam Khomeini Insurance, Oil Ministry Insurance)

- Private insurances (Dana, Iran, Asia, Alborz…)

The governmental insurances companies pay 90% of inpatient and 70% of outpatient costs. The rest must be paid by the patient. Private insurances however reimburse up to 100% of the costs depending on the contract. Rural inhabitants (less than 10000 people) are covered by a different scheme. They are able to use the PHC facilities free of charge.

Medical Representatives

Medical Representatives exist in Iran; they are, however, weak and play a secondary role.

3.1.2.3 Medical Education

After a Madjlis (parliament) decision in 1986 the health ministry was entrusted with medical education.

Since 1993 all local healthcare services (PHC, Primary health care) are subordinated to the medical universities which are also responsible for all the health services as well as hygienic control.

With a successful entrance examination with excellent grade the students are allotted to the medical universities based on a prescribed process. After 4 semesters of basic study, medical students have to pass "Pathophysiology" course. This is followed by a practical semester called „Pysi opatologie" and 2 years of work in hospitals. After this there is another examination. Once it is passed the student is called "Intern". In this stage the students are allowed to examine the patients on their own responsibility, but are not permitted to prescribe medicine. After these 2 years the students are recognized as Medical Practitioner. Afterwards they are able to gain a specialist title for a special medical qualification.

3.2 The Hospital

The hospital (Krankenhaus, Bimarestan) is an appellation for „an institution in which sick persons are examined and undergo medical treatments". [BROCKHAUS2006]

A special form of a hospital is the hospice which is used exclusively for palliative care of mortally ill patients. Also included are hospitals that are predominantly at the disposal of a specific group of persons like hospitals and wards in prisons and army. Additionally in Iran there are hospitals which belong to specific ministries (e.g. oil ministry) and are used only for their employees.

Influential factors that affect the building, administration, and maintenance of hospitals are the legal boundary conditions, demographical development, general economical conclusions, the level of research and development, the state of the information technology as well as the quality requirements. The quality requirements differ in different countries, as can be seen in the evaluation of the project data (ref. chapter 5).

3.2.1 Hospitals in Germany

3.2.1.1 History of hospitals in Germany

In the middle ages the first charitable welfare facilities came into being in the course of Christianization, with the ideal of charity and mercy. In foreign lodgings (Xenodochien) accommodation and care were given to pilgrims, travelers and poor people. This had not occurred in the western culture before. A milestone in the development were the religious order communities which spread out in Europe after 529 and devoted themselves especially during the crusades (from 1096 to 1270) to nursing. The first German hospitals were established in the baroque time (1566-96). Already in 1601 surgeons, chemists and trained orderlies were employed and different areas were provided in the hospitals, like examination rooms, doctor's rooms, insulated rooms for seriously ill patients and rooms for devices. The Augustinians provided the foundation for qualified, regular nursing. At the famous "Hotel Dieux" in Paris 1000 sisters and novices took care of 2800 patients daily at the end of 17th century. The term Hospital came into use in the ninth decade

Figure 3-6: A hospital hall of the 16th century, etching around 1560

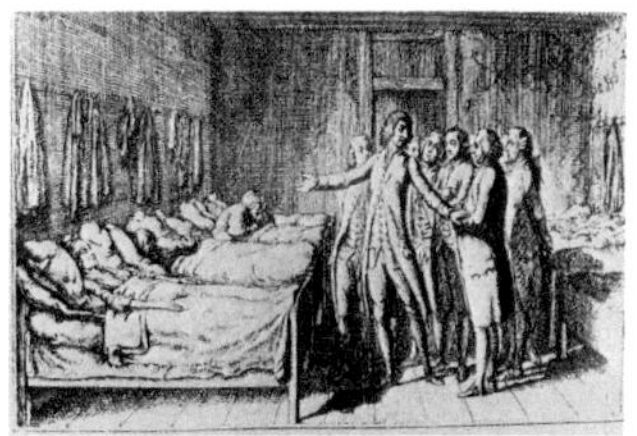

Figure 3-7: Sickroom of the Charite. Etching of Daniel Chodowiecki, 1783

of the 18th century. This was also the time of great efforts under Maria Theresia and Joseph II. In 1780 a great medical centre was founded for 2000 patients. With the German idealism (Kant: "Dignity of the Human Being ", moral philosophy) the development of the hospital from state institution to central institution of the urban public welfare took place in the Biedermeier period time to provide welfare for the poor. Inspired by progress of the physicians (discovery of anesthesia in 1846), hospitals were established as independent institutions. Through the researchers' curiosity and the steadily growing social engagement, the poor hospital developed to the „civil hospital for all classes ".

General hospitals and institutions established themselves between 1785 and 1845. They were founded and sponsored by the state and contained departments for surgery and internal medicine as well as smaller additional stations for patients with contagious diseases. The supervision was performed by an academically qualified doctor. Sanitary-hygienic achievements (permanent air ventilation, immediate feces removal, washing and bath facilities as well as eco- friendly location) and the high level of the hospital hygiene established new standards in this time, because separation was performed between different kinds of diseases [MURKEN1988, S.59].

About 1817 the first attempts were made to establish private hospitals which were independent of the state and "patients institutes" in which ambulant medical help was given free of charge. Also the first teaching hospitals with hardly more than 20-30 beds (1780 in Freiburg) started their activities. Nursing started to become a respectable activity that opened ways for women to professional emancipation.

Figure 3-8: Laundry, 1848

Operation halls as well as catering and laundry moved economically to the foreground and were located in preferential places in the centre of the hospital complex. The fact that the Charite in 1848 had its own laundry [MURKEN1988 P.93] shows how seriously the hygiene of the bedclothes and the patient's clothing were treated. At the same time the cleaning of the sanitary arrangements, the floors and the quick hygienically flawless removal of the wastes where emphasized to a greater extent than ever before. With the rise of Germany as an industrial state and the strengthening of the socialist labor movements under Otto von Bismarck the exemplary, social insurance system (1883), accident insurance system (1884), Invalidity and old age insurance law (1889) came into being.

Robert Koch in 1882 discovered the cholera and tuberculosis bacteria. This led to architectural and structural change of the hospitals. The room air was cleaned to avoid "air driven infection. New results in the area of bacteriology and hygiene gained importance in the fight against contact driven infection" [MURKEN1988, S.143]. The "disinfection institution" founded in 1873 allowed for the first time in Germany the disinfection of infected clothes, laundry, bedclothes and other objects by dry heat. In 1972 the economic position of the hospitals in Germany was put on a reasonable base with the" law of the economic protection of the hospitals (KHG)". In this model the investment costs were covered with public money, the running operating expenses, however, were claimed over the daily hospital and nursing rates of the health insurance schemes (dual financing system). This allowed a gradated clinical care also in rural regions.

Since 1886 attention was paid to the concentration of certain functional processes. Related areas were placed together, because long ways were seen as a disadvantage. With the advent of reinforced concrete construction and lifts, the technical and structural demands arose to stack the hospital stations on top of each other instead of placing them next to each other horizontally. The paths for patients, visitors, staff, laundry, catering were separated. Advantages of the "bed high rises" are the short paths, lighter installation of pipelines and wiring for the energy and water supply as well as more favorable separation and combination of fields of work and departments.

With the invention of antibiotics (1935) and penicillin (1944) natural remedial factors were gradually renounced. Since the 50's and 60's three basic types have been distinguished for hospitals with less than 200 beds according to H. U. Meyer:

1. Horizontal type: Functional equipment and nursing area at one level

2. Mixed type: Allocation of functional equipment and nursing area at two levels

3. Vertical type: Functional equipment under the nursing area

3.2.1.2 Hospital Management

A partial segment of the extensive health care service is the hospital that must fulfill the patients' demands as a customer adequately. In the framework of the existing legal boundary conditions the hospital is obliged change from pure administrative competence to business management competence, and to develop into an economically oriented enterprise with increased automation. In the general hospital sector the numbers of facilities and beds have decreased during the past years as shown in fig. 3-9. At the same time there has been an increase in the number of patients (16.539.398 in 2005) [The NORTH LB 2002, p.18]

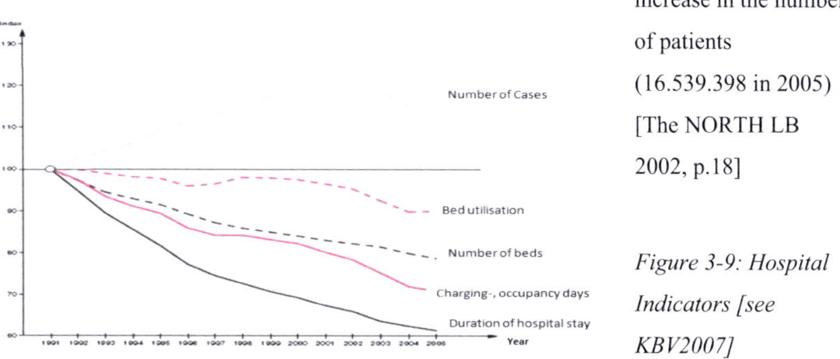

Figure 3-9: Hospital Indicators [see KBV2007]

With 510,767 beds in 2,014 hospitals in 2006 there has been a decline by 13.057 beds. Figure 3-9 shows the distribution of the hospitals, number of beds, stationary and ambulant OPs in 2005 vs hospital size.

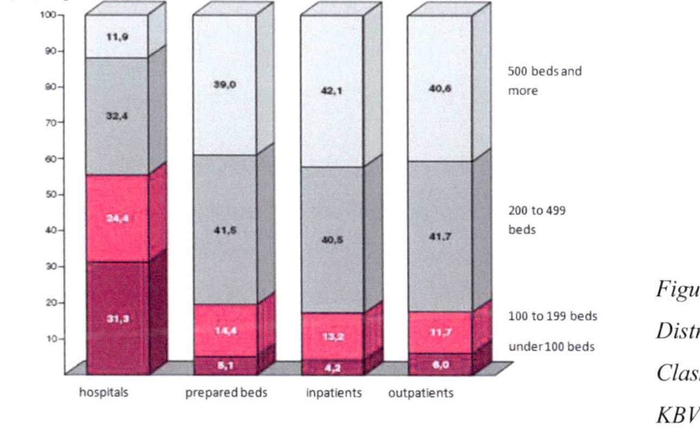

Figure 3-10: Distribution vs. Classes 2005 [see KBV2007]

Due to the steadily decreasing resting time (average resting time in general hospitals was 7.9 days in 2006) the number of the counting / occupation days is also decreasing. It dropped in comparison to 2005 (143.2 Mil.) by 0.69 percent to 142.3 mil. charging / occupancy days [BASIC DATA the hospital statistical Federal Office 2006].

Hospital Sponsors

The sponsorship of public, free-charitable and private hospitals differs. Hospitals which are operated by the territorial entities (federal government, state, scenery federation, association of municipalities, district, municipalities), by unions of such institutions (e.g., work groups, purposive associations) or by social insurers (e.g. land assurance institution, professional association) or are financially supported mainly by these entities belong to the public hospitals. Free charitable hospitals are those which are sponsored by church municipalities, ecclesiastical and worldly unions, cooperatives as well as by endowments. [BMG2006]

In Germany about one third of the hospitals are in the legal sponsorship of states and municipalities, about one third in the property of free welfare federations and about one third in private hand (often in the hand of doctors). About 60% of the beds however stand in the public hospitals (of the states and municipalities), one third of them in free charitable medical centers (in the possession of free welfare federations) and only the rest of about 5% in private hospitals. Private sponsors head mainly small special hospitals, while the big emergency hospitals are mostly public. The trend is in the direction of private hospitals. The percentage of private clinics in the general hospitals rose from 15.9% in 1992 to 22.3% in 2000. Many private hospitals are a part of bigger hospital chains, e.g., of the Rhönklinik AG or the Helios AG.

Hospital Financing

The financing system is a dual one. The federal states are responsible for the financing of the hospital infrastructure and follow this purpose by installation of hospital plans and provision of funds. The medical funds unions have an obligation to guarantee the ambulant care. The hospital financing law makes it obligatory to prepare the hospital plans with participation of the state hospital society and the federal sates (Bundesländer) of the health insurance schemes. Until 1972 there was a dual financing from states and health insurance schemes.

Since 2003 a monistic financing system exists, the accounting system is oriented to case rates in the form of DRG's. [BMG2006]

The *prime costs cover principle* was abolished by the health structure law in 1993, so was the guarantee that all expenses of the hospital institution are taken over. Since that time the hospital can only claim medically adequate hospital and nursing charges. There are uniform case rates as well as additional special fees for every hospital of a federal state. 20-30% of the budgets of a hospital are covered by these case rates and special fees, the rest by the base and departmental hospital and nursing charges.

Cost structure of a hospital

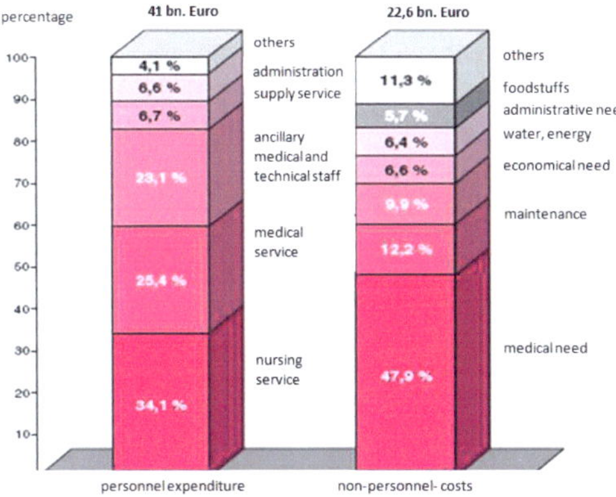

The hospital expenses in 2005 were about 63.6 bn. Euros. It can be recognized from the figure 3-11 that the expenses in a hospital are primarily personnel expenditures. The admission in a hospital normally depends on the referral of a registered doctor (mostly of the family doctor). Without the referral only emergencies are admitted.

Figure 3-11: Cost Structure in Hospitals 2005 [see KBV2007]

3.2.2 Hospitals in Iran

3.2.2.1 History of hospitals in Iran

The oldest discovery of a medical intervention in Iran was made in the "burnt city" in Balutchestan. An extravagantly formed artificial eyeball was discovered in the left eye socket of a 4800 year old skeleton. A large number of practical medical instructions have been recorded in the

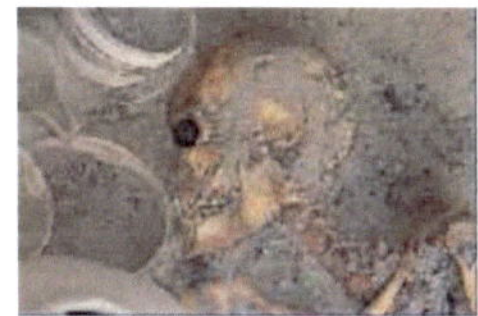
Figure 3-12: 4800 years old skeleton with artificial eye

library of Asshurbanipal in Niniveh in Assyria (668 BC). The majority of them deal with prescriptions for treatment of patients; however there are also clay tablets with the title "Procedures". One of the largest and oldest collections of such treatments is known as "Procedures of medical diagnosis and prognosis" (100 BC). The causes of diseases were related to different kinds of entity like ghosts, gods and demons. This opinion made a specific entity responsible for a specific symptom. However, organic failures and general abnormal behaviour of patients were examined as well. The integrated attitude and procedure led to a clear allocation of duties and treatments that resulted in two types of physicians. The task of the first one, called Aschipu was the identification of the causes for the symptoms and their treatment on the spiritual level. The responsible entity was diagnosed here and the treatment (spells, ointment and herbs) was performed accordingly. Patients with purely physical symptoms went to the second type of physician;

Figure 3-13: Visit at home

called Asu. Due to his activity he was also called "physiologist". The patients were referred by the Ashipu or came directly (injuries). The Asu used other herbs and ointments than the Ashipu. Additional abilities of Mesopotamian physiologists are based on the Hammurabi's (1700 BC) where the first operations are mentioned. The responsibility and liability of the physiologists towards the patients as well as the reimbursement – which varied based on the position and financial status of the patient - are exactly specified here. The location of the treatment was at that time mainly the house of the patient. His relatives were the caring staff. The places and temples of the healing creatures and goddesses were specially rivers, preferential points for treatment. The largest Persian dynasty of the Achemendians (550 BC.) contributed an enormous amount of cultural and scientific achievements. The medical libraries of Babylon present a foundation of the knowledge about cures. Moreover, there existed an active exchange with Greek and Egyptian physiologists. With the famous university „the royal cure school of Jondishapur a centre of medical science was developed which was unique until that time. Studies and developments at this centre exerted their influences worldwide. This kind of combined medical school and care is supposed to be typical for the later Islamic hospitals. With the Islamisation in the 7th century the east was connected to the west. A wave of translations of every scientific treatment found the way over Spain also to Europe and facilitated an exchange from Spain to Indonesia.

The new seat of authority and consequently also the spiritual centre was located in Baghdad from 754 ad The first hospital (Bimarestan) was also built here based on the Jodishapur model by the famous Persian physiologist Razi [864-925 AD]. One of the famous Razi's methods to find a location for the building of a hospital was the following: He distributed pieces of meat in specific places in the town. The hospital was then built at that place where the meat remained unspoiled for the longest period of time.

In the course of the following centuries there were further developments which were mainly driven by Greek and Persian physiologists and scientists. Well-known names in this connection are Galen (Blood circulation) and Hippocrates (Disputation of mental disease-causes), the brilliant Persian scientist Razi (distillation of alcohol), or Abu Ali Sina / Avicenna [980-1037 AD] who wrote over 100 books about medical subjects. His medical Canon presents the most important and fundamental work of medical science.

The purely logical and rational medical contemplation of the Greeks was weakened by the increased Islamic and Arabic view of the world in the 13th century, so that superstition assumed a significant role in fighting a disease. The religious leaders strongly rejected the lessons of anatomy and therefore there was no further significant development until the 19th century. Then however, in the second half of the 19th century, there was a strong political and ideological transformation in Iran.

Through European translations modern science and Western ideas found their way into the orient and aroused a movement of modernization.

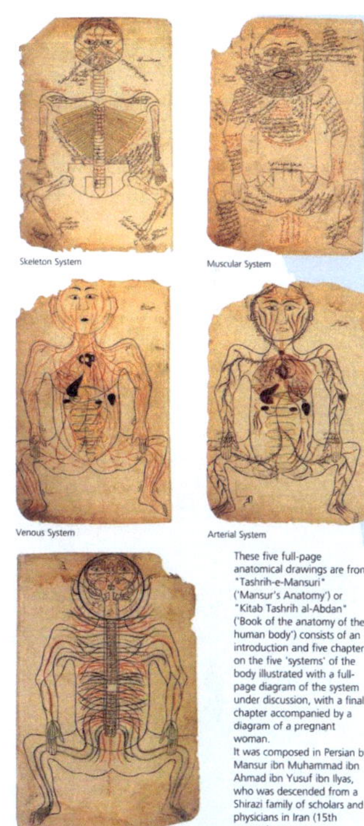

Figure 3-14: System of a human being 15th Century

The "Dar ol Fonoon" University, where international professors taught, was very famous in the time of Gajjars and Amirkabir. Its medical faculty was the unique training centre before the foundation of Tehran University. Especially physiologists who returned from Europe played a big role as they brought their knowledge of modern medicine to Iran.

Fees were charged for pilgrimages to Iraq and tombs in holy cities in order to have a budget for medical care. Centres were built with these funds especially in the north of the country, where many diseases (e.g. Cholera) found their way into the land. In addition to the religious and state-financed facilities, foreign institutions were also active; so hospitals were built by the Portuguese in 1622 on the Hormoz Island in the Persian Gulf, by the Russians in 1845 on the Ashuradeh Island in the Caspian Sea, by the Britons in 1876 in Isfahan and by the Americans in 1905 in Orumieh. Some of these hospitals became university clinics and received financial support from the government. In 1892 the doctors with more than 10 years of experience were permitted to be registered and received state-owned offices.

The first state-owned hospital was Sina hospital which was developed by Dar ol Fonoon. It was founded in 1894 by the students of Dar ol Fonoon and had a budget of 1200 Toomans. The 30 hospital beds were extended to 88 beds during the Second World War and supplied medical care for 4000 patients annually. In 1918 the first health ministry was founded, which was, however, closed after 3 months. Diseases like Typhus spread across the country and the hospitals were faced with large problems due to lack of resources so that the state hospital Sina were reduced to 25 beds and many hospitals were closed. The Health ministry was re-established in 1920 and combined with the health association under a new name. 3 years later a new institute was founded, which took over the tasks of the Health Ministry and stemmed the spreading of diseases with the help of quarantine. The number of state-owned hospital beds was increased in 1926 in Tehran to 130 and Iran owned 4 city-owned clinics, to 70 beds in Mashhad.

Religious institutions ran various hospitals with up to 500 beds. About 7500 people were vaccinated in 1921, and 905 Doctors were working in Iran one year later in 1922. The institute for pharmacy was founded in 1918 which was made subordinate to Dar ol Fonoon. These times were overshadowed by pest and typhoid epidemics. Until 1953 the hospitals were led by Doctors who were performing medical care and scarcely followed organizational tasks. In this year 137 projects were implemented in health care which was however not realizable. This was also the foundation year of "The institute for hospital management"

One of the first projects in the area of health and hygiene is the project "education of Behdars". These Behdars had to have at least A-level grades and were sent to the small towns and villages of the country. This project was stopped at the beginning of the 40s.

3.2.2.2 Hospital Management

This section is more detailed than the corresponding section on German hospitals, because it concerns the research object OPIK Iran. The hospital system of Iran will be discussed in this course with the help of the example of the Medical University of Teheran.

Hospital System

The hospitals were led after the revolution for some time by the Behdarie Ostan- the health authorities of the provinces-. Since 1986, however, they are under the control of the Ministry of Health and Medical Education (Vezarate Behdasht Darman va Amuzeshe Pezeshki).

According to the decision of the Madjlis (parliament) of the Islamic republic Iran in 1986 the health ministry was put in charge of medical education. Consequently all medical universities were under the management of the ministry and were administratively separated from the remaining universities.

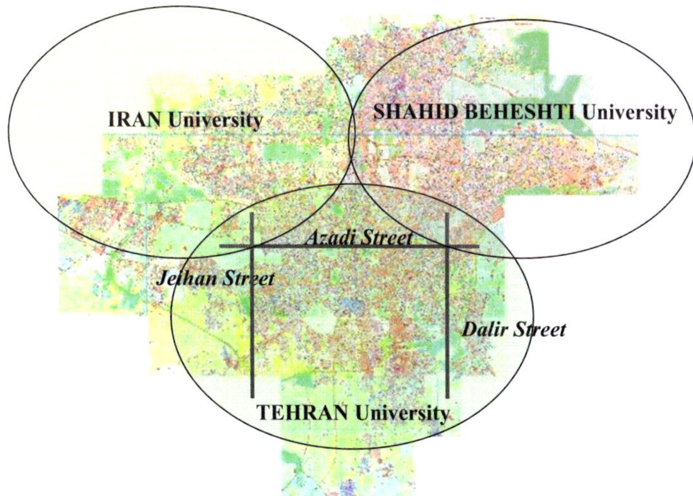

Figure 3-15: Areas of Responsibility of the Universities in Tehran [AUTHOR2007]

Since 1993 all local health services (PHC) have also been under the leadership of the medical universities which are consequently responsible for all health services as well as hygiene controls. The region of Teheran with about 12 million inhabitants is covered by three universities as shown in picture 3-15.

 The "Iran University", in the north and the west, the "Shahid Beheshti University", in the east and partly also in the north and the "University of Teheran" whose area of responsibility covers the south of Teheran and the southern areas of the Teheran province with Eslamshahr and Shahre Ray. The University of Teheran is used as example for the construction and workflow of hospital organisation and management.

District Health centers	Urban Health centers	Rural Health centers	Health posts	Health Houses
Eslamshahr	9	7	13	19
Ray	17	6	15	28
Djonoub	28	-	22	-
Total	**54**	**13**	**50**	**47**

Table 3-1: PHC- Centers of the University of Tehran [TUMS2005]/ [AUTHOR2007]]

Table 3-1 shows the PHC-centers for in the area of responsibility of the University of Teheran. All together the university of Teheran is responsible for the following health institutions: 15 governmental, 19 private, charitable hospitals, hospitals of the Tamine Ejtemai ("national assurance") and other organizations or hospitals of the ministries, a total of about 6500 beds, 200 pharmacies, 147 labs, daily medical centers and doctor offices.

The Medical University of Tehran

The first modern medical university in Iran was founded in Teheran in 1934, based on the values of the Dar-ol-Fonoon which had established a foundation in the history of the medical universities more than 150 years ago. This university with eight faculties, 16 university hospitals, 25 research centers in Teheran and 15 additional centers in 10 provinces, as well as eight national scientific centers, is regarded as a "mother of the universities" and is a symbol for "higher education" in Iran. At the moment 10.000 students study in 1300 institutes and it must be mentioned that only the best of the "Concours" (central exam for the universities) are admitted here. The medical university is subdivided into 6 departments as shown in the organization chart (see fig. 3-5):

1. Representative education (Moavenate omure Amuzeshi)
2. Representative research (Moavenate Tahghighat)
3. Representative students and culture (Moavenate Daneshjui-Farhangi)
4. Representative public health (Moavenate Salamat)
5. Representative drugs and food (Moavenate Daru-Ghaza) since 2006
6. Representative support (Moavenate Poshtibani)

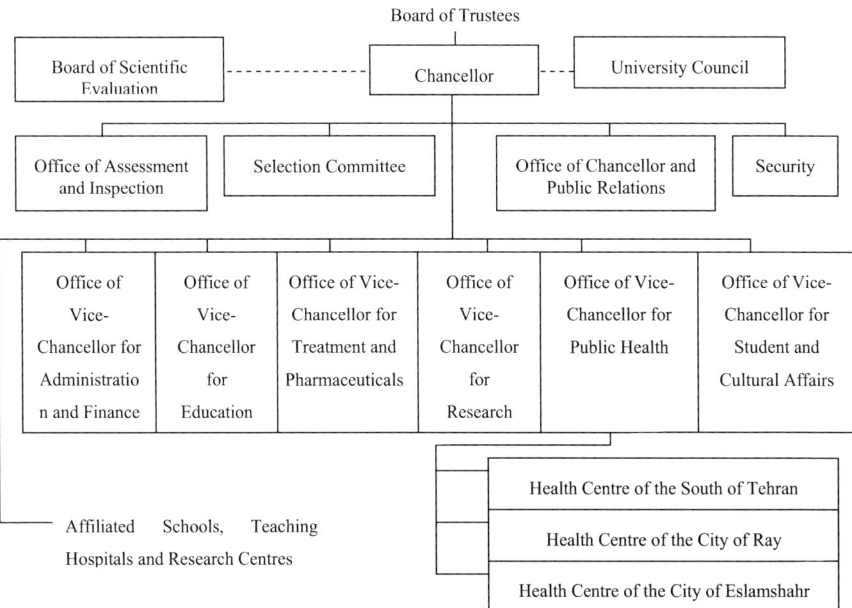

Figure 3-16: Organisation chart of the University of Teheran [TUMS2005]

The „Representative for public health" of the University of Teheran - Moavenate Salamate Daneshgahe Teheran

In 2004 the representatives "hygiene" and "cure" were combined to form a representative "public health". The representation health which will be discussed in the scope of this thesis is split into the following groups and units:

- Group for dental hygiene and oral hygiene
- Group for hygiene in schools
- Group food
- Group hygiene and family planning
- Group of preventive measures, countermeasures to the disease generation
- Group environmental hygiene and environmental specialist
- Unit for development of the Health Houses
- Unit medicine technology
- Unit hygiene training
- Administration / accounting

The first department is the hygiene area whose major tasks are the control of all hygiene services, quality controls as well as the solution of all public hygiene problems in the whole area of responsibility. The second department is the cure area. This area is responsible for the assessment and control of all health facilities and is in charge of the licence assignment of the doctors and all other employees of the health sector.

Once a year an announced control of the hospitals takes place by the employees of the department of hospital assessment of the representative of the public health (see figure 3-17).

The department that consists of experts from different disciplines (catering, laundry, medicine technology, hospital management, hygiene and cleaning) assesses with the help of an interrogative catalogue for the hospitals of the university. With the help of this assessment the hospitals are classified in 5 classes (1 is very good... 5 is unsatisfactory). The hospital is also budgeted and gains its reputation according to this classification. Hence, the ranking plays a very important role for the whole hospital, mainly because their yearly budget is depending on that rank.

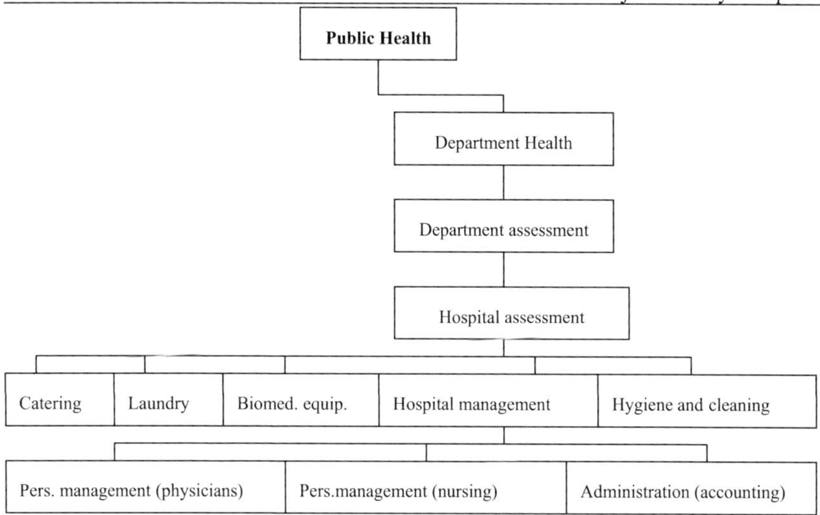

Figure 3-17: Organisation chart of the hospital assessment [AUTHOR2007]

Hospital Sponsorship

The following hospital sponsors can be mentioned in Iran. They are under the care of the respective universities:

- State Hospitals
    - Dependent on the health ministry
        - Educational hospitals
            - State hospitals (Dolati)
        - Non educational hospitals
            - Tamine Ejtemai- social security hospitals („social security organization")
            - Ministry Hospitals (e.g. oil ministry)
    - Independent of the health ministry
        - Army hospitals (Bimarestane Artesh)
            - Pasdaran-hospitals (revolutionary guard)
            - Sepah –hospitals (army of the guardians of the Islamic Revolution)
        - Police hospitals (Niruye Entezami)
        - Municipal hospitals (Bimarestane Shahrdari)
        - Central bank hospitals (Bimarestane Banke Markazi)
- Non State hospitals
    - Private hospitals (Khosusi)
    - Charitable hospitals (Kheyrie)

Hospital financing

The hospital planning is carried out by the ministry or by investors. The entire expenses for the state hospitals (also medicine technology and personnel expenditure) are provided by the state. Construction is financed from the budget of the hospital (calculated by the ministry or the university), from subsidies of the ministry or from donations.

The construction ministry, which owns a special department for hospital building, is mostly involved in the construction. In some cases private contractors are also put in charge of hospital construction. However, in both cases the technical office of the university or the ministry takes over the controlling function. This office must also confirm all steps of the construction before they are executed.

4 The research project OPIK

4.1 OPIK-Germany

The initial intention of the OPIK project in 2001 was the study of hospital statistics. This analysis showed that the trend in Germany's health care system leads to less medical centres, less beds and more patients with shorter resting time. (See also chapter 2)

To ensure the "survival "of the hospitals, especially in view of the huge cost pressure, a detailed knowledge of the processes in hospitals is one of the most important premises. This relevance of the process quality as well as the Benchmarking as a "best practice method" stood in conjunction with the successful introduction of the DRG. The adoption of these "Diagnosis Related Groups" in 2003 was directly linked with the process orientation of the facility management in hospitals [LENNERTS2002a].

Thus the research project OPIK has originated from the necessity of a detailed analysis of processes in hospitals. Indeed, a large number of process analyses from different sources already existed, but none presented a comprehensive standard for the clinic scenery in Germany.

This was the setting for the planning of the research project "optimization and analysis of processes in hospitals" OPIK established by the professorship of Facility Management at the University of Karlsruhe (TH) in cooperation with the professional union for hospital technology (Fachvereinigung für Hospitaltechnik (e.V.) (FKT)).

The research project is financed by industrial partners/ service providers and supported with their know-how. The cooperation with the industry and hospitals was chosen in order to be able to apply learning through cooperation with business partners, to fulfil their requirements and to guarantee a quick market penetration of the research results. The heterogeneous character of the participating partners was the guarantee for an interdisciplinary exchange of specialist knowledge which created a unique connection between practice and research in the Health Care System of Germany. [LENNERTS2002b]

The special area of conflict that results from the different perceptions of the participants affected the research results positively because the different positions of the hospitals, service providers and science could be taken into consideration.

Figure 4-1: Partners of the OPIK- Project (2007)

The approach can be divided into four phases. In the first phase the processes were described, visualised and standardised by the panel of experts (participating medical centres). Next the cost- and quality factors were determined theoretically for every process step and related interfaces to the "primary field" in the hospital were described. In the third phase data for the cost and quality factors were raised in the participant partner hospitals. Finally the data collection that formed the core of the process analysis was evaluated in the fourth phase.

The standardisation of the process results was implemented in three steps:

1. Analysis and optimisation
2. Versification
3. Development of a standard process

In regular workshops results were discussed with the partners and the process standardisation was sped up. Studies and censuses in the different hospitals were conducted by the scientific staff and supported by research activities and theses. [LENNERTS2002c]

Since 2001 it was possible to analyse the following processes in the research project OPIK:

- *Wartung Medizintechnik- Maintenance of medical equipment -* (نگهداشت تجهیزات پزشکی)
- *Instandsetzung und Wartung technischer Anlagen- Maintenance of technical facilities* - (نگهداشت تاسیسات فنی)
- *Medikamentenlogistik- Logistic of pharmaceuticals -* (پشتیبانی دارو)
- *Logistik medizinischer Produkte- Logistic of medical products-* (پشتیبانی محصولات پزشکی)

- *Reinigungsmanagement- Cleaning Management-* (مدیریت نظافت کاخداری)
- *Sterilisierung– Sterilisation-* (استریلیزه)
- *Abfallbeseitigung- Waste Management-* (مدیریت پسماندهای جامد)
- *Abwasserreinigung- Waste Water Management-* (مدیریت پسماندهای مایع)
- *Organisation- Organization-* (مدیریت ساختار سازمانی)
- *Hausmeisterdienste- Service Management-* (سرویس دهی)
- *Energiemanagement- Energy Management-* (مدیریت انرژی)
- *Reparaturdienste- Repair Management-* (تعمیرات ساختمانی)
- *Catering- Catering-* (مدیریت خوراک)
- *Wäscherei- Laundry Management-* (مدیریت رخت شویی)
- *Telekommunikationsmanagement- Telecommunication Management-* (مدیریت مخابرات)
- *IT-Management- IT- Management-* (مدیریت رایانه ای)

Analysed processes in this PhD-thesis

4.2 OPIK- Iran

The project OPIK-Iran started in February 2006 with the support of the Public Health Institute (Moavenate Salamat) of the Medical University of Tehran.

All process partners (service enterprises, subcontractors, management, representatives of the Health Institute, the Medical University of Tehran and the Ministry of Health and Medical Education MOHME) were invited to the kick-off meeting in Tehran. At this juncture the different perceptions, the unequal level of knowledge and the different use and appliance of the processes became apparent to the responsible persons and different hospitals.

Figure 4-2: Workshop- OPIK-Iran

Special workshops with specialists were arranged for three selected processes (maintenance of medical equipment, maintenance of technical facilities, laundry) to determine the standard for the Iranian process and the framework. Therefore 3 hospitals of Tehran University were chosen as "pilot hospitals".

The project was split into two parts, similar to the German parent project, a theoretical and a practical one. The theoretical part analysed the current status of the processes, which were developed in respective workshops. The practical part, the data collection in the pilot hospitals, was accomplished with the help of 10 students of the Medical University of Tehran, because most of the needed data was not documented and the information had first to be registered.

After the questionnaires (see Appendix) had been translated and adapted to the existing boundary conditions (legislative regulations, cultural assimilation, etc.) the data of 20 days was collected. To make the values comparable with each other, a data collection period (20 days) equal to that in the German parent project was chosen.

4.2.1   The project hospitals[17]

The project hospitals are introduced briefly. All of them are teaching hospitals of the medical University of Tehran, located in the center of the capital. [18]

4.2.1.1 The Imam-Khomeini Complex– Mojtamae Emam Khomeini[19]

The „Imam-Khomeini complex", famous as the „1000 bed hospital" (largest hospital in Iran) includes three self managed hospitals

- the Imam Khomeini Hospital
- the Vali Asr Hospital
- the Cancer Institute and
- the Medical Imaging Center.

Figure 4-3: Imam Khomeini Complex

---

[17] All information is from [MOHME2006]

[18] Detailed Information about the hospitals you can find in table 5-1

[19] The Imam Khomeini Complex joined this project, because the laundry of the whole complex is washed in one laundry centre (so the laundry of the Vali-Asr Hospital is washed there too).

The complex covers a tract of around 250.000 square meters and serves around 550.000 inpatients and outpatients every year. With various medical research centres, such as reproductive health, science and technology in medicine, neurology and Iranian Tissue Bank, etc. this complex ranks among the most important health centres in the Middle East.

4.2.1.2 The Vali-Asr Hospital – Bimarestane Vali-Asr

The general hospital known under the name "Dr. Eghbal Hospital" was build in 1975 with the budget of the Oil Ministry (Vezarate Naft).

The total area of the hospital is 20.000 m² on a ground of 28.000 m². The bed capacity is around 350 beds. The staffs include 48 scientific *(Heyate Elmi)* and 212 nursing members. The hospital is situated in the northern part of the Imam Khomeini Complex and contains a research centre as well as a large number of laboratories.

Figure 4-4: Vali-Asr Hospital

4.2.1.3 The Shariati Hospital – Bimarestan Shariati

This general hospital was founded in 1974 on a total area of 72.000 m². 105 scientific and 475 nursing staff members care for 136.000 inpatients and outpatients every year. This teaching and therapeutic hospital with different active research centres and wards is one of the important medical education and research centres in Iran.

For the first time in Iranian medical history heart- and bone marrow transplantations were performed in this hospital.

Figure 4-5: Shariati Hospital

4.2.1.4 The Tebie - Children's Hospital - Bimarestane Tebie Kudakan

The children's hospital was established in 1968 and first named after its founder Dr. Ahari. It is the first paediatrics hospital in Iran that cares only for children and is even today the most comprehensive centre for children in Iran.

It is built on a total area of 38.000 m² and offers 245 beds.

68 scientific members and 124 members of the nursing staff serve annually around 62.000 children.

The specialities of this children's hospital are the immunological, allergy and asthma research centres that are located in that facility.

Figure 4-6: Tebie Kudakan Hospital

Table 5-1 (see page 61) gives a resume of the most important data of the pilot hospitals.

4.2.2 The processes of the project

The selection of the processes was guided by the comparison of the project countries. Therefore technical processes were given a higher priority in order to weaken the extraneous influences, like political, organizational or cultural parameters. Moreover German constructions and machines enjoy a good reputation in Iran. This is one cause why a large number of the technical facilities in Iran are German. That might make a comparison particularly interesting, because the processes in similar technical facilities should be similar or even the same.

4.2.2.1 State of the medical equipment

The first medical equipment engineering department in Iran was founded in the Imam Khomeini hospital in Tehran in 1992, consisting of 35 experienced and suitably educated persons. When in 1996 the division of medical equipment engineering was established under the direction of the Public Health Institute (Moavenate Salamat) nine former employees of the Imam Khomeini hospital started the activities. The remaining employees were reallocated to other hospitals. So today ten of the fifteen hospitals belonging to the administrative district of the Medical University of Tehran exhibit such a department.

Figure 4-7: Medical equipment (Iran)

However, this does not reflect the situation of the country. Around 10% [20] of the hospitals have a medical equipment department. At other hospitals, the duties and responsibilities for the medical equipment are borne by the technical department (Tasisat), the supervision over food and pharmaceuticals (Ghaza va Daru) or a staff member of the medical equipment department of the medical university. This person controls the hospital and acts as a technical adviser for the hospital management.

Two of the three pilot hospitals have the medical equipment engineering that is shown in a dashed box in figure 4-8.

According to a study of the University of Iran only 50% of the available capacities of the medical equipments are used. [ALFAGHED2004]

Reasons for the unused capacities are:

- the lack of specialised staff

- unawareness of the users concerning the important functions of the costly equipment

- available equipment is not put into operation

- wrong use of equipment leads to failures

- spares for the repair of equipment are rare and in most instances must be ordered in bigger cities or even abroad.

---

[20] According to MOHME (Ministry of Health and Medical Education)

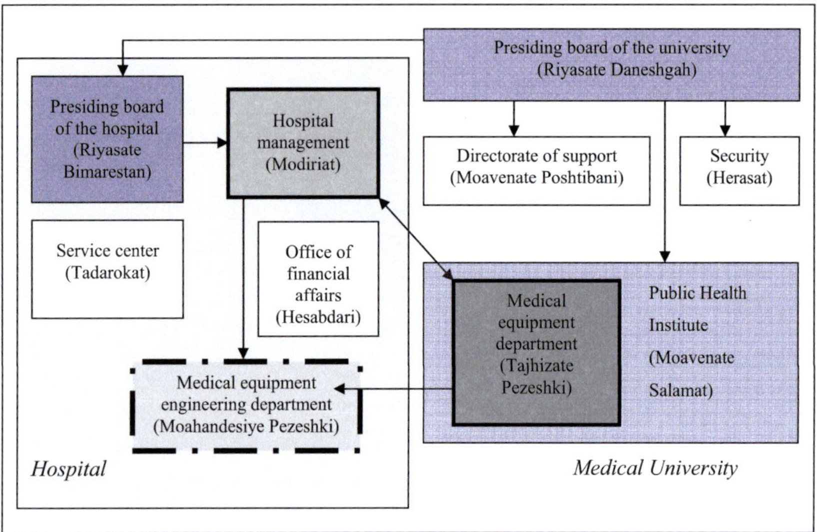

Figure 4-8: Relation of the medical equipment engineering department [AUTHOR2007]

Nevertheless, over the last years the state of the medical equipment in the country has continuously been improved, e.g. degree programs like biomedical engineering and medicine technology (Mohandesie Pezeshki) are offered in many universities and are strongly en vogue the last years. Also the Ministry of Health and Medical Education acted and instituted a committee for medical technology that compiles legislative proposals and basics for this topic.

4.2.2.2 State of the technical facilities

With the construction of modern hospitals, technical facilities also found their way into these buildings and the technical department developed as the "physical heart" of every hospital. Electricity, air, heat, cooling, gas, freshwater and the effluent are managed by this division. Many technical departments also assume the repair of doors, beds, elevators or "janitor services" (e.g. exchange of electric bulbs).

The implementation of these activities, for which the technical department is responsible, is frequently entrusted to subcontractors who are responsible for the services agreed to by contract. These primarily private companies are on the hospital site. They mainly carry out minor repairs or advice the hospital management on technical defects.

One of the pilot hospitals maintains a technical office that counsels the hospital management as an internal technical expert.

Figure 4-9: Technical Department (Iran)

Generally the technical department is controlled directly by the management (Modiriat) or by the technical office (Daftare Fani) of the medical university.

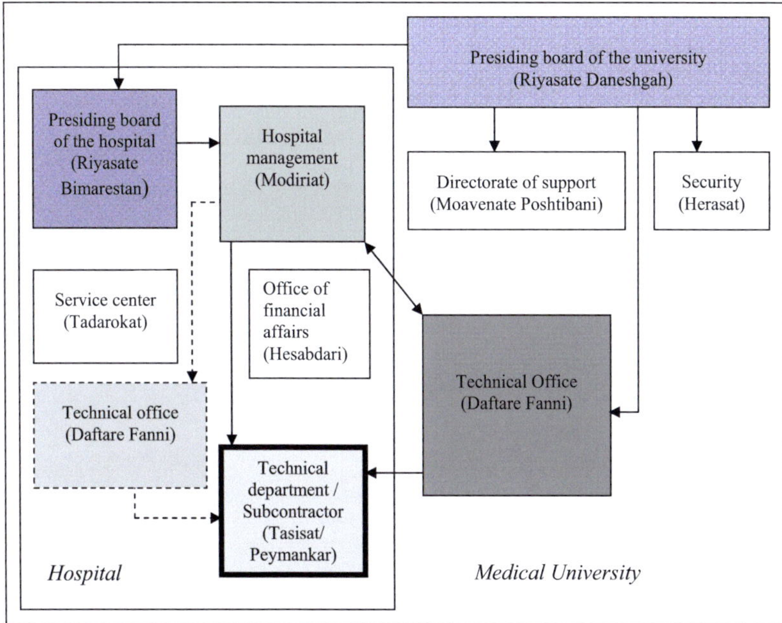

Figure 4-10: Relation of the technical department [AUTHOR2007]

The technical office of the university (Daftare Fani) is responsible for all technical areas of the university. Except for the construction or modification of all public health facilities, this office is responsible for technical inspections as the skilled agent of the medical university.

Figure 4-10 shows all parties who are involved in the process of the "maintenance of technical facilities" at different levels, the university level and the hospital level. The dashed line shows the hospital with the special arrangement as described above. The arrows show the "command" flows in and between the different levels.

As mentioned in table 5-1 (page 61) the analysed pilot hospitals are older objects possessing old technology with accordingly old installations. Therefore most of the machines are overage and need more attention, repair, spare parts and time. As described at length in chapter 5, maintenance is done rarely, because the employees of the technical department are mostly involved with the repair or have not been instructed to perform maintenance work.

4.2.2.3 State of the laundry

In the course of privatisation the trend is to entrust the laundry to subcontractors. The state at the pilot hospitals confirms this trend; one laundry has already been outsourced.

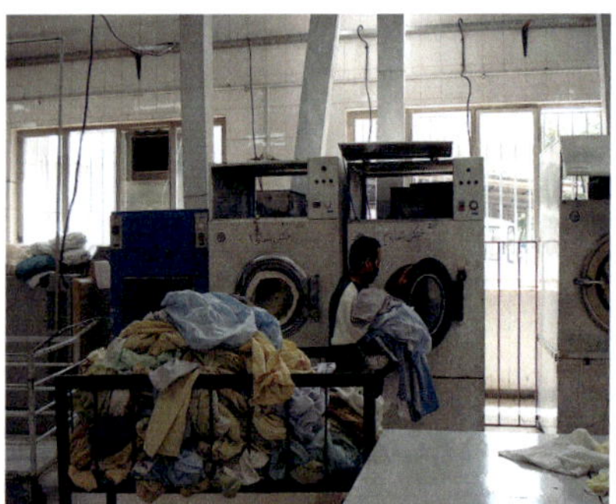

Figure 4-11: View in the laundry (Iran)

Interesting is that contrary to the German pendant Iranian laundries only wash sheet laundry and lab coats. Only in some special cases the clothing of the nursing staff is also cleaned. Generally, the staff themselves must take care of the cleanness of their work clothes, what is controlled by their supervisor. The responsibility of the cleanness of the laundry in the department is carried by the nursing staff. Intern this is controlled, by the hygiene department if one exists, otherwise by the management. (cf. dashed box in figure 4- 12)

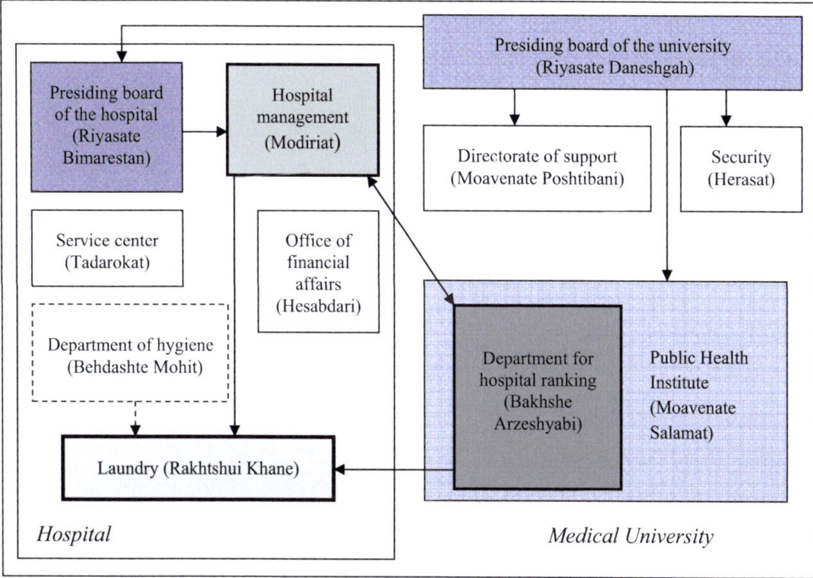

Figure 4-12: Relation of the laundry [AUTHOR2007]

Externally the hospital laundry is controlled by the department of hospital ranking that reports to the Health Institute of the university.

Annually this department furnishes an opinion in which the hospitals of their administrative area (cf. chapter 3) are ranked. At this point the hospitals are classified in "categories" graded from 1-5. These ranking forms the basis for the budgeting as well as the accounting of services rendered. The significance of the laundry is given more emphasis in the expert report than the two other processes (technical facilities and medical equipment).

5 Evaluation and relevance of the project results

This chapter represents the results of investigation of the processes *maintenance of medical equipment, maintenance and repair of technical facilities* and *laundry management* of the OPIK- Project in Iran and in Germany.

The processes which have been investigated are described, evaluated and compared with each other. Similarities, common features and differences are analysed. Especially the question of the differences is discussed, and the causes for these differences as well as the influences of various parameters are considered.

First the general questionnaires (cf. Annex A-1) which were filled in by all three Iranian project hospitals are evaluated. Next the process description is represented and explained. The scope of applications, the customers and the aims of the process are treated and the process responsibilities are defined. After that the details of the life cycle of the processes in both countries are described and compared with each other. Then the processes are analysed, the process data are evaluated and their significance is presented. Finally the expertise which compares the processes is summarised in the resume.

The German evaluation of the OPIK-Project is based on the results of Mr. Jochen Abel (medical equipment), Mr. Uwe Pfründer (technical facilities) and Mrs. Karin Diez (laundry management) and their workgroups [LENNERTS2001-2007d].

5.1 General results

Table 5-1 summarises once more the most important data of the pilot hospitals in Iran. The data are valid for the Iranian year 1384[21] (this corresponds to march 2005-march 2006).

The results of the questionnaires (cf. Appendices-A1 and -A2) are given below. Besides organizational, technical and logistical questions, problems of the appreciation of facility management and optimisation potentials are treated in order to give an overview of the existing state of the hospitals.

---

[21] The Iranian calendar is a solar calendar, also called "Dschalali" calendar. The term also known beyond Iranian borders "solar calendar after the emigration" [hidschri schamsi] goes back to the fact that the calendar begins with the emigration of the prophet Mohammad. The actual Iranian calendar is an advancement of the Nouruz-Nameh-calendar (Nouruz-Nameh = book for New Year) of Omar Khayyam. It was implemented on March 31st 1925.The first day of the Iranian calendar is determined by the astronomical beginning of spring. This occurs in the Gregorian calendar between March 19th and March 21st.

Hospital	Tebie Kudakan	Vali-Asr	Shariati
Founding year	1968	1975	1974
Ranking	1	1	1
Number of hospital beds	245	365	830
Number of inpatients 1384	62.000	12.011	136.000
Bed occupancy [%]	82%	75%	86,7%
Number of employees	600	566	1.070
Number of departments	12	11	38
Number of operation rooms	2	9	14
Number of sterilization departments	2	1	n.s.[22]
Area of the site [m²]	3.600	28.000	72.000
Area of the hospital [m²]	n.s.	20.000	33.247
Annual budget [23] [Milliarden Tooman]	1.2	4,4	n.s.
Number of computers	150	95	250
Number of TV	40	40	n.s.
Heat consumption [kWh]	n.s.	n.s.	n.s.
Electricity consumption per month[kWh]	253.125	903.740	n.s.
Water consumption per month [l]	4.179.000	2.776.658	n.s.
Waste production per month [t]	160.000	n.s.	n.s.
Value of the medical equipment [Tooman]	800 Mio.	n.s.	n.s.
Motor pool (number of ambulances)	2	5	n.s.

Table 5-1: Overview of the general questionnaire [AUTHOR2007]

---

[22] Not specified

[23] 1Euro is equivalent to around 1200 Tooman (2006)

The evaluation of the questions concerning management issues (cf. Appendix-A3) of the three hospitals leads to the following conclusions:

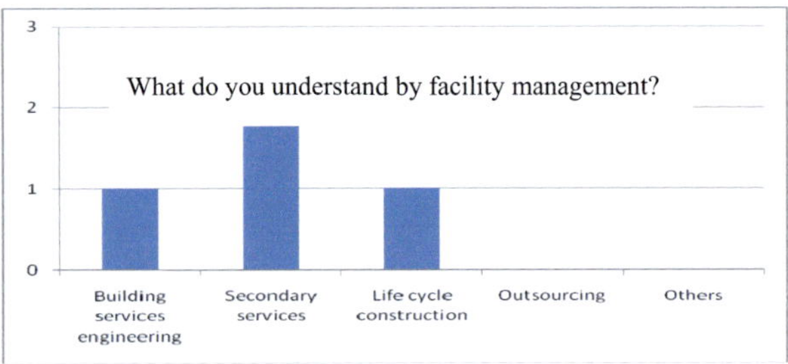

Figure 5-1: Concept of facility management [AUTHOR2007]

Facility management is predominantly regarded as a secondary service in the project hospital as can be seen in figure 5-1, while one hospital 1 took the answers secondary service and life cycle construction, hospital 2 decided building service and hospital 3 building services engineering.

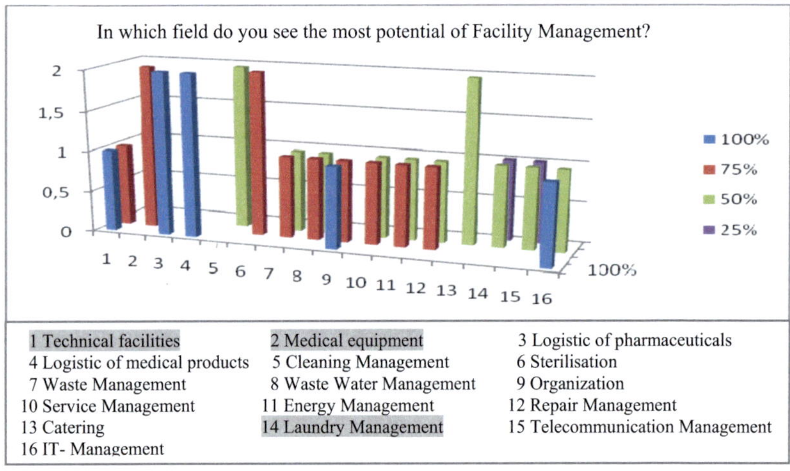

Figure 5-2: Potential of facility management [AUTHOR2007]

The answer of the potentials of facility management in their hospitals is only answered by two hospitals. As can be seen in figure 5-2 these are logistics of pharmaceuticals and medical products that are chosen by both hospitals. One 100% is given by one hospital for maintenance of technical facilities, organization and IT-management. Least potential is seen in laundry and telecommunication management.

All three hospitals have already experiences with process analyses, however, these were carried out primarily for nursing services (see table 5-2).

	medical service	nursing service	administration	facility management
Number of hospitals	1 of 2 hospitals	2 hospitals	1 of 2 hospitals	1 of 2 hospitals
Occurred process	medical training	training of the nursing staff	patient admission	maintenance of medical equipment

Table 5-2: Earlier management analysis experiences [AUTHOR2007]

The ranking of the skills expected of the facility management staff was headed by leadership skills and technical know-how, whereas commercial know-how takes the last rank. Potential for optimisation is seen in the maintenance of technical facilities and medical equipment.

5.2 Maintenance process of medical equipment

5.2.1 Process description

Application area of the process

The process *maintenance of medical equipment* contributes to the flawless supply of medical equipment in hospitals. In Germany the process is applied for all technical devices that fall under the law on medical products (Medizinproduktegesetz (MPG)). In Iran such a law is presently (2007) being drafted by a commission of the Ministry of Health and Medical Education (MOHME).

Definition of the customer

Customers of the process „life cycle of medical equipment" are several types of persons. Passive customer of this process is, primarily, the patient for whose healing medical equipment is employed. The patient does not intervene actively in the process; he is a beneficiary of suitable appliances.

Primary customers of the process are the medical and the nursing service that need medical equipment for their own processes.

The customer in this process can be involved as a user (medical and nursing service), as a guarantor (e.g., equipment management officer) or as an orderer (e.g., department that needs special medical equipment).

Aims of the process

Aims through the eyes of the customer

From the point of view of the patient the guarantees of optimum support of the healing process is of primary importance. Important for the recovery process is the trust in the capacity of the technique. From the point of view of the hospital operator the preparation of the processes and the minimisation of the liability risk remain paramount.

Aims through the eyes of the organisation

By application of the process, maintenance according the guidelines is guaranteed in Germany. In Iran the subject of maintenance is not yet top ranked. Presently the aim of the operators is still the ability to guarantee the services of the hospital, which primarily depends on the repair of the available equipment.

Aims through the eyes of the employees

The process should support the employees in performing the maintenance services at the required quality level.

Process responsibility

The organisation unit that is assigned to the medical equipment is responsible for the care, advancement, observance and control of the processes. In Germany the following laws, regulations and requirements must be observed:

- law of medical products (Medizinproduktegesetz (MPG))

- regulations for medical product operation (MPBetreibV)

- radiation control regulation (Röntgenverordnung (RöV))

- regulation of medical products (Medizinprodukteverordnung (MPV))

- radiation Protection Ordinance (Strahlenschutzverordnung (Str.SchV))

- regulation of detection, assessment and averting of risks with medical products (Verordnung über die Erfassung, Bewertung und Abwehr von Risiken bei Medizinprodukten)

- requirements concerning hygiene in connection with the processing of medical products (Anforderungen an die Hygiene bei der Aufbereitung von Medizinprodukten)

As described in chapter 4 only around 10% of the hospitals in Iran have a department for medical equipments. Laws and regulations do not yet exist officially. The Ministry of Health and Medical Education has appointed a committee for medical technology that is compiling legislative proposals and basics.

Activator of the process

Activator of the process life cycle of medical equipment is the procurement of a medical instrument.

Description of the process

Germany

Procurement

In the sub-process *procurement* the agreement between the parties is very important. Essentially four parties are involved in this process; these are in general the device user, the financing unit, the supplier and the department of medical equipment. For process purposes the financing of the maintenance costs for this sub-process is very important. During the process the device-specific maintenance strategy is determined in agreement with the financing unit.

Installation

The medical equipment is installed in accordance with the regulation of medical product operation (MPBetreibV).

Operation

The *operation* is based upon the law of medical products (Medizinproduktegesetz (MPG). Responsible for this process is the operator in collaboration with the user. Parts of the operation (e.g., training, checking according to the radiation control regulation (Röntgenverordnung (RöV)) can be delegated.

Maintenance and inspection

Maintenance and inspections are generally amenable to planning. According to the legal guidelines (e.g. radiation control regulation (Röntgenverordnung (RöV)) inspections are on-site. To guarantee a smooth execution of the process, it is of great importance to inform customers about the performance of maintenance and inspection work in time. This makes it possible for the customers to change appointments.

An essential component of maintenance and inspection is the inspection analysis. Only a careful analysis of the inspection results assures timely activation of the maintenance orders and makes it possible to adapt the service intervals to safety related and commercial aspects. The optimization of the individual sequences concerning the pure service provision is determined in the contract deviations analysis.

Corrective maintenance

Corrective maintenance is initiated by a notification of a maintenance requirement. The notification concerns a service requirement and is issued for the purpose of service management.

During investigation of a claim it is attempted on the one hand to find the cause of the damage, while it is checked on the other hand whether the claim concerns a subject that must be registered. Should it concern such a case, e.g. §3 of the MPBetreib (requirement of medical product operation) or §42 RöV (radiation control regulation (Röntgenverordnung)), the appropriate authorities and the executive board must be informed.

The economic efficiency of corrective maintenance must be verified before it is carried out. The cause analysis prevents repeated repairs that become necessary if the real cause of the damage is not removed. Damage can be caused by design errors and by handling errors. If not all causes are removable, the basic causes should be identified in agreement with the user. This decision must be based on an analysis of profitability. The causes must be documented. If corrective maintenance is not profitable, the device must be shut down or the purchase of a replacement must be initiated. A functional check according the legal guidelines (e.g.

engineered safety control (STK Sicherheitstechnische Kontrolle)) is part of the resumption of service.

Shut-down

Every individual device of medical equipment must be shut down in an adequate and orderly manner. It is important to ascertain whether it was necessary to obtain a permit for the operation of the device or not. If there is an obligation to obtain a permit the law of medical products (Medizinproduktegesetz (MPG) § 25 "General disclosure duty ") must be applied.

The respective regulations for shut-down have to be considered (see also Law for Promoting Closed Substance Cycle Waste Management and Ensuring Environmentally Compatible Waste Disposal[24].

Furthermore the repurchase by the producer and the resale or utilisation by a third party can be negotiated. Particular attention should be devoted to a check of the maintenance contract to avoid unnecessary payments.

Iran

Procurement

According to the regulations of the Ministry of Health and Medical Education, procurement is the task of "purchasing committees "(Komitee-ye-Kharid) (see also chapter 4.2) that consist of the hospital director, the hospital management, the department of financial affairs and a representative of the technical staff, as well as the head or leader of the medical equipment department if such a department exists. Purchasing does not follow a strategy or annual plan of the hospital or the university. Most of the purchasing is the result of short term demand particularly by request of the doctors. In comparison to all other steps of the process, procurement is subjected to the strictest laws, which are designed to support the domestic biomedical technology market. Before purchase of a device, it must be determined whether such a device is produced in the country. The second-hand import of medical equipment is officially forbidden. Because such import nevertheless occurs, big problems arise particularly with respect to guarantee, maintenance and repair contracts.

---

[24] (Gesetz zur Förderung der Kreislaufwirtschaft und Sicherung der umweltverträglichen Beseitigung von Abfällen (KrW-/AbfG))

Installation

The medical equipment is installed by the medical equipment department or by a subcontractor who is selected on the basis of submitted tenders. For bigger and costly devices this tendering is frequently associated with a time consuming, expensive und primarily bureaucratic process.

Figure 5-3: Medical equipment department (Iran)

Operation

A law for the operation of medical equipment is being prepared. At present the operation is guided by experience or training. It must, however, be pointed out that training is very rare and does not take place for all users. Especially the devices in the teaching hospitals suffer from the high fluctuation of the staff (residents).

Maintenance and inspection

For some very expensive devices service contracts are concluded with commercial enterprises in the country. If the servicing company does not have a representation in the country, a similar local company is selected from a list of advertisers. For this special purpose, the university department of medical equipment publishes the "yellow pages for biomedical technology". In addition, the Ministry of Health and Medical Education forges a classification and registration of maintenance and servicing companies.

73% of the companies for maintenance and servicing in the pilot hospitals of the OPIK-project were subcontractors, but no maintenance was performed for approximately 90% of the devices.

Repair

The corrective maintenance and repair of the devices is performed by the internal departments. If a device is repaired by an external company, this is done outside the hospital if possible. The cause of damage is usually not determined. Compared with the German process, the analysis of economic efficiency does not have a high priority.

Shut-down

Before a piece of equipment is shut down, the requirements of the hospital itself and of the other hospitals of the university are determined. If a device is no longer usable, it is either disassembled in the hospital or stored in the "spare part camp" or sold on "biomedical bazaars".

Comparison of the Processes in Germany and Iran

The flow of the overall process and the accompanying partial processes are shown in the accompanying flowcharts of Appendix-A8. The most remarkable differences are summarized in the following table.

Process step	Germany	Iran
Procurement	• device-specific maintenance strategy is determined in agreement with the financing unit • Costs of maintenance are calculated	• No strategy or annual plan • short term demand • particularly by request of the doctors
Installation	• according to the rules of medical product operation (MPBetreibV)	• follows the operating instructions of the manufacturer
Operation	• law on medical products (Medizinproduktegesetz (MPG))	• operation follows experiences and training • a law is in development
Maintenance and inspection	• training, check according the radiation control regulation (Röntgenverordnung (RöV)) • Inspection analysis • Analysis of the maintenance	• only takes place for few, expensive types of equipment
Corrective maintenance	• economic efficiency • cause analysis	• no cause analysis
Shut-down	• repurchase by the producer • regulations for shut-down	• spare part camp • biomedical and scrap metal bazaars • transmission, sales to hospitals that can use it

Table 5-3: Comparison of processes for medical equipment [AUTHOR2007]

5.2.2 Characteristic variables of the processes

The characteristic variables of the process steps are shown in Appendix- A10. The decisive factors for the total cost of the maintenance [DIN31051] in Germany are the partial processes "maintenance and inspection" and "corrective maintenance". The cost allocation between the processes can vary and to some extent reflects the maintenance strategy. As already mentioned, maintenance is considered to be of less importance in Iran. This is the reason why purchasing still plays the decisive role.

Cost factors*:

- Costs of maintenance according to the replacement value of the installed equipment

- Purchase costs according to inpatient days

Quality factors*:

- Number of the registered interventions with medical equipment

[* Operation figures OPIK-Germany and Iran]

5.2.3 Interfaces of the process

The interfaces of the process in the project countries were investigated as well and are presented in Appendix-A9. Evident in the comparison in Iran there is a higher bureaucratic complexity that primarily leads to loss of time.

5.2.4 Process data

Scope of the process data

For the purposes of data collection, the process was divided in maintenance/ inspection and repair. In Germany 24 reports for maintenance / inspection and 187 reports for repair were collected and entered in the data base (Appendices-A6 and -A7). This corresponds a performance ratio of 40,8 % for maintenance / inspection and 51,7% for repairs. The allocation to the respective hospitals can be seen in table 5-4 and 5-5.

Germany	KH 1	KH 2	KH 3	KH 4	KH 5	KH 6
Registered days – debit [WD[25]]	20	20	20	20	20	20
Registered days – actual [WD]	19	10	20	0	0	0
Registered days –WD [%]	95%	50%	100%	0%	0%	0%

Table 5-4: performance ratio for maintenance / inspection (Germany) [LENNERTS2001-2007d]

Germany	KH 1	KH 2	KH 3	KH 4	KH 5	KH 6
Registered days – debit [WD]	20	20	20	20	20	20
Registered days – actual [WD]	19	10	20	13	0	0
Registered days –WD [%]	95%	50%	100%	65%	0%	0%

Table 5-5: performance ratio for repair (Germany) [LENNERTS2001-2007d]

38 reports for maintenance / inspection and 197 reports for repair were gathered from the Iranian pilot hospitals. For maintenance / inspection this corresponds a performance ratio of 23,3 %; for repair the performance ratio is 100%.

Nevertheless, it must be mentioned that 2 of the 3 project hospitals did not conduct maintenance.

Iran	KH 1	KH 2	KH 3
Registered days – debit [WD]	20	20	20
Registered days – actual [WD]	0	14	0
Registered days –WD [%]	0%	70%	0%

Table 5-6: performance ratio for maintenance / inspection (Iran) [AUTHOR2007]

Iran	KH 1	KH 2	KH 3
Registered days – debit [WD]	20	20	20
Registered days – actual [WD]	20	20	20
Registered days –WD [%]	100%	100%	100%

Table 5-7: performance ratio for repair (Iran) [AUTHOR2007]

---

[25] WD- working days

5.2.5   Process analysis

Maintenance and repair analysis

Work orders for the maintenance of medical equipment were grouped according to device types. Unfortunately the number of the service reports does not suffice for representative results. The following figures show box-plots of the total transit time in hours for different types of devices. The number of the service reports for each device type varies from 10 to 30. Considering the standard deviation and the expected total number of orders for each type of device per year, the number of service reports should be at least twice as high.

The results of the evaluation are shown by means of the box-plots. The accompanying table also shows the data in absolute figures. The minimum and maximum values represent the orders with the lowest and the highest total transit time in hours or total costs. The quantiles specify thresholds which section the range of the values. In the case of the maintenance of monitors (see figure 5-3) this means that 5% of the orders needed less than a 30 minute transit period. 25% of the orders needed at most a 60 minute transit period. The 75% and 95% quantiles are interpreted in a similar manner. The median represents the factor that divides the range into two equal parts. In other words the median specifies the transit period or cost that is exceeded in 50% of the cases. The lines right and left of the box represent the dispersion of the orders. In this way the box-plot shows if a normal distribution or a left or right incline is present. With the knowledge of the median transit period and median costs an excellent process- orientated input base for the evaluation of performances is available. The box-plots are also very suitable for the calculation of the working hours of the medical equipment staff.

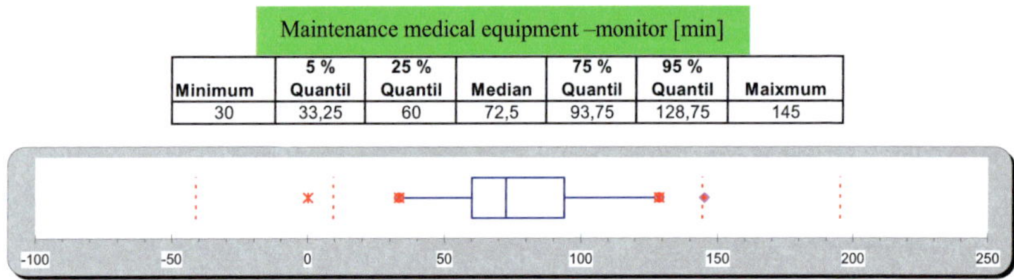

Maintenance medical equipment –monitor [min]						
Minimum	5 % Quantil	25 % Quantil	Median	75 % Quantil	95 % Quantil	Maixmum
30	33,25	60	72,5	93,75	128,75	145

Figure 5-4: Box-plot maintenance medical equipment- monitor [LENNERTS2001-2007d]

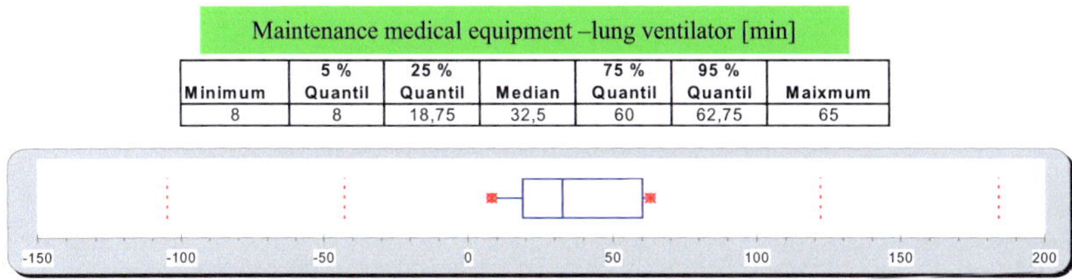

	5 % Quantil	25 % Quantil		75 % Quantil	95 % Quantil	
Minimum			Median			Maixmum
8	8	18,75	32,5	60	62,75	65

Figure 5-5: Box-plot maintenance medical equipment- lung ventilator [LENNERTS2001-2007d]

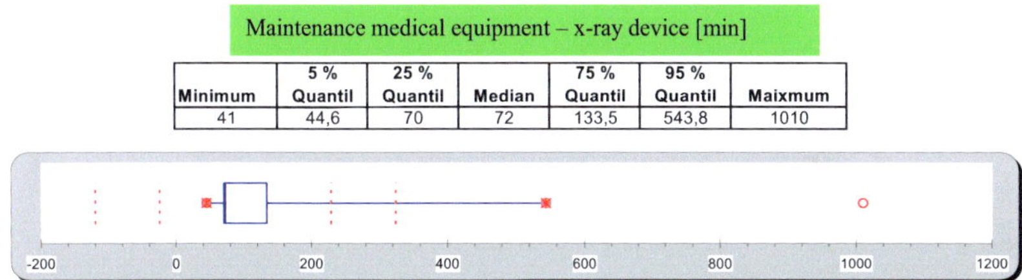

	5 % Quantil	25 % Quantil		75 % Quantil	95 % Quantil	
Minimum			Median			Maixmum
41	44,6	70	72	133,5	543,8	1010

Figure 5-6: Box-plot maintenance medical equipment- x-ray device [LENNERTS2001-2007d]

In the Iranian hospitals the maintenance reports were not grouped according to device types but analysed by hospital.

Maintenance: Only one of the hospitals conducts any maintenance. 16 of the 36 delivered questionnaires contained data for the maintenance period. The average value was 10 minutes, as can be seen in Figure 5-7. Too few data were given for the statistical variation of the "quality", so that no conclusions could be draw.

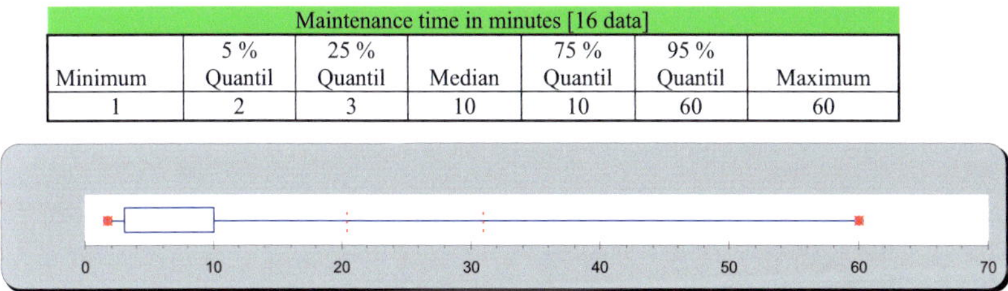

	5 % Quantil	25 % Quantil		75 % Quantil	95 % Quantil	
Minimum			Median			Maximum
1	2	3	10	10	60	60

Figure 5-7: Maintenance time [AUTHOR2007]

Repair: For the repair only repair time and repair costs (spare parts) data were collected.

Hospital 1

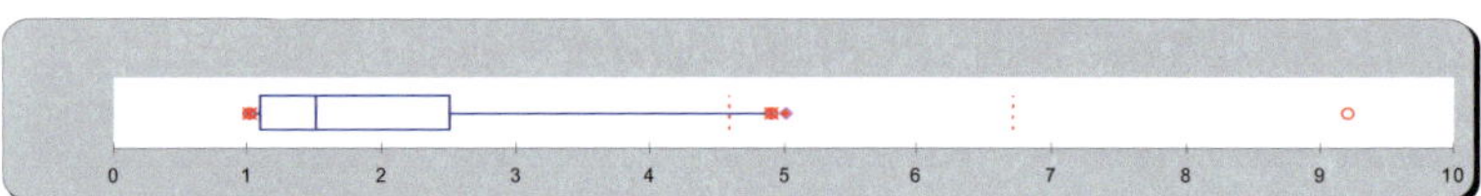

Hospital 1: repair costs [thousand Tooman] [64 data]						
Minimum	5 % Quantil	25 % Quantil	Median	75 % Quantil	95 % Quantil	Maximum
0	1	1	2	3	5	9

Figure 5-8: Hospital 1-repair costs [AUTHOR2007]

Hospital 1: repair time in days [44 data]						
Minimum	5 % Quantil	25 % Quantil	Median	75 % Quantil	95 % Quantil	Maximum
1	1	1	1	2	3	7

Figure 5-9: Hospital 1-repair time [AUTHOR2007]

The following statements about the maintenance quality were given by the hospital:

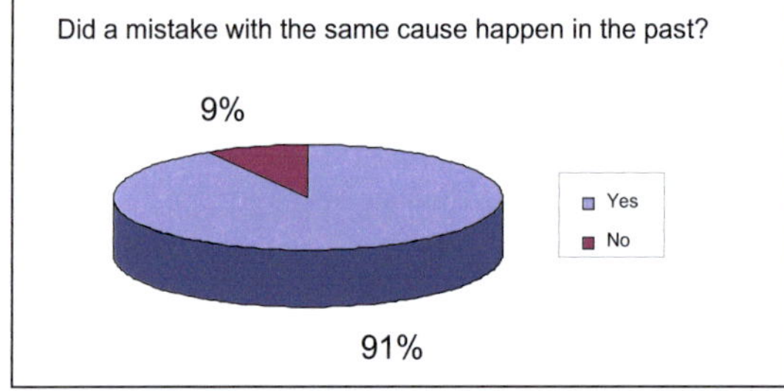

Figure 5-10:

Hospital 1- quality question1 [AUTHOR2007]

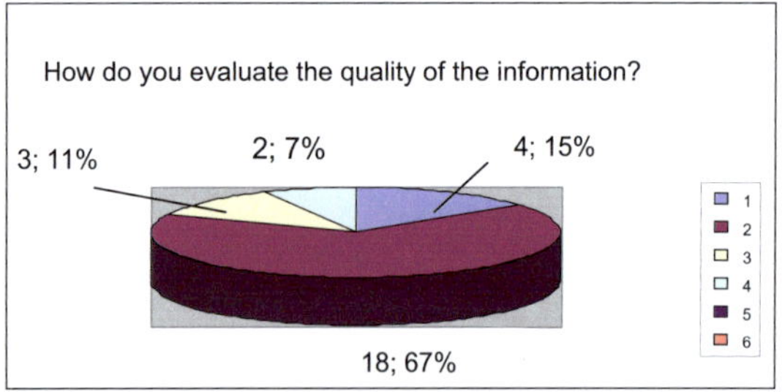

Figure 5-11:

Hospital 1- quality question 2 [AUTHOR2007]

Figure 5-10 shows that for more than 90% of the mistakes, the same mistake had already happened in the past. The question "How many trials were required to complete the order successfully?" was answered two on the average. The quality of the information given by the user was judged with good by 67%.

Hospital 2:

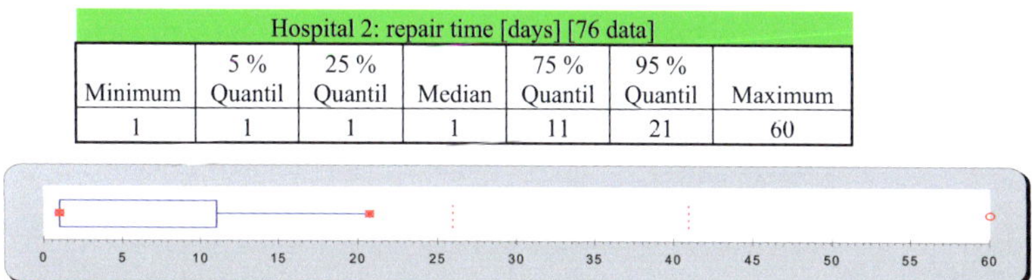

Hospital 2: repair time [days] [76 data]						
Minimum	5 % Quantil	25 % Quantil	Median	75 % Quantil	95 % Quantil	Maximum
1	1	1	1	11	21	60

Figure 5-12: Hospital 2 repair time [AUTHOR2007]

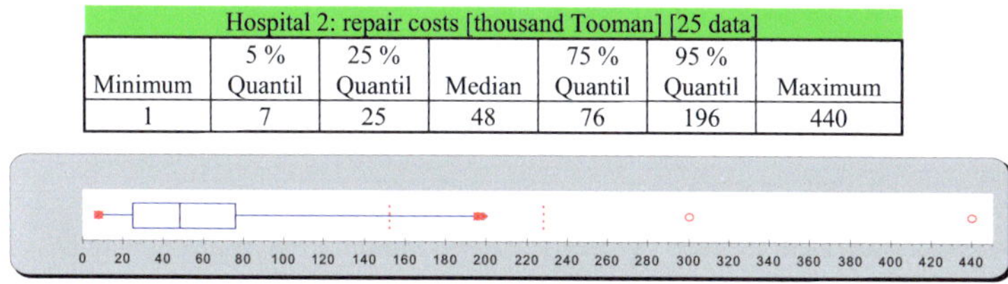

Hospital 2: repair costs [thousand Tooman] [25 data]						
Minimum	5 % Quantil	25 % Quantil	Median	75 % Quantil	95 % Quantil	Maximum
1	7	25	48	76	196	440

Figure5-13: Hospital 2 repair costs [AUTHOR2007]

Hospital 3:

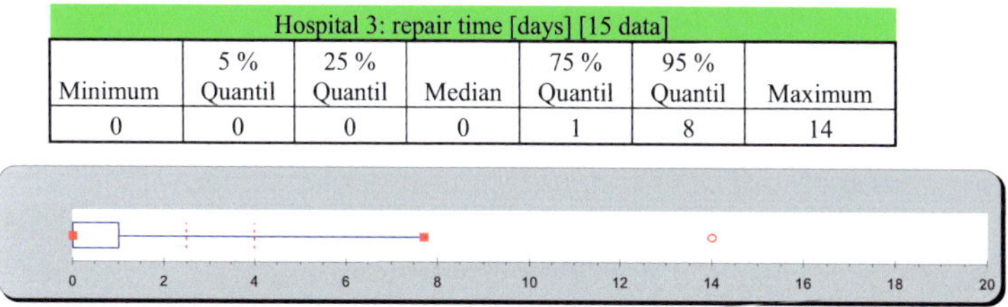

Hospital 3: repair time [days] [15 data]						
Minimum	5 % Quantil	25 % Quantil	Median	75 % Quantil	95 % Quantil	Maximum
0	0	0	0	1	8	14

Figure5-14: Hospital 3 repair time [AUTHOR2007]

The result of the analyses shows that the repair time is not measured in minutes as in Germany, but in days. The average repair time for the pilot hospitals is one day. That shows that the importance of time in Iran is not yet as high as in Germany.

Hospital 3: repair costs [thousand Tooman] [8 data]						
Minimum	5 % Quantil	25 % Quantil	Median	75 % Quantil	95 % Quantil	Maximum
1	1	3	15	38	93	160

Figure 5-15: Hospital 3 repair costs [AUTHOR2007]

Table 5-8 summarizes once again the results of the analysis of the hospitals of the OPIK-Iran project.

	Hospital 1	Hospital 2	Hospital 3
Maintenance time [min]	-	10	-
Repair time [days]	1	1	0 (within a day)
Repair costs [Thousand Tooman]	2	48	15

Table 5-8: Results of the data analyses [AUTHOR2007]

Process flow

The process steps were analysed with respect to labour time, information time and transit time. The result for Germany shown in figure 5-16 reveals that the effort for the activities related to the identification and removal of the causes is low. Claims investigation followed by maintenance and reconnection are rated highly.

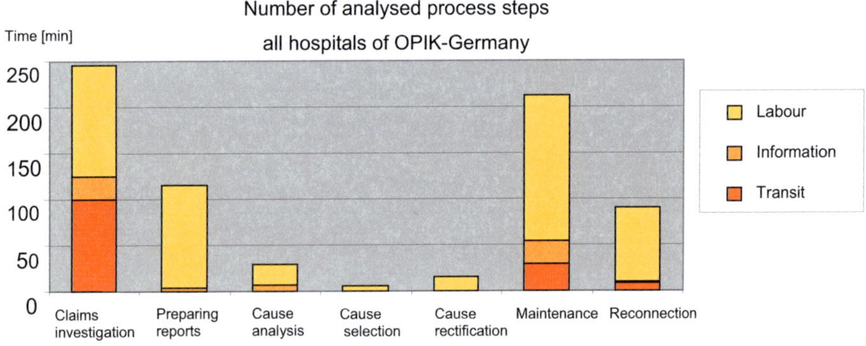

Figure 5-16: Number of the analysed process steps [LENNERTS2001-2007d]

The information exchange accumulates for claim investigation and the maintenance. In general the information exchange is very low and does not exist in all process steps. For optimal process flow purposes the information exchange in the process step maintenance should be reduced and shifted towards inspection and reconnection. It is assumed that information exchange during maintenance disturbs the flow of the primary processes as well as the secondary processes. It is most interesting that the time required for the activities in this process step, as shown in the cause analysis (figure 5-17) is disproportionately high. The process step reconnection is performed on a small scale. This is very amazing because an inspection of the functions after the repair is prescribed in the regulation of medical product operation (Medizinprodukte-Betreiberverordnung (MPBetreibV)).

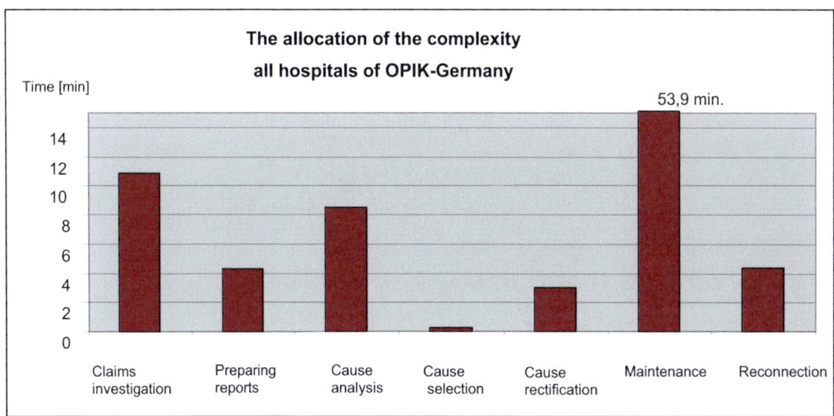

Figure 5-17: Allocation of the complexity [LENNERTS2001-2007d]

In the Iranian hospitals very few data are available for the documentation. The same applies to the cause analysis. The only data that led to significant results could be collected for maintenance and repair (see table 5-8).

Distribution of the expenditures

Germany

A cause analysis requires a large amount of work. The average duration of the process flow without the time required for the analysis is more than 60 minutes. A damage analysis is performed mainly if it concerns a complex or elaborate order. Repair plus analysis on the average requires approximately 80 minutes. In comparison the average time without analysis lies around 45 minutes. Hence, it can be assumed that causes analysis for recurring damage is either not performed at all, or only insufficiently. Experience shows that extensive available data for medical equipment are not used for analysis. The results of the data collection also reflect this situation. At this point in time, the process should be automated to improve the recognition of the causes of recurring damage and device defects.

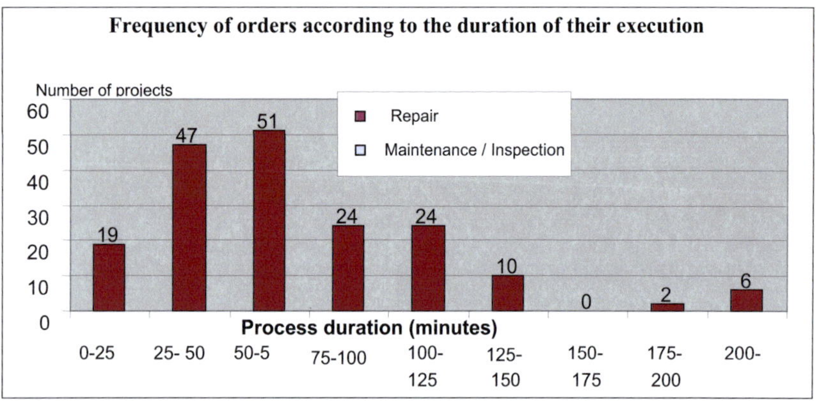

Figure 5-18: Frequency of the orders [LENNERTS2001-2007d]

If the expenditures related to work- , information- and transit time are considered, it becomes apparent that transit- and information time requirements are low (see figure 5-18). The time is needed for the transport of the medical equipment between its location and the garage. The information time requirement can probably be interpreted by simple communication, because the client and the contractor have a similar level of know-how.

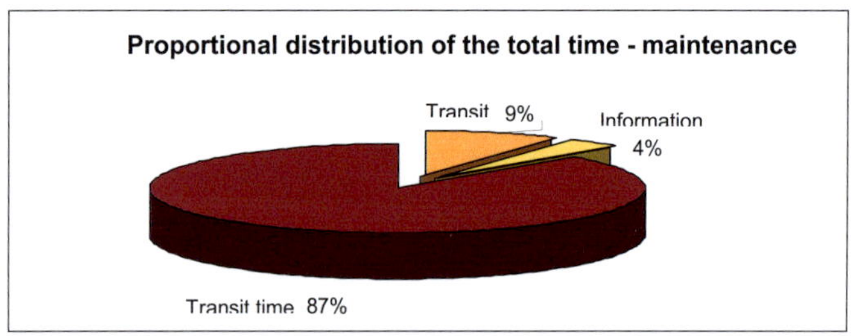

Figure 5-19: Total time – maintenance [LENNERTS2001-2007d]

IRAN

Because it was not possible to gather a sufficient amount of data for the total times, questionnaires for maintenance and repair were prepared and a survey was conducted to obtain the information. All together 23 questionnaires were filled out by the customers of the medical equipment, of these 19 by nursing staff, one by a doctor and 3 by other hospital staff (see figures below). In this context the grading begins with 1 (very good / very important) and cnds with 6 (very bad/ not important).

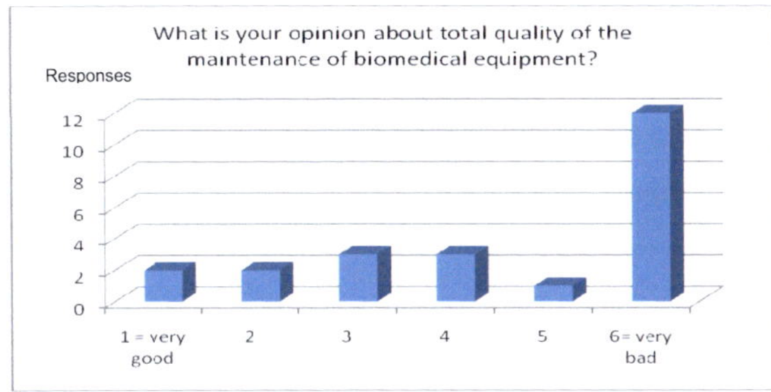

Figure 5-20:

Result question 1

Figure 5-21:

Result question 2

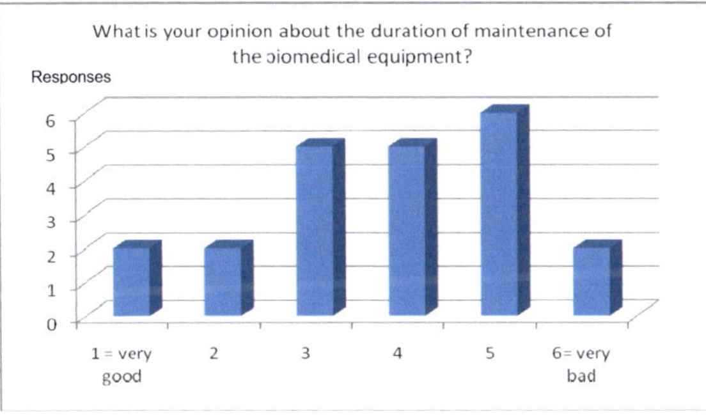

Figure 5-22:

Result question 3

The answers to question1 point at the discontentment of the customers with the quality of the maintenance. The customers are also dissatisfied with the high training needs and the duration of the maintenance as can be seen in figure 5-21.

5.2.6 Summary medical equipment

Germany can draw on long standing experience provided by a medical branch of technology that is leading worldwide. Numerous unions and federations have been founded that initiate developments in different areas. Laws and regulations have been established that apply to everyday transit time processes and have proved to be suitable. Regular controls by different authorities (public health office …), guarantee the optimal use of the medical equipment.

Maintenance and repair (spare parts) can be carried out at low cost because the companies and suppliers are local. Planning (Maintenance and servicing planning at the time of procurement), analysis (causes analysis...) and the documentation takes an important position.

In Iran, however, laws and regulations are only beginning to be introduced. A large part of the hospitals do not even have a department for medical equipment. The main budget for the medical equipment is used for the purchase of the devices. Maintenance is very rare, and is primarily restricted to very expensive devices. The lack of specialist staff and a low level of training and documentation in addition to a lack of control of the equipment influence the medical equipment technology in particular. Interesting is that in Iran much more departments are involved and integrated in the single process steps, what stretches the single process steps and expends a lot of time and consequently money.

These aspects can be seen in the comparison, too. The processes in Iran are not so detailed or even not existing, for example the maintenance that is only operated in one hospital. Another indicator is the repair time that is calculated in Iran in days while the measure in Germany is minutes. That once again shows the position and the worth of time on the one hand and on the other hand the still low rank of the medical equipment in the Iranian hospitals compared with the situation in Germany.

5.3 Maintenance and servicing process for technical facilities

5.3.1 Process description

Area of application of the process

The process of *maintenance and servicing technical facilities* contributes to the flawless supply of technical services in hospitals. In Germany the process is applied for all technical devices that do not fall under the law of medical products (Medizinproduktegesetz (MPG)). In Iran such a differentiation is not used. There are regulations for each hospital that are controlled by the technical office of the university, but there are no laws for the country as a whole.

Definition of the customer

Various types of persons are customers of the process „life cycle of maintenance and servicing technical facilities". Passive customer of this process is, primarily, the patient for whose healing a suitable environment including the technical equipment should provide a healthy, friendly and hygienic room climate. The patient does not intervene actively in the process; he is a beneficiary of suitable equipment and facilities.

Primary customers of the process are the medical service and the nursing service that need medical equipment for their own processes. Responsible for the correct function of the technical facilities are the appropriate qualified units (e.g. technical department, technical chief, external service or subcontractor). The customers in this process can be involved as users (medical and nursing service), as guarantors or as orderers. They can be direct or indirect users.

Aim of the process

Aims through the eyes of the customer

For patient the reliable functioning of the technical facilities is important. From the point of view of the hospital operator the preparation of the processes and the minimisation of the liability risk are of paramount importance.

Aims through the eyes of the organisation

The process is applied in Germany with the aim to guarantee maintenance according the guidelines, technical certifications and norms. The process should be transparent.

Aims through the eyes of the employees

The process should assist the employees to perform the maintenance services at the required level of quality. Important is the transparent implementation of the process.

Process responsibility

The organisational unit that controls the technical facilities and construction is responsible for the care, propagation, observation and control of the processes.

In Germany the following laws, regulations and requirements must be observed:

- DIN 31051 Instandhaltung (maintenance)
- VDI Richtlinie Instandhaltung (guideline maintenance)
- VDI 6022 Hygiene-Anforderungen an Raumlufttechnische Anlagen und Geräte (hygiene requirements to air conditioning plants and devices)
- VBG 4 Elektrische Anlagen und Betriebsmittel (electrical facilities and equipment)
- Strahlenschutzverordnung Str.SchV (Radiation Protection Ordinance)
- Auflagen Genehmigungsbehörde (requirements of the authorising agency)
- DIN VDE Normen -norms of the VDE (association of the electronics, the electrical and information technology engineers)
- Prüfungen nach BGV A2/GUV 2.10 -Prüfung ortsveränderlicher Geräte (Inspection of mobile devices)
- DIN 6280 –Stromerzeugungsaggregate (current generating sets)
- DIN 4675 -Brandmeldeanlagen (fire alarm systems)
- DIN 13080 - Gliederung des Hospitales in Funktionsbereiche und -stellen (classification of the hospital into function areas and function sites)
- VDS Richtlinien -Richtlinien zur Brandschadensanierung (guidelines for restoration after damage due to fire)
- RKI Richtlinien - Richtlinie für Hospitalhygiene und Infektionsprävention (recommendation of hospital hygiene and infection prevention)
- DIN 1946 Belüftung von Krankenhäusern (air ventilation of hospitals)

Activator of the process

Activator of the process maintenance and servicing of technical facilities is the procurement.

Description of the process

Germany

Procurement

In the sub-process *procurement* the agreement between the parties is very important. Essentially four parties are involved in this process; these are in general the device user, the financing unit, the supplier and the department of technical facilities or the management unit. For process purposes the financing of the maintenance costs of this sub-process is very important. During the process the device-specific maintenance strategy is determined in agreement with the financing unit.

Installation

The *installation* of technical facilities takes place in accordance with the norms and regulations of the respective technical facilities.

Operation

The *operation* is based upon the regulations of DIN 31051 Instandhaltung (maintenance) and the manufacturer's data. Responsible for this process is the operator. Parts of the operation can be delegated.

Maintenance and inspection

Maintenance and inspection are generally amenable to planning. Some of the inspections are checks according to the legal specifications (see DIN 31051). In order to guarantee a smooth flow of the process it is very important to inform customers in time about the execution of maintenance and inspection work. This enables the customers to change appointments.

An essential component of maintenance and inspection is the inspection analysis. Only a careful analysis of the results of the inspections assures timely issue of the maintenance orders. The maintenance intervals can then be adjusted to safety related and commercial aspects. The degree of optimization of a specific sequence of service provision is determined in a contract deviation analysis. This analysis investigates the optimisation of the individual process steps concerning the service provision.

Corrective maintenance

A corrective maintenance is triggered by a service request. The request describes a service requirement for the purpose of service management (e.g. service demand, etc.). It is a matter of service demand in terms of a service management. During the claim investigation, it is attempted on the one hand to find the cause of loss, while on the other hand it is determined whether the loss concerns a subject that must be registered. Should it concern such a case, the responsible authorities and the executive board must be informed. Before the corrective maintenance is performed its economical efficiency must be verified. All steps must be documented. Causes for the need of the repair must be documented. If a corrective maintenance is not profitable, the shut-down of the equipment or a replacement purchase must be initiated. One aspect of the reconnection of the equipment is the functional check according the legal guidelines (e.g. engineered safety control (STK Sicherheitstechnische Kontrolle)).

Shut-down

Every individual device of technical facilities must be shut down in an adequate and orderly manner. It is important to determine whether there was an obligation to obtain a permit for the operation of the device or not. Similar regulations for shut-down must also be observed. (See also the Law for Promoting Closed Cycle Waste Material Management and Ensuring Environmentally Compatible Waste Disposal (Gesetz zur Förderung der Kreislaufwirtschaft und Sicherung der umweltverträglichen Beseitigung von Abfällen (KrW-/AbfG)).

Furthermore the repurchase by the producer and the resale or utilisation by a third party can be negotiated. Particular attention should be paid to the check of the maintenance contract to avoid unnecessary payments.

Iran

Procurement

The procurement is adjusted to the value of the facilities of the different parties. For great purchases the directorate of support (Moavenate Poshtibani) and the security (Herasat) are involved, because they mainly provide the financing. The coordination is mainly provided by the technical office (دفتر فنى) of the University that controls the construction of the hospitals and is responsible for technical facilities, purchase, inspection and maintenance. In general they control the hospitals once a month. In some hospitals some employees of the technical office are head of the technical department, too. They are responsible for the coordination and the control of the subcontractors. Procurement normally does not follow a strategic plan, because such plans do not exist or are not implemented.

Installation

The installation of technical facilities is mostly treated separately because the procurement is concerned "only" with the purchase of the equipment, not with the planning for the installation, maintenance or servicing. In rare cases the maintenance enterprise is authorised to do the installation. For any tendering the "advertising committee" must hold sessions which lead to interminable processes

Operation

The *operation* is based upon the manufacturer's data (if available and foreign language competence is present) because no laws or regulations exist.

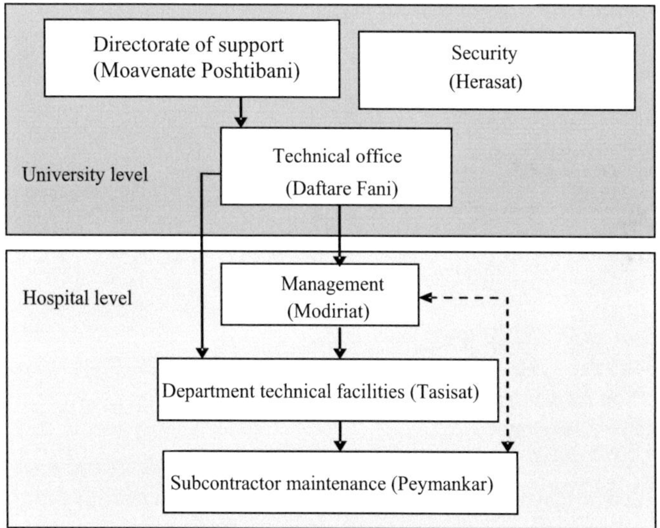

Figure 5-23: The hierarchy of the corrective maintenance [AUTHOR2007]

Maintenance and inspection

If maintenance is performed at all, it is only performed for very expensive equipment. The maintenance starts after the expiration of the guarantee period for the device. If maintenance is performed it is strongly influenced by the recommendations of the technical office to the management. The advertising committee decides whether a maintenance contract should be concluded.

Figure 5-24: Technical Office (Iran)

Corrective maintenance and repair

The hierarchy of the corrective maintenance process can be gathered from figure 5-23.

This process varies according to the strategy of the particular hospital. The process usually flows as follows: The staff of the in-house subcontractor carries out a daily "stock-taking tour" through the hospital and identifies all devices in need of repair. In addition, hospital staff may request repairs. With the resulting repair list the subcontractor returns to the office and takes the needed tools and spare parts (if available). If spare parts are not available a procurement application must be filed that must at least be approved by the management (e.g. also for spare bulbs). If the spare parts are expensive, the Directorate of support (Moavenate Poshtibani) of the hospital must also approve the application.

Figure 5-25: Plant room (Iran)

Shut-down

Facilities can only be shut down or disposed with the permission of all involved parties at the university and hospital level. Finally the device is mostly kept in the camp for "gutting" or it is sold on the "scrap metal bazaars".

Comparison of the processes in the Germany and Iran

The entire flow of the overall process and the accompanying partial processes is shown in the accompanying flowcharts of Appendix-A13. The most significant differences are summarized in the following table:

Process step	Germany	Iran
Procurement	• device-specific maintenance strategy is determined in agreement with the financing unit • Costs of maintenance are calculated	• No strategy or annual plan
Implementation	• Laws and the requirements of DIN 31051 and manufacturer's data	• Mostly manufacturer's data
Operation	• Laws and the requirements of DIN 31051	• Mostly manufacturer's data
Maintenance and inspection	• training check according the regulations(DIN 31051) • Inspection analysis • Analysis of the maintenance	• only occurred for a few, expensive devices
Corrective maintenance and repair	• economical efficiency • cause analysis	• no cause analysis
Closure	• resale to third party • repurchase by the producer • regulations for shut-down (Law for Promoting Closed Cycle Waste Material Management and Ensuring Environmentally Compatible Waste Disposal and regulation for electronic scrap metal - Entsorgungsverordnung (KrW-/AbfG und Elektronikschrottverordnung))	• spare part camp • scrap metal bazaars • transmission sales to hospitals that can use it

Table 5-9: Comparison process of technical facilities [AUTHOR2007]

5.3.2 Characteristic variables of the process

The process steps "maintenance and inspection" and "corrective maintenance" are decisive for the total maintenance cost. The cost allocation between the processes can vary and to some extent reflects the maintenance strategy.

Cost factors*:

- Costs of maintenance according to the replacement value of the installed equipment

- Purchase costs according to inpatient days

Quality factors*:

- Number of the registered interventions with the technical facilities

- Operation figures OPIK-Germany and Iran

[* Operation figures OPIK-Germany and Iran]

The characteristic variables of the various process steps are shown in Appendix- A15. The significance of the characteristic variables depends on the process steps. Eye catching is that there is no planning phase in Iran. As mentioned before, maintenance is considered to be of less importance in Iran.

5.3.3 Interfaces of the process

The interfaces of the process in the project countries were identified as well and are presented in Appendix-A14. Remarkable in the comparison is a higher bureaucratic complexity in Iran that primarily leads to loss of time and consequently money.

5.3.4 Process data

Scope of the process data

For data collection purposes, the process was divided into maintenance/inspection and repair [DIN31051]. 322 reports for maintenance/inspection and 654 reports for repair were collected and entered in the data base in Germany. The allocation to the various hospitals can be seen in table 5-10 and 5-11.

Rendered reports

H 1	H 2	H 3	H 4	H 5	H 6	H 7
45	161	7	80	5*	10	14

Table 5-10: Data sheet - maintenance/ inspection of technical facilities [LENNERTS2001-2007d]

Rendered reports

KH 1	KH 2	KH 3	KH 4	KH 5	KH 6	KH 7
57	320	14	146	73*	23	21

Table 5-11: Data sheet – repair of technical facilities (Germany) [LENNERTS2001-2007d]

* This hospital filled another 250 Data sheet for maintenance/inspection and repair that could not be considered due to formal faults.

The Iranian pilot hospitals were grouped into maintenance/ inspection and repair like the German pendant. In general maintenance is not performed in these hospitals. Two of the hospitals therefore did not supply any data. Hospital 1 delivered 8 reports that were not evaluated because of their small number.

For the process step repair 544 reports for technical facilities were collected 20 days.

Rendered reports

H 1	H 2	H 3
186	168	170

Table 5-12: Data sheet – repair technical facilities (Iran) [AUTHOR2007]

The appropriate data sheet can be found in Appendices- A11 and -A12.

As mentioned before, a sufficient number of data sheet were collected for technical facilities, but so many fields were left empty that only an evaluation for repair time and quality was possible.

5.3.5 Process analysis

Maintenance and repair analysis

By detailed consideration of the individual hospitals different strategies can be observed. Thus the relatively low rate of maintenance orders in hospital 9 can be explained by higher costs for the extern subcontractor.

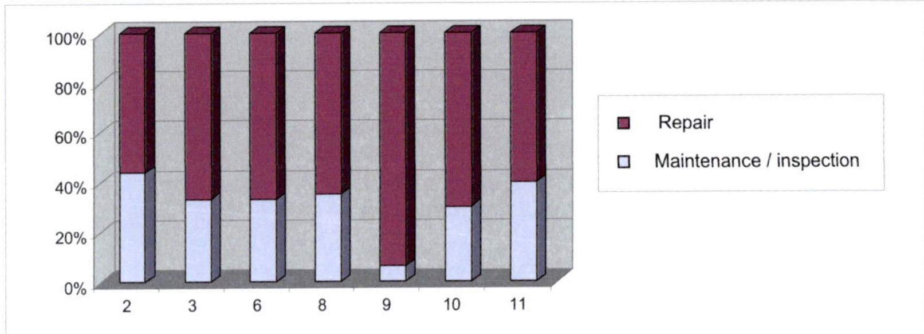

Figure 5-26: Subdivision maintenance/ inspection and repair [LENNERTS2001-2007d]

As mentioned before, maintenance and inspection are only performed in Iranian hospitals for a few devices, which lead to contracts with external subcontractors.

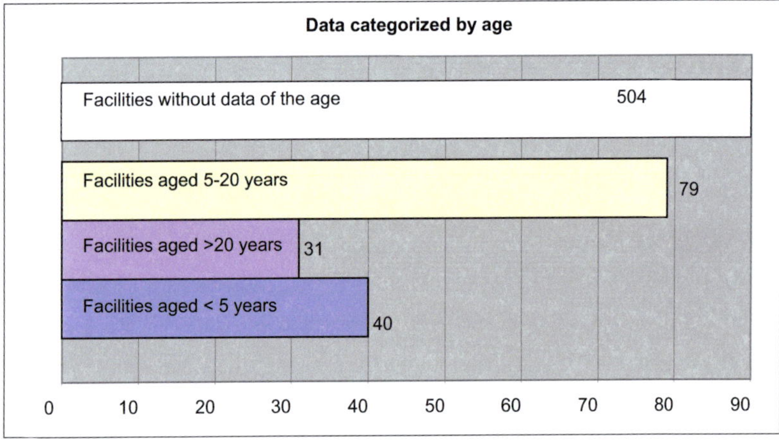

Figure 5-27: Data categorized by age [LENNERTS2001-2007d]

Repair

If the age distribution of the technical facilities in the German hospitals is considered, it is observed that most of the facilities have an age of 5-20 years. It must be mentioned that only 23% of the hospitals replied to this question.

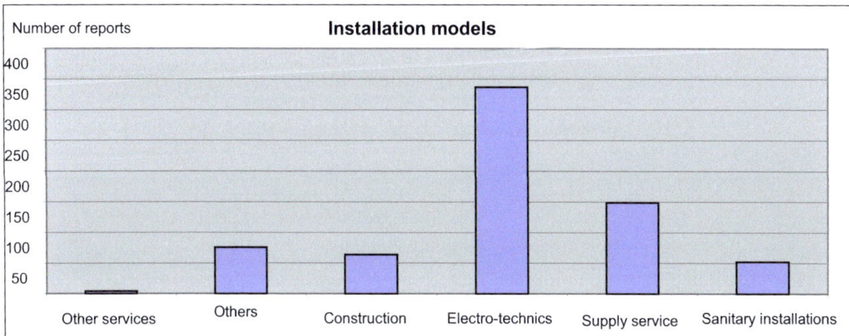

Figure 5-28: Allocation of the installation models in all hospitals [LENNERTS2001-2007d]

Due to a low number of replies the evaluation of the age structure of equipment at the Iranian hospitals could not be carried out. Nevertheless, it can generally be said that most facilities stem from the formative years of the hospitals (see table 5-1) in the seventies or even before then. The same applies to the installation models; here also there were too few data for evaluation. However in the German hospitals the repair orders were allocate as shown in figure 5-29 to the following installation models. The largest part of the orders concerned electro-technics and supply services.

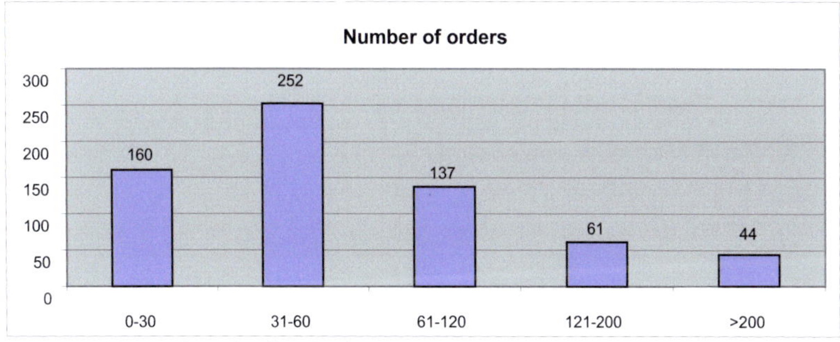

Figure 5-29: Number of orders by duration (Germany) [LENNERTS2001-2007d]

Transit periods (repair of technical facilities)

The orders for repair of the German technical were analysed by duration and described in box-plots. The results are shown in the figure below. While the German facilities are grouped by installation model the Iranian results are reported by hospital because they don't make the division in facility groups.

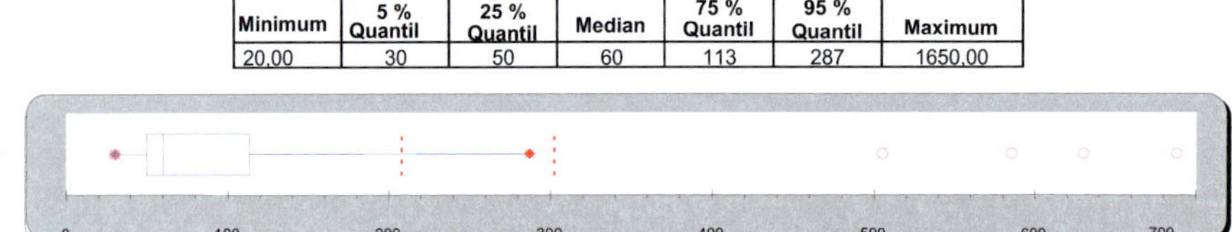

Duration repair of supply service [minutes]						
Minimum	5 % Quantil	25 % Quantil	Median	75 % Quantil	95 % Quantil	Maximum
20,00	30	50	60	113	287	1650,00

Figure 5-30: Box-plot supply service (Germany) [LENNERTS2001-2007d]

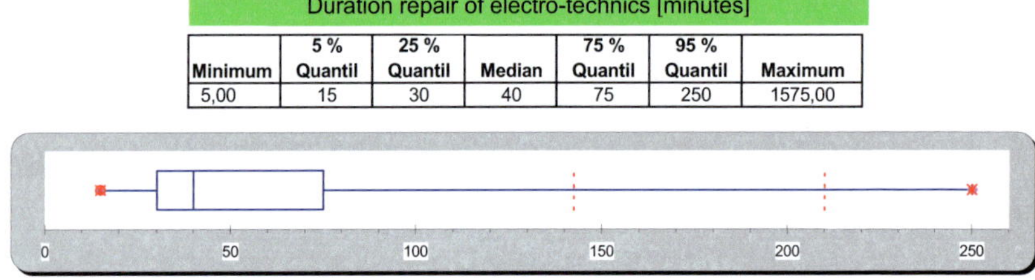

Duration repair of electro-technics [minutes]						
Minimum	5 % Quantil	25 % Quantil	Median	75 % Quantil	95 % Quantil	Maximum
5,00	15	30	40	75	250	1575,00

Figure 5-31: Box-plot electro-technics (Germany) [LENNERTS2001-2007d]

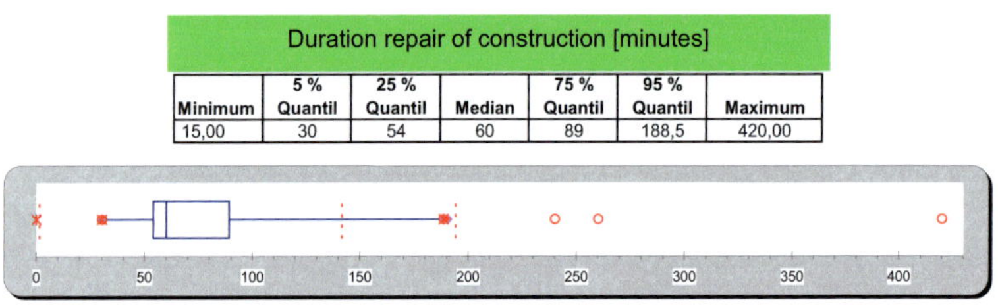

Duration repair of construction [minutes]						
Minimum	5 % Quantil	25 % Quantil	Median	75 % Quantil	95 % Quantil	Maximum
15,00	30	54	60	89	188,5	420,00

Figure 5-32: Box-plot repair construction (Germany) [LENNERTS2001-2007d]

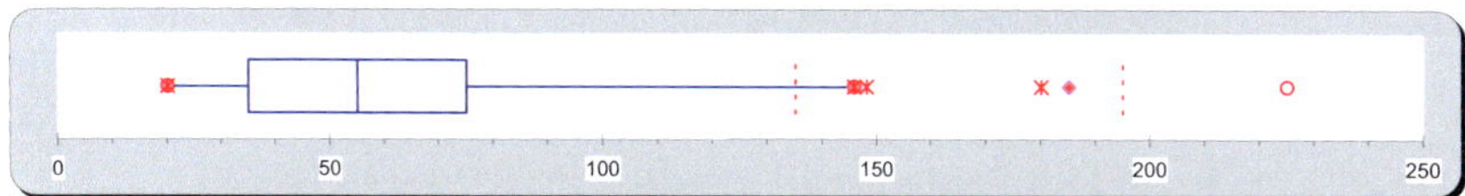

Duration repair of "others" [minutes]						
Minimum	5 % Quantil	25 % Quantil	Median	75 % Quantil	95 % Quantil	Maximum
10,00	20	35	55	75	145,75	225,00

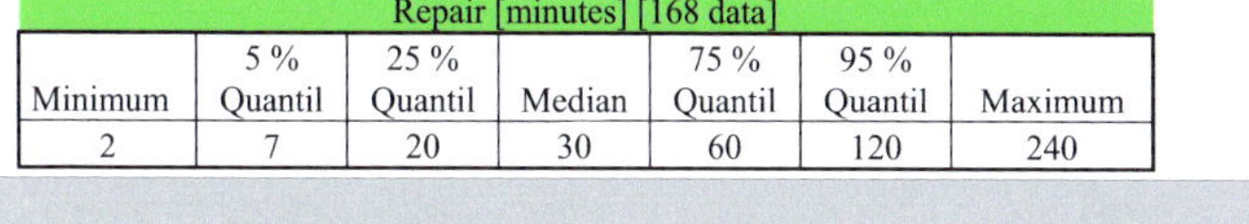

Figure 5-33: Box-Plot repair „others"(Germany) [LENNERTS2001-2007d]

The results for the duration of repair in the Iranian pilot hospitals are presented below.

Hospital 1:

Repair [minutes] [168 data]						
Minimum	5 % Quantil	25 % Quantil	Median	75 % Quantil	95 % Quantil	Maximum
2	7	20	30	60	120	240

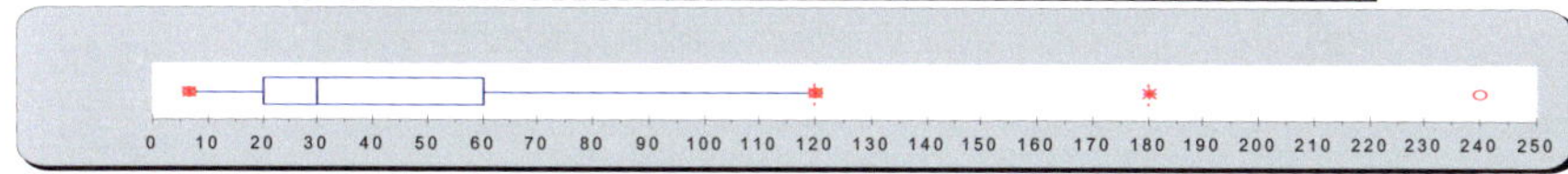

Figure 5-34: Box-plot repair hospital 1 (Iran)

Hospital 2:

Repair [minutes] [143 data]						
Minimum	5 % Quantil	25 % Quantil	Median	75 % Quantil	95 % Quantil	Maximum
3	5	10	15	25	50	220

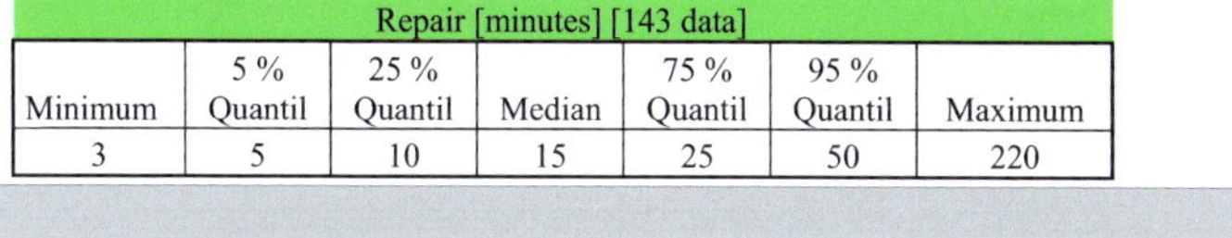

Figure 5-35: Box-plot repair hospital 2 (Iran)

Hospital 3:

Repair [minutes] [132 data]						
Minimum	5 % Quantil	25 % Quantil	Median	75 % Quantil	95 % Quantil	Maximum
1	4	10	20	40	180	300

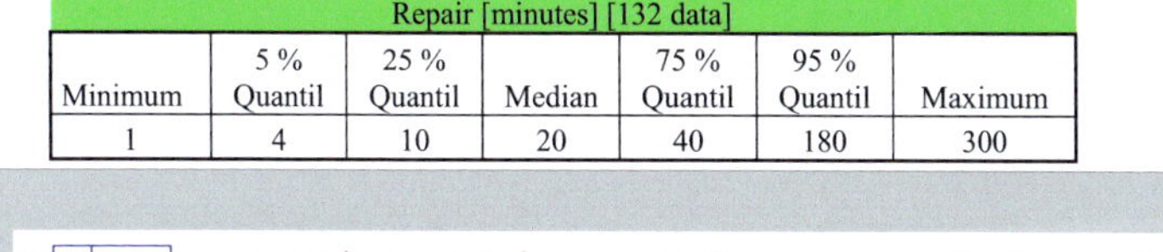

Figure 5-36: Box-plot repair hospital 3 (Iran)

The evaluation of the duration of repair in Germany is 54 minutes per order. The Iranian technicians need an average time of 22 minutes. In comparison with the German duration this is approximately 30 minutes less. The difference could be due to the documentation work in Germany that is more comprehensive and hence takes more time.

Because of the lack of documentation of the repaired facilities in the Iranian project hospitals no statement could be made; it could only be said that during the data evaluation the frequency of the repair of chillers and ventilation was conspicuous.

Process flow

The analysis of the process flow in Germany reveals that activities in conjunction with the recognition and removal of causes have a small volume (see figure 5-37). The time frame of the activities in this process step is especially high in the cause analysis. The process step decommissioning likewise has a small volume. This can be explained by the circumstance that there are no regulations for the decommissioning of small devices in Germany.

The information exchange shows accumulations in the area of claim investigation and repair. Overall it can be said that information exchange is only performed in a few processes. For an optimal process flow the information exchange should be low during the process step repair and shifted either to the check or the decommissioning. It is assumed that information exchange during repair disturbs the flow in the primary as well as the secondary process.

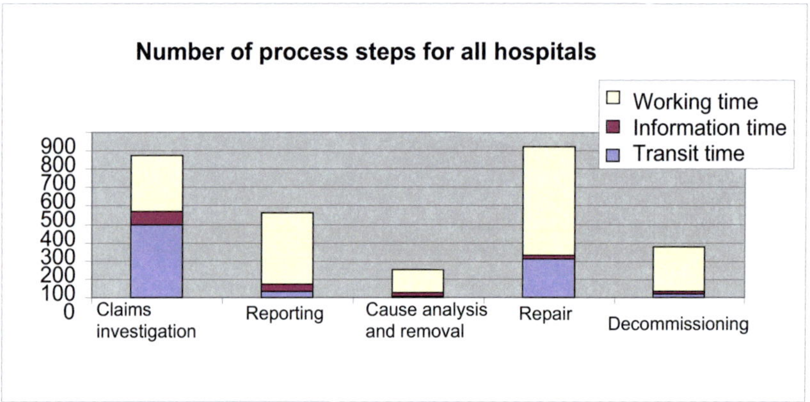

Figure 5-37: Number for process steps in all hospitals (Germany) [LENNERTS2001-2007d]

Absolutely considered a cause analysis requires a large amount of work. A comparison of work-, information- and transit time reveals that the most time is needed for work. The observed transit times are attributed to the coupling processes that take place. After completion of an order the next order is started immediately. The less information time can be explained by simple communication or that even no communication happens.

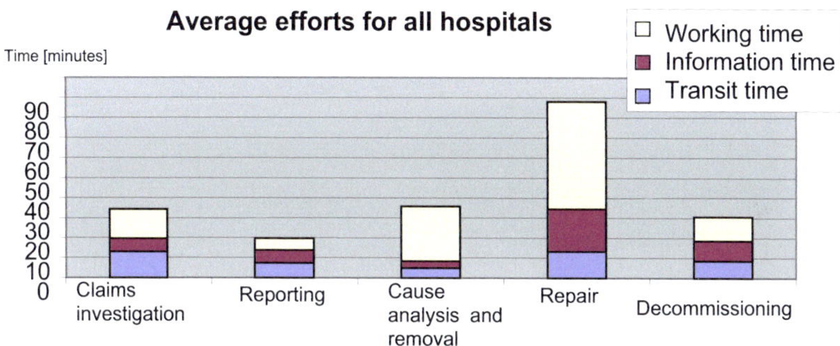

Figure 5-38: Average efforts for all hospitals (Germany) [LENNERTS2001-2007d]

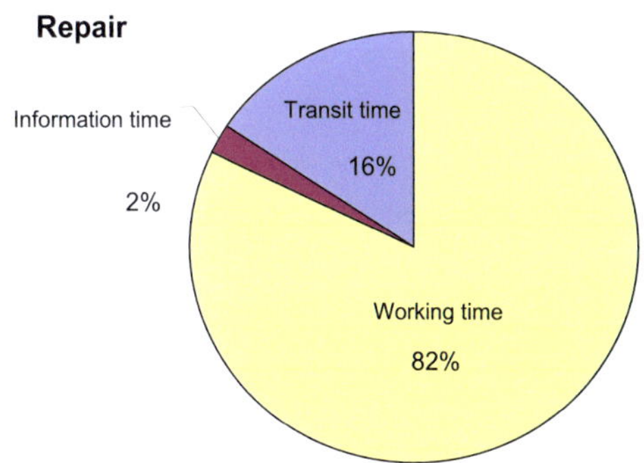

Figure 5-39: Percentile of the total time repair [LENNERTS2001-2007d]

For lack of data total times for repair in the Iranian hospitals could not be evaluated. Therefore questionnaires on the quality of the repairs were distributed. They led to the following results.

Hospital 1:

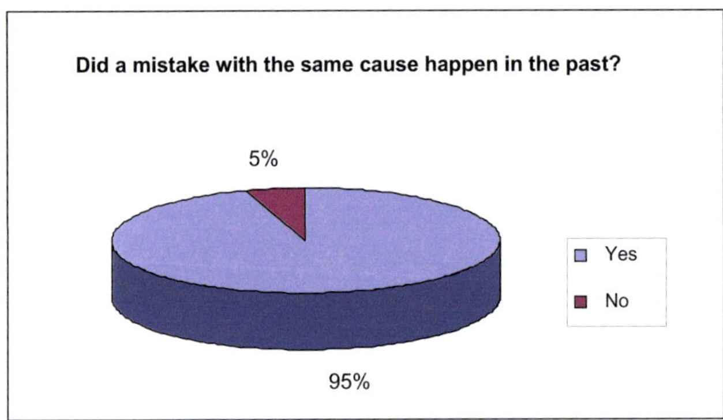

Figure 5-40: Result quality hospital 1 [AUTHOR2007]

The evaluation of the data leads to similar conclusions for all hospitals. In two of the three hospitals mistakes with same causes happened in all cases. The results of the evaluation show that on the average 98% of the facilities were repaired repeatedly. This situation is the consequence of the missing process step cause analysis. The implementation of this process step with qualified staff would definitely reduce the error and repair rate considerably.

The small volume of the information that could be gathered is the result of poor documentation and of the position that technical facilities have in the Iranian hospitals.

5.3.6 Summary of the technical facilities

Similar to the process of medical equipment the evaluated process for technical facilities is affected by the highly advanced position of technical equipment and facilities in Germany. The importance of technique beginning with numerous apprenticeships, study courses and degrees continues with unions and federations for every specific field. The numerous regulations and laws that are controlled very strictly by different authorities point no less to the significance of this field.

In addition, maintenance and repair (spare parts) can be carried out with less cost because the companies and suppliers are local. Planning (Maintenance and servicing planning commencing already with the procurement), analysis (causes analysis...) and the documentation play a decisive role.

In Iran, however, laws and regulations are introduced at the university and hospital level. The main part of the budget is used for the purchase of facilities. Maintenance is very rare, and is performed primarily for very expensive facilities. In the course of privatisation the management tries to settle the low priced operating company what leads to a very high fluctuation. In addition the high fluctuation of the management, lack of specialist staff, a low level of documentation, a small budget besides and the lack of control of the mostly elder facilities influence the situation of the technical facilities in particular.

5.4 Process laundry management

5.4.1 Process description and parameters

Application area of the process

The process *laundry management* contributes to the flawless supply of fresh laundry (ward-clothes, scraps and work clothes) to the cost centre (ward, bed central, operating room) and the cleaning and the disposal of that laundry after it has been used.

Aim of the process

Aims through the eyes of the customer/ cost centre

The aim of the management process is to assure that the functionality and quality of the laundry are suitable for the medical staff. The process flow depends on the installation and observance of a hygiene plan. From the point of view of the patient as an indirect customer, the aim of the process is to assure that the laundry cannot cause infections, and that the recovery process is supported in an optimal manner.

Aims through the eyes of the organisation

The aim of the management process is to minimize the logistic load of transit routes and the space requirement of the storage. The laundry consumption per cost centre becomes transparent. Investment and maintenance costs can be minimised by the detection of unsuitable laundry.

Aims through the eyes of the employees

The employees are supported in the performance of the process by explicit work orders. They are also protected against the danger of infections.

Process responsibility

The responsibility for the care and development of the process as well as control and compliance rests with the organisation unit „laundry supply". This unit has different compositions in the various hospitals that were studied.

Activator of the process

Activator of the process is the need for laundry. This need exists constantly, while the amount of the used laundry fluctuates with the degree to which the capacity of the hospital is used.

Description of the process

Germany

Requirements planning

The basic parameter for the requirements planning of the laundry supply is the current or the average laundry consumption of the individual cost centre. It is useful to distinguish between three types of cost centres: ward-clothes, scraps and work clothes. The consumption of station laundry depends on the size (number of beds), the kind and the extent of the ward. The number of surgery rooms and the kind of surgery performed determine the laundry consumption of these facilities. It is necessary to distinguish between the uses of reusable or non-returnable laundry. Work clothes are separated into pool supply and personal work clothes. Important for the requirements planning is the number of employees and the in-house clothing standard or requirements.

Procurement

In the sub-process *procurement* the coordination between the parties is very important. Essentially four parties are involved in this process. In general these are the user, represented by the head of nursing staff or domestic economy, the financing unit (according to the ownership of the laundry (own or rental laundry) hospital or subcontractor), the supplier and the laundry facility. Personal work clothes must be fitted to the user.

Allocation

Laundry is normally allocated cyclically according to the volume required by the individual hospital. Logistically the process can be separated into two parts; the in-house logistics, including the determination of the required volume, and the external logistics. The consignment sale can occur either internally or externally. Depending on the supply model, the work steps are performed either by the customer (represented by the matron or the domestic economy) or by the laundry facility. The responsibilities and the interfaces are distributed accordingly. Exceptional cases, e.g. emergency, can cause an extraordinary need for laundry that must be covered by direct allocation.

Utilisation

Responsible for the use of laundry are the cost centres. The duration of use depends on the hygiene plan of the hospital and on the in-house standards.

Interfaces (to the laundry / sewing shop, matron and domestic unit) exist only if the quality of the laundry does not satisfy the quality requirements of the cost centre, or a supply gap occurs in an exceptional case.

Dirty washing accumulation / - collection

Dirty washing is accumulated cyclically according to the laundry volume of the different hospitals. Logistical potential for savings exists in the implementation combined with the process of allocation of the fresh laundry. Precondition is the sub-process dirty washing collection; the responsibility therefore rests with the customer or the cost centre.

Similar to the allocation process, the collection of washing is separated into in-house and external logistics. Depending on the supply model, the laundry is thus collected by the customer (sometimes represented by the matron or the domestic unit) or by the laundry facility. The responsibilities and interfaces are assigned in a similar fashion.

Cleaning

The laundry is cleaned after the cyclic accumulation / - collection. The laundry facility is responsible for this process. The flow is prescribed by guidelines. The process step ends with the temporary storage of the laundry in the picking stock. Depending on the supply model the customer / cost centre or the laundry facility is responsible.

Repair

The process of the repair is activated by a reclamation that is issued by a cost centre or a quality control of the laundry. The sewing shop, that according to the supply model is part of the laundry or the matron/domestic unit, is responsible for the process step. It is important that the interface between the cost centres and the sewing shop is clearly defined (e.g. use of a separate laundry bag for defect laundry or marking of defect laundry, so that the defect laundry can be removed from the use-cleaning-cycle quickly).

The interface between sewing shop, cost centre, the matron/domestic unit and the laundry facility is of particular importance for the damage check, which makes it possible to remove the cause of the damage. Generally, the repair arrangements are checked of their efficiency, whereas the repair costs are compared with the costs of the replacement of a piece of laundry. After the repair, the piece of laundry is placed back into the cycle of utilisation.

Shut-down

A piece of laundry is sorted out if repairing is not cost effective. The responsibility lies at the sewing shop. The records of proper waste management are given to the persons in charge (e.g. the matron/domestic unit or the laundry facility) to be used for the requirements planning by the sewing shop.

Iran

Requirements planning

The basic parameter for the requirements planning of the laundry management is experience. The costs are not charged to the cost centres. Accounts are regulated directly with the management. In general the analysed hospitals differentiate between ward clothes and scraps.

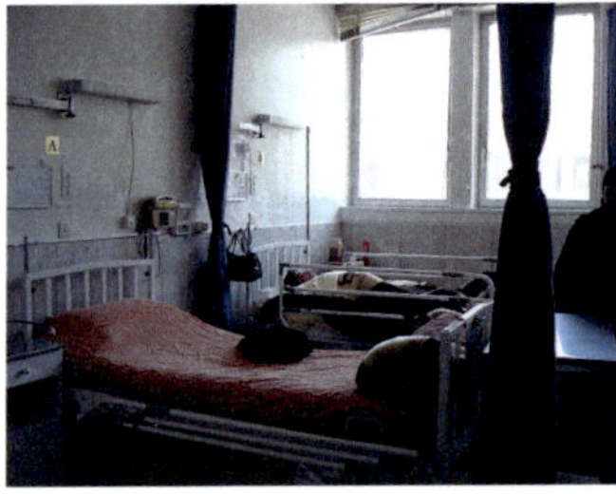
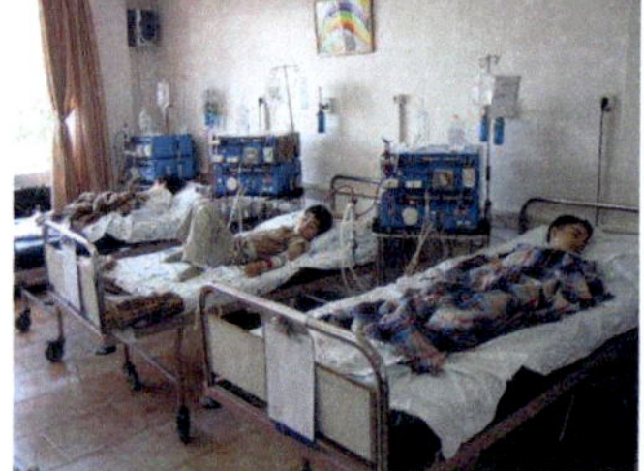

Figure 5-41: View to a station (Iran)

The ward clothes are provided only for the doctors and the nursing staff, but in some hospitals the nursing staffs itself is responsible for clean clothes. Other working clothes (e.g. technical staff) are washed in exceptional cases. Mostly the working clothes are provided by a subcontractor (Peymankar) who is responsible for the cleaning, too.

Procurement

An annual plan is drawn up by the management, the wards and the laundry facility which assigns a certain amount of laundry to each department/ward. A small part of this laundry is ordered; most of the required laundry is produced in the in-house sewing shop. In this case, procurement consists mainly in the purchase of rolls of fabrics (this is the reason why most data are given in meters).

Production

The sewing shop is responsible for the production of the laundry (mostly bed linen). The shop is authorized by the supervisor of the laundry facility.

Figure 5-42: Production (Iran)

The rolls of fabrics are ordered at the bazaars by the management or the supervisor of the laundry facility. Different fabric colours are chosen for different wards/ departments. All of the laundry is labelled with a hospital/ward stamp to prevent theft (see Figure 5-43). Rental of laundry, as it is practised in Germany, is not known in Iran yet.

Figure 5-43: Identification mark of the fabrics (Iran)

Allocation

The laundry is normally allocated once a year. Occasionally laundry is also distributed if the demand exists and the replacement stocks are available. The process is concentrated almost exclusively on in-house logistics. Mostly an employee of the laundry team or of the ward is responsible for the pickup and delivery service. Pickup is only internal. The working steps are mostly performed by the laundry facility and its management.

Utilisation

The different wards/ stations are responsible for the use of laundry. The duration of use depends to the hygiene plan and fluctuates considerably between the different hospitals. Interfaces (to the laundry/ sewing shop, management, stock, financing unit) are defined for communication if the quality of the laundry does not fulfil the requirements, supply gaps appear or unusual requirements are made by the ward/ stations unusually high consumption of laundry).

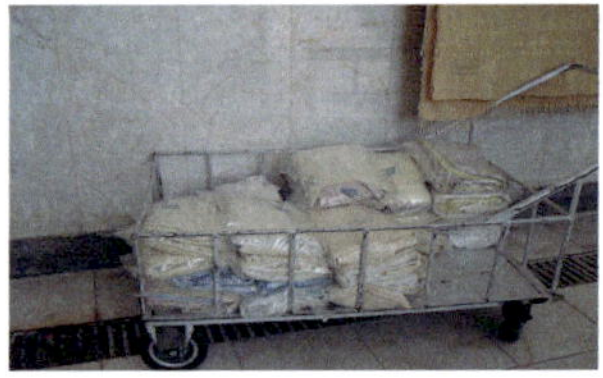

Figure 5-44: Transfer service (Iran)

Dirty washing accumulation / - collection

Dirty washing is accumulated cyclically (generally daily) according to the laundry volume of the ward. The washing in most cases is accumulated in the ward in tons and is then brought directly to the laundry by an employee of the ward staff or by an unskilled worker (Khadame). One of the pilot hospitals has a central collection center as can be seen in figure 5-46. In all ward halls are loops were the dirty washing is forwarded to the collection center and can be collected or sorted by an employee.

Figure 5-45: Dirty washing ton (Iran)

Figure 5-46: Dirty washing collection (Iran)

Cleaning

The washing occurs referring to the cyclically dirty washing accumulation and collection. The responsibility lies in the hand of the laundry. The procedure is given though there are no regulations. The process step ends with the temporary storage in the picking stock that and is still in the responsibility of the laundry.

Figure 5-47: Cleaning and picking (Iran)

Repair

The process of the repair is activated by a reclamation that is issued by a cost unit or the quality control of the laundry facility. The final decision whether the laundry is repaired is taken by the supervisor of the laundry facility. The sewing shop that is part of the laundry facility is responsible for this process step.

Figure 5-48: Sewing shop (Iran)

The cause of damage is not investigated further, except for obvious problems (e.g. constantly recurring deficiencies, suspicion of malicious destruction). After the repair, the piece of laundry is returned to the cycle of utilisation.

Shut-down

A piece of laundry is sorted out if it has defects whose repair is not cost effective.

The supervisor of the laundry is responsible for this process. Before the useless laundry is disposed it is checked whether it can be of further use (cleaning cloth). Records of proper waste management must not be kept.

Comparison of the processes in Germany and Iran

The flow of the overall process and the accompanying partial processes is shown in the accompanying flowcharts of Appendix-A17. The most remarkable differences are summarized in the table below.

Process step	Germany	Iran

Requirements	• Need planning of the individual cost centres
	  o ward-clothes
	  o scraps
	  o and work clothes
		• No cost centres, directly setteled with the management
		  o ward-clothes
		  o scraps
		  o and work clothes (very rare)

Procurement	• Agreement between user, financing unit, supplier and the laundry	• According to the annual plan agreement between laundry, ward and management
Production	• Service is rendered externally	• Production in the in-house sewing shop from bought rolls of fabrics
Allocation	• According to the supply model in-house or external	• Either by the ward staff or by the laundry facility
Utilisation	• Cost centres are responsible	• Wards are responsible
Dirty washing accumulation/ – collection	See allocation	
Cleaning	• flow is predetermined by laws or guidelines	• flow is predetermined but laws or guidelines are not applied
Repair	• efficiency test	• decision is taken by laundry supervisor

Shut-down	• Records of proper waste management must be kept.	• decision is taken by laundry supervisor
		• No records needed

Table 5-13: Comparison process laundry management [AUTHOR2007]

5.4.2 Characteristic variables of the process

The process of laundry management must supply the primary processes with laundry without interfering with the logistics. The costs must be minimized. The process consists of an inner (cleaning – use) and an outer (need planning, procurement – shut-down) cycle.

The process step of fresh laundry allocation is of high logistical importance in the inner cycle. In addition, the consumption of laundry per cost centre can be measured so that this process step delivers data that can be used for the requirement planning (outer cycle).

In the Iranian hospitals the laundry could be recorded by wards. But because the laundry facility negotiates directly with the management, recording does not take place.

The process step cleaning is the core sub-process of the entire process. It is characterized by:

- Judicative guidelines
- Time factors
- Transport duration
- Duration of the cleaning procedure/ conditioning
- Duration of the storage

The characteristic variables of the individual process steps are presented in Appendix- A17.

5.4.3 Interfaces of the process

The interfaces of the process are presented in Appendix- A18. In Germany legislation and guidelines are obligatory. This is not the case for the Iranian laundry facilities. There are in-house hygiene controls and an inspection from the university, but there are no official guidelines. Hence, optimisation potential in Germany is given in the area of equipment technology, -dimensions and –efficiency. In the outer cycle the procurement is of special importance. The annual purchase of laundry must be adjusted to the existing stock of laundry.

5.4.4 Process data and analysis

Scope of the process data

In Germany the evaluation of the process factors was based on a questionnaire (see Appendix-A19). A report was prepared in the workshops and send to all pilot hospitals. This German "parent" questionnaire could not be adapted one to one, because other important factors existed in the Iranian hospital. Therefore the report was translated, modified, adapted to the Iranian concept for the OPIK-Iran project and given to the management or to the supervisor of the laundry facility.

Classification of the sample

In Germany eleven hospitals delivered data for the evaluation. Figure 5-50 shows the particle size distribution of the participating hospitals. It can be seen that each size range is represented by at least two hospitals. The majority of the studied houses are middle sized (200 to 1000 beds).

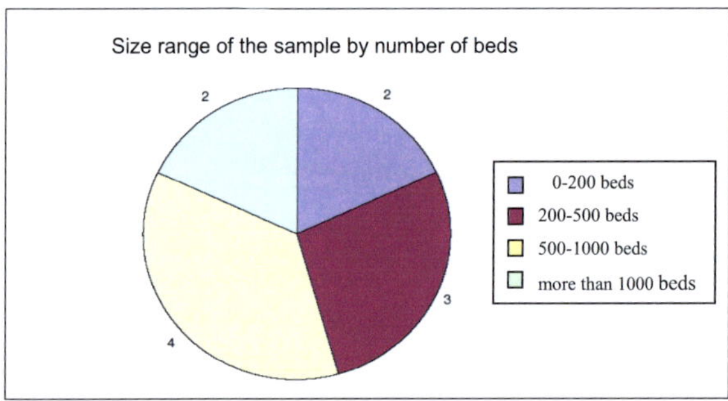

Figure 5-49: Size range of the hospitals (Germany) [LENNERTS2001-2007d]

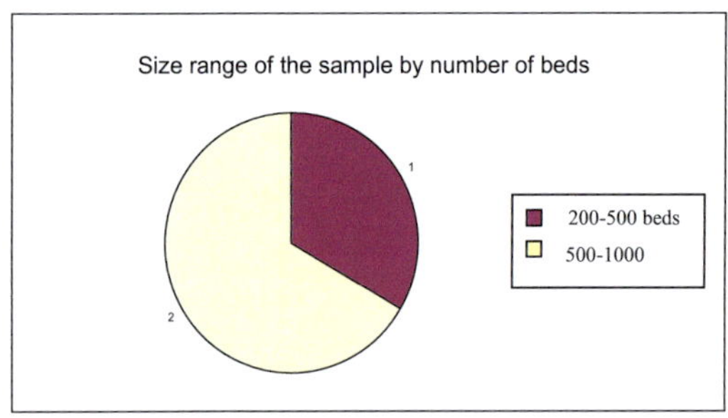

Figure 5-50: Size range of the hospitals (Iran) [AUTHOR2007]

In Iran all three hospitals provided data. Their size range can be seen in figure 5-51.

The models of supply of the two project countries can be seen in figures 5-52 and 5-53. According to the figure most German hospitals have external laundry and in-house logistics. In three houses the whole process is performed in-house, two hospitals have made the process completely external.

This distribution reflects the general status of care by the hospitals in Germany; it can be observed that the number of houses with laundries of their own is becoming less and less. At the same time the tendency goes towards replacing the total process including the in-house logistics.

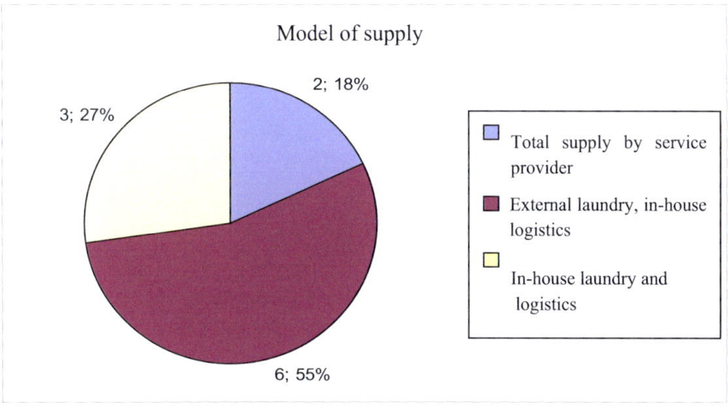

Figure 5-51: Model of supply (Germany) [LENNERTS2001-2007d]

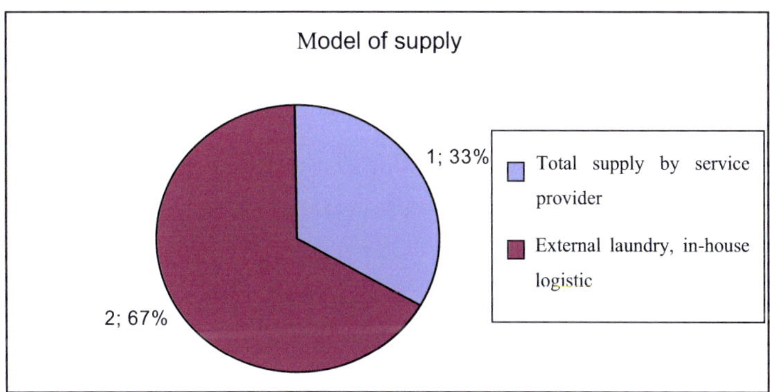

Figure 5-52: Model of supply (Iran) [AUTHOR2007]

A recommendation by the Iranian Ministry of Health and Medical Education indicates that hospitals with more than 50 beds should have a laundry of their own. The distribution in figure 5-52 can evoke a wrong image because of the third hospital. This hospital is using one common laundry together with a whole hospital complex; however the logistics lie in the hand of the individual hospital and is procured by internal staff members.

Collection

The laundry can be collected in-house or externally. Figure 5-53 shows that 73% of the pilot hospitals are collecting in-house, only 2 houses are collecting externally and one house is collecting both in-house and externally.

The demand of space in the pilot hospitals varies significantly. On the average it is around 18m² per 100t of collection of laundry.

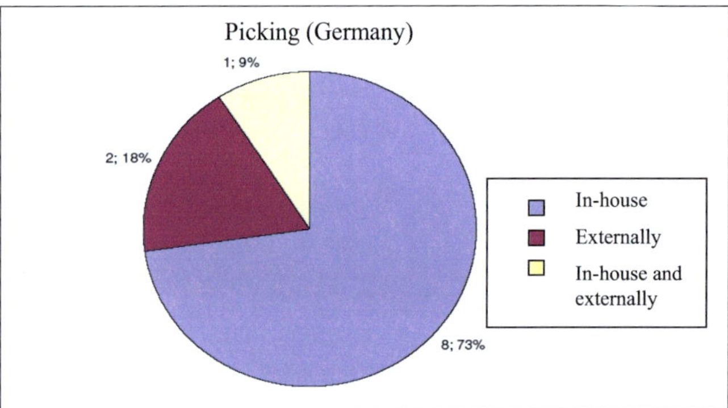

Figure 5-53: Collection (Germany) [LENNERTS2001-2007d]

80% of the collection in the analysed Iranian hospitals is external, 20% a mixture between in-house and external collection. On the basis of the data of the Institute of Public Health is assumed that around 80% of the collection is done in-house.

Stocking system

Stocking system for work clothes

Work clothes can be stocked in for different ways; by an open entry in safe deposit boxes, by a manual release through the laundry stock; delivery in lockers or by automats.

Figure 5-54 shows which systems – also in combination – are used in the German hospitals. The open entry in safe deposit boxes is used most frequently. The manual release is nearly used as little as possible.

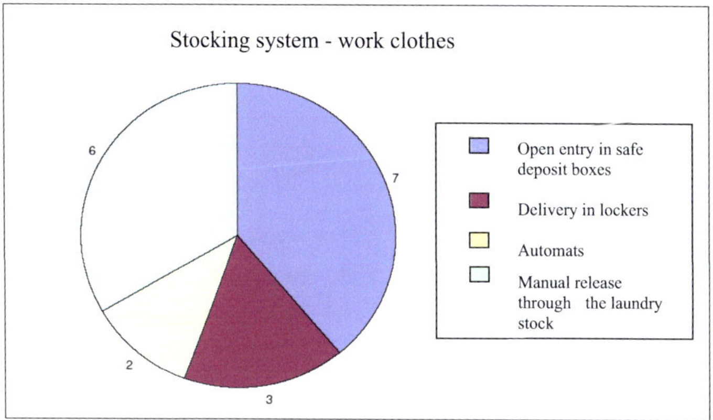

Figure 5-54: Stocking system –work clothes (Germany) [LENNERTS2001-2007d]

As mentioned in chapter 4.3.2, most of the Iranian hospitals provide only lab coats and ward clothes. In some houses these are only sewed and the ward clothes are not even washed. The physicians, on the other hand, can have their lab coats washed in nearly all laundries. The clothes of the service providers (Peymankar) like the cleaning staff; technicians etc. must be cleaned by themselves. Up to the present there have not been any requests to have these clothes washed in the hospital laundries. In Iran delivery in lockers or by automats are not common. The work clothes frequently are released through the laundry stock; in some cases safe deposit boxes are used.

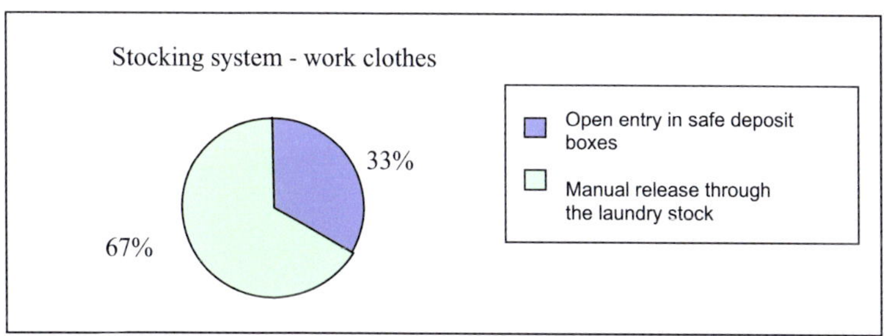

Figure 5-55: Stocking system –work clothes (Iran) [AUTHOR2007]

Stocking system for ward clothes

Ward clothes are stocked in two different ways; either in built-in cabinets or in laundry wagons that are used as cabinets. The advantage of the laundry wagons used as cabinets is that they are suitable for the logistics of a local point of delivery where „ first-in/first-out" is a good strategy.

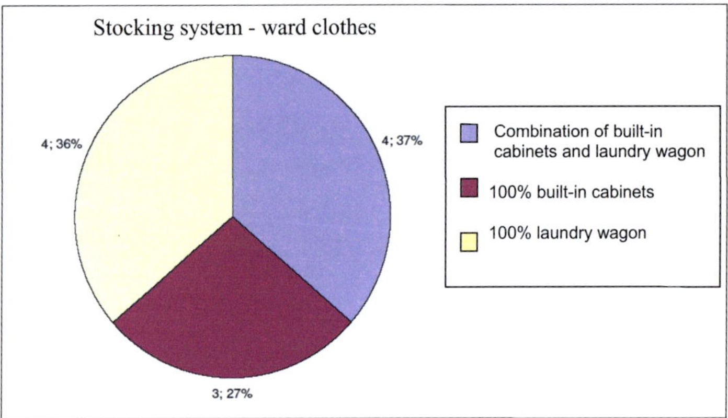

Figure 5-56: Stocking system- ward clothes (Germany) [LENNERTS2001-2007d]

The principle of the delivery is that the remaining fresh laundry is placed on top of the new delivery, so that the remaining stock is withdrawn first. The emptied laundry wagon can be used as a means of transport for the dirty laundry by linking the fresh-laundry delivery and dirty- laundry collection.

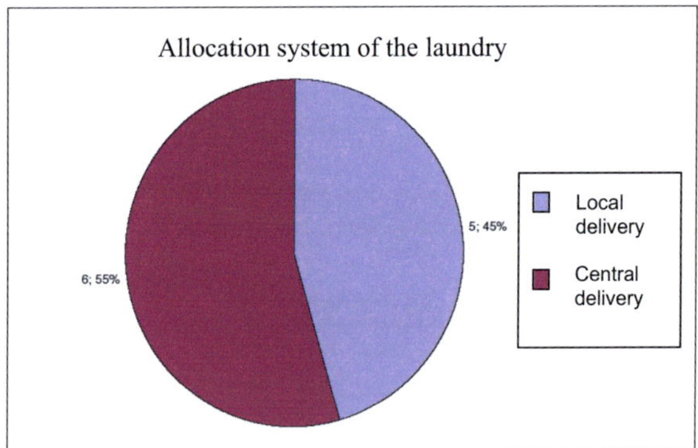

Figure 5-57: Allocation system (Germany) [LENNERTS2001-2007d]

Figure 5-57 shows that five houses allocate the laundry locally in a star-like pattern around the point of delivery; in six houses laundry is delivered in a central-ring pattern. A disadvantage of the central delivery pattern is that the volume of delivery per point of delivery cannot be determined conveniently for cost centres-related account. Figure 5-58 shows that only three of the houses logistically link the fresh laundry delivery with the dirty laundry collection, although this would be the logical consequence of a delivery based on a star-like supply pattern.

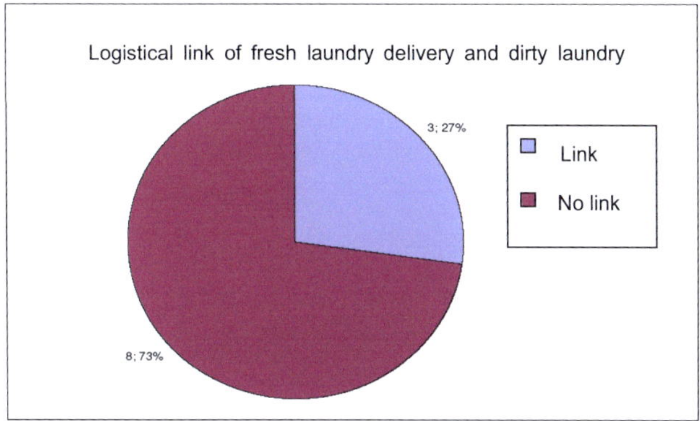

Figure 5-58: Logistical link of fresh laundry delivery and dirty laundry (Germany) [LENNERTS2001-2007d]

Figure 5-58 shows that three houses stock their ward clothes in built-in cabinets. One half of the remaining eight houses have only laundry wagons; the other half is using a combination of both systems.

In the Iranian hospitals built-in cabinets are used exclusively. The delivery of the laundry is 100% decentralised. All wards have employees that are responsible for the laundry delivery service. Nevertheless, because the laundry is not paid by the ward it is not meaningful to speak about individual cost centres. The fresh- and dirty laundry are linked in all pilot hospitals; it must be noted that contrary to the German system the terms fresh delivery and dirty collection are used. The differences of the two systems become apparent here once again. In Germany the cleaning belongs to the activities of the laundry facility while in Iran cleaning is the work of the staff members.

Logistics

The duration of the in-house logistics is a very important cost factor because employees are involved. Figure 5-59 shows the duration of the logistics per hospital, accounting for the parameters number of buildings, points of delivery weight handled by the logistics and collection rhythm. In the following the influence of these parameters on the duration of the logistics is investigated. Because of the low personal costs in Iran, particularly in the state hospitals, the in-house logistics do not play the very significant role that they do in Germany. However, it must be pointed out that there are differences in the logistic systems. On account of this disparity, a direct comparison cannot be made for Iran.

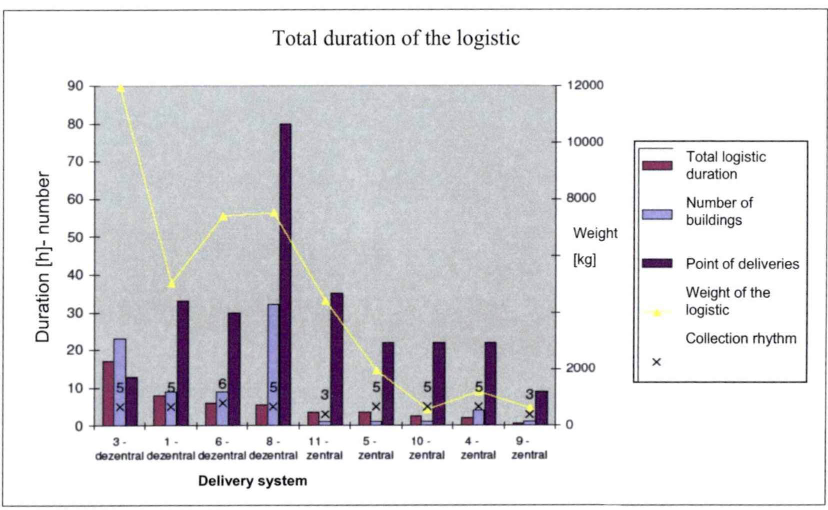

Figure 5-59: Total duration of the logistics (Germany) [LENNERTS2001-2007d]

Collection rhythm

The weekly delivery- and collection rhythm in Germany can be observed exactly in figure 5-60. In most of the hospitals the fresh laundry is delivered five times a week; the dirty laundry collection has the same rhythm. In three houses the dirty laundry is collected six times a week. However, in two hospitals the laundry is delivered and collected only three times a week. These differences between the houses are increased by the different sizes of the pilot hospitals.

It is, however, remarkable that the differences are not consistent. One of the houses with less than 200 beds has five deliveries per week. On the other hand a house in the range of 200 – 500 beds has only three deliveries per week. The same applies to the bottom up deviation; two houses in the range of 500-1000 beds have their laundry collected six times a week while a house with more than 1000 beds gets by with five collections. This shows that there must still be potential for optimisation.

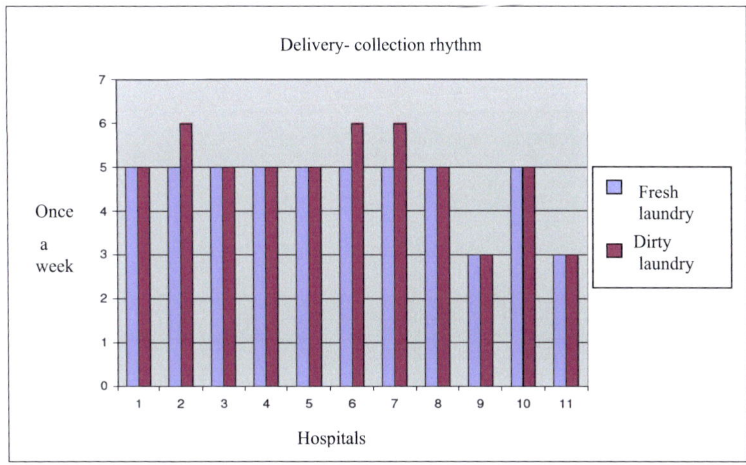

Figure 5-60: Logistic rhythm (Germany) [LENNERTS2001-2007d]

The logistic rhythm influences the delivery and collection weight directly. In general it is assumed that the delivery weight per delivery is on the average the same; before the weekend or delivery breaks, however, the fresh laundry delivery weight is higher and, corresponding to this, after a delivery break (at the start of the week) the dirty collection weight is higher. In the following, the duration of the logistics depends on the logistic weight per point of delivery.

In the analysed Iranian hospitals the average weekly delivery and collection rhythm is five days. As shown in figure 5-61 the rhythm varies between two and seven days; the two days can be allocated to the smallest hospital.

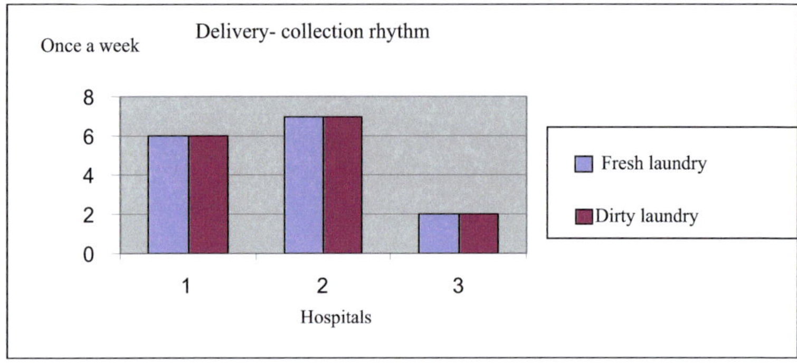

Figure 5-61:Total duration logistic (Iran) [AUTHOR2007]

Ward clothes

Figure 5-62 shows that two of the eleven houses have rental laundry instead of private hospital laundry. In the case of private hospital laundry it must be taken into account that in addition to the capital expenditure for the laundry itself there are costs for extra employees and space.

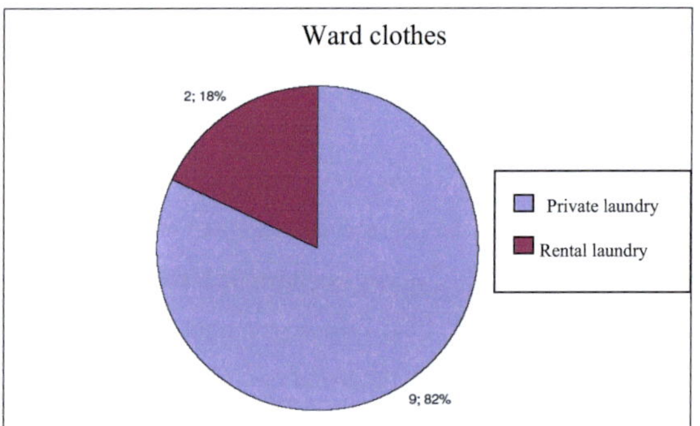

Figure 5-62: Ward clothes (Germany) [LENNERTS2001-2007d]

The ward clothes of the OPIK-Iran project are largely produced in the in-house sewing shops and are 100% privately owned by the hospital. Rental laundry that is used in most German hospitals has not found its way into the Iranian market.

Figure 5-63: Ward clothes (Iran)

In order to prevent that staff members or patients take the laundry with them, all of the laundry is marked with the hospital logo. The colour of the laundry indicates the ward to which it belongs.

Scraps

Figure 5-64 shows that the use of scraps in Germany is divided into two nearly equal parts. One part is the exclusive use of single use scraps; the other part is a combination of a single use and a reusable laundry.

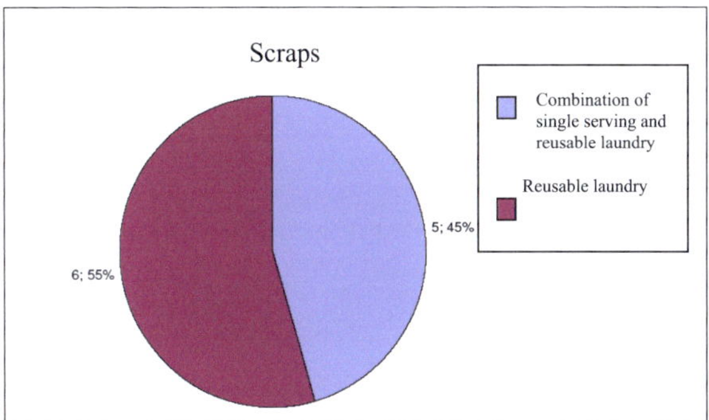

Figure 5-64: Scraps (Germany) [LENNERTS2001-2007d]

The ownership of the reusable laundry of the 5 pilot hospitals is shown in figure 5-65. Three houses are proprietor of the reusable laundry; in one house a part of the laundry is rented and one house uses only rented laundry. The greatest part of the scraps consists of reusable laundry.

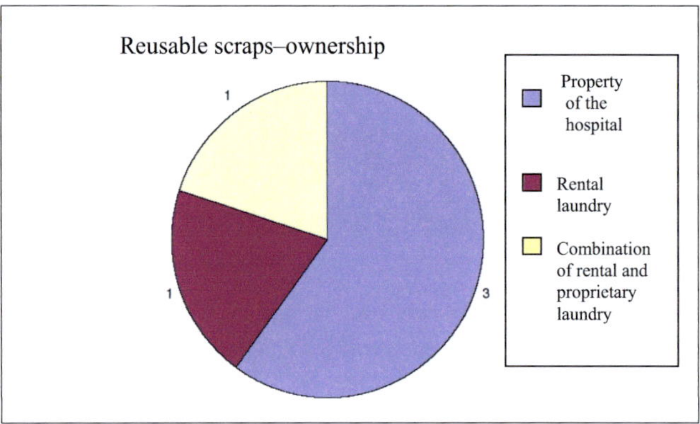

Figure 5-65: Reusable scraps (Germany) [LENNERTS2001-2007d]

One of the three Iranian pilot hospitals is using a combination of single use and reusable laundry. Two thirds use only reusable laundry that is cleaned in the in-house laundry. The proprietor is again the hospital.

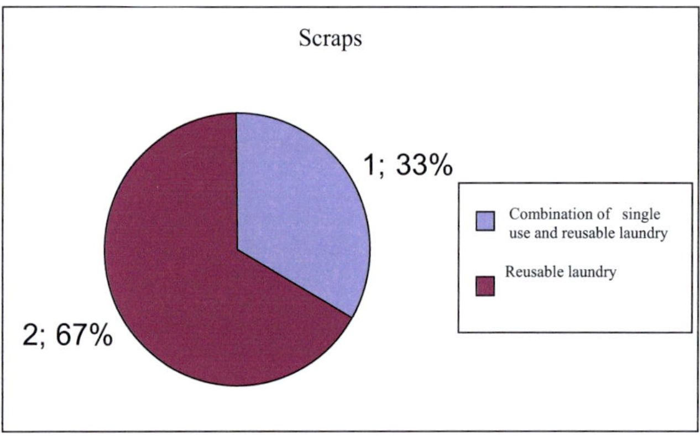

Figure 5-66: Scraps (Iran) [AUTHOR2007]

Work clothes

Figure 5-67 shows that the largest part of the work clothes is property of the hospitals; only two of the hospitals use only rental clothes.

The laundry usage depends decisively on the accoutrement of the employees. The number of pieces of work clothes that must be provided is a cost factor. Special personal work clothes must be provided for every individual employee. Neutral pool clothes have the advantage that the clothes can be shared. In case of illness or vacation the clothes of an employee are unused in his locker; stock quantity can be adjusted to the average attendances and usages. On the other hand, personal work clothes have the advantage that the individual person is responsible; this tends to prevent that the rhythm of collection of dirty laundry and delivery of fresh laundry is missed. As mentioned above, the work clothes which are produced in the in-house sewing shops in Iran are provided only for physicians and the nurse staff. All other employees receive their clothes from the service provider for whom they work, or they themselves are responsible for their work clothes. A trend that becomes apparent in many hospitals is that the nursing staff must pay for their work clothes. That means that the sewing shop produces the clothes and sells them to the nursing staff.

But in general the work clothes are the property of the hospital.

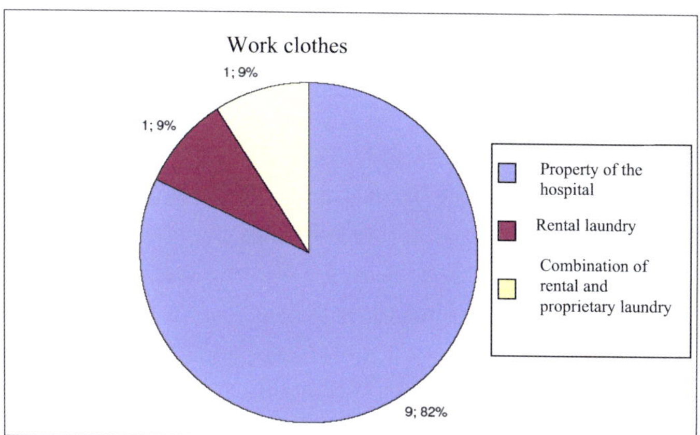

Figure 5-67: Work clothes (Germany) [LENNERTS2001-2007d]

Space requirements for the laundry supply in hospitals

Figure 5-68 shows the space which is provided per bed by each of the hospitals. It should first be observed that only the space for the in-house sewing shop can be completely economised, for instance by using only a rental supply (see house 2 and house 5). All other spaces must exist in certain variations.

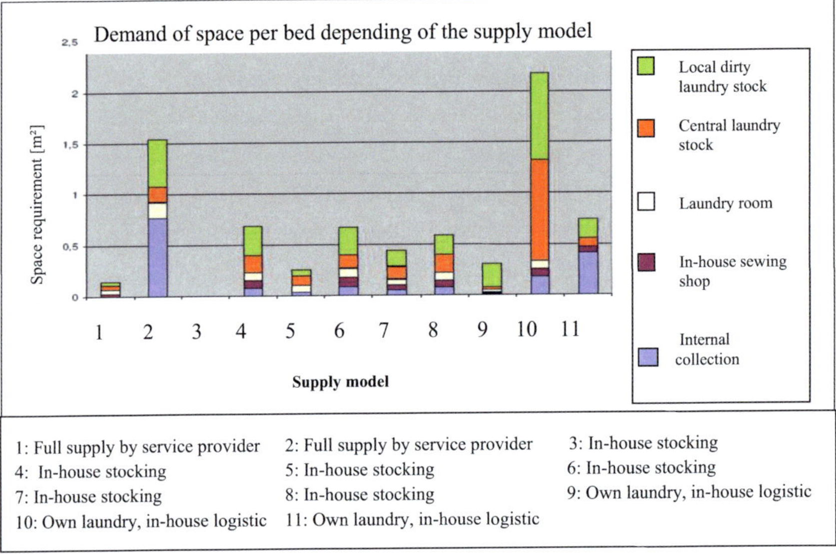

Figure 5-68: Space required per bed (Germany) [LENNERTS2001-2007d]

In figure 5-69 the hospitals are ordered by their supply model. This model does not seem to influence the space requirement. Optimising potential is apparent for the dirty laundry spaces in house three. For local dirty laundry stock at the collection station eventually a combination with a waste management space could be ideal. Astonishing is that houses two and eight a have large space for the delivery of stock, especially because house 2 has no in-house collection. The spaces of house two belong to the in-house laundry that is no longer used and will be torn down.

On the average a space requirement of around 0,5 m² per planed bed can be assumed for the laundry management in hospitals.

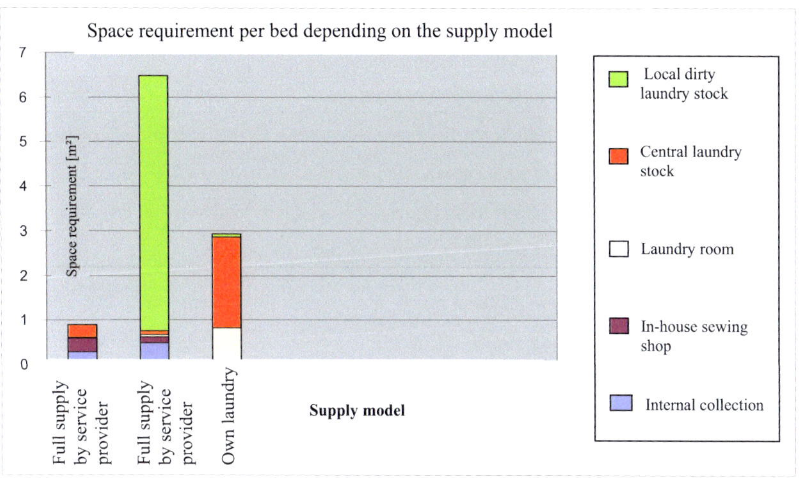

Figure 5-69: Space required per bed (Iran) [AUTHOR2007]

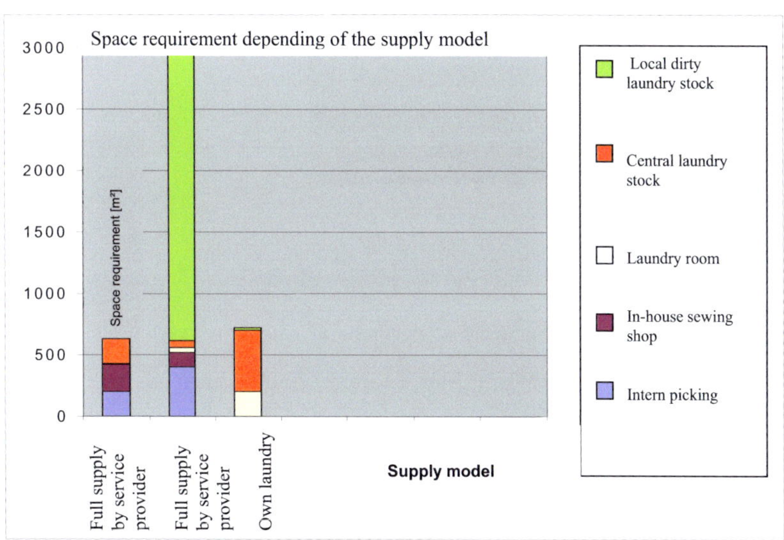

Figure 5-70: Space requirement depending on the supply model (Iran) [AUTHOR2007]

Figure 5-70 shows the demand of space per planed bed for the supply model of the OPIK Iran project. The extreme differences between the different pilot hospitals are apparent. While the local dirty laundry stock (green) requires very little space for hospital three, no such space is provided for hospital one. In hospital two this space equals two thirds of the total space. The central dirty laundry stock in hospital three occupies the most space after the laundry room (white) although neither an in-house collection nor a sewing shop exist in that hospital.

Comparison with figure 5-69 reveals a similar allocation of the spaces. Again the large amount of space for the dirty laundry stock in hospital two is apparent. The space of approximately 3000 m² (as can be seen in figure 5-71) could certainly be used better for other purposes.

Figure 5-71: Laundry space (Iran)

Employees required for the laundry management in hospitals

Figure 5-72 shows the number of in-house employees required per 100 t laundry supply in the hospitals, sorted by supply model. As expected, the numbers in house one and two are the lowest (these are both full supplier).

House one needs employees only for the sewing shop, because in-house laundry is used. In spite of total supply, the hospital uses employees of its own; this model is usual when available employees are providing subcontractors or adopting a part of the work to avoid exemption. The three houses with in-house laundry facilities are also shown in the figure too. It is remarkable that houses that have in-house logistics but have outsourced the washing process need around 0,4 employees in the laundry, but no employee for the administration.

(House ten is comparable, because of an exceptionally low usage of laundry). The employee distribution is reversed in houses that are doing 100% of the process in-house. Hospital three and six need comparatively more staff for the collection and delivery service. The need of employees for the sewing shop is relatively constant for houses with in-house laundry and lies at 0,3 employees per 100t of laundry usage. Only two of three hospitals have specified their employee requirements, although it is realistic to assume that these are low for all hospitals.

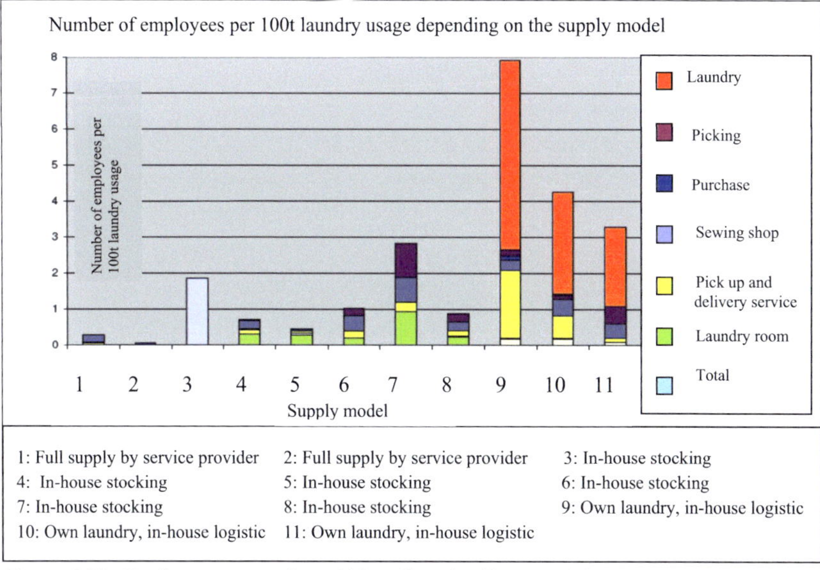

1: Full supply by service provider   2: Full supply by service provider   3: In-house stocking
4: In-house stocking   5: In-house stocking   6: In-house stocking
7: In-house stocking   8: In-house stocking   9: Own laundry, in-house logistic
10: Own laundry, in-house logistic   11: Own laundry, in-house logistic

Figure 5-72: Employees required per 100 t of laundry (Germany) [LENNERTS2001-2007d]

As can be seen in figure 5-73, two of the Iranian hospitals (hospitals 1 and 2) have 15 and 14 persons employed in the laundry. Both hospitals have outsourced their laundry to "Sherkati"-staff (company staff). The Sherkati-staff has a direct contract with the hospital, but they are no public servants. They are given a special contract and work under other conditions than the public servants hospital staff (see table 5-14). The laundry staff of hospital three consists entirely of public servants employees; their number is relatively small. Compared with the employee structure in Germany, it is remarkable that the breakdown is simpler. They are assigned to the laundry facility and the sewing shop. The management is mostly in the hand of the supervisor of the laundry facility. Only hospital three divides the four employees between the laundry facility, pickup and delivery service, laundry collection and sewing shop.

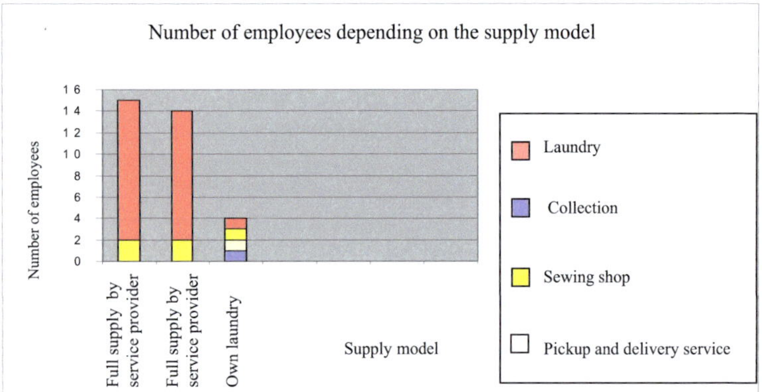

Figure 5-73: Employees required per 100 t of laundry (Iran) [AUTHOR2007]

Costs

The costs of the laundry belonging to the hospital consist of different factors. It must be pointed out that the space for the laundry and maintenance of the laundry are cost factors too. These factors are usually forgotten. Table 5-14 shows extracts of the "types of space usage" (Raumnutzungsarten) according to DIN 277. The spaces in the hospital can be related to these types of usage: local dirty laundry stock at the ward, central dirty laundry stock, laundry room, collection and supply space as main usable area 4, and the sewing shop as main usable area 3.

"types of space usage " (Raumnutzungsarten) according to DIN 277		
Main usable area 3	30000	Production, hand- and machine work, experiments
(HNF3)	32000	Garage
Main usable area 4	40000	Stock, allocation, sell
(HNF4)	41000	Stock room
	44000	Assumption and emission room

Table 5-14: Types of space usage (Raumnutzungsarten) according to DIN 277

According to the "types of space usage" (Raumnutzungsarten) and the dimensions of the spaces, the space costs for 12 months were calculated and related to the total annual laundry consumption in kg for every house.

Workspace	Tariff allocation (Age 40 years old, married,1 child)
Purchase	BAT[26] 3
Administration	BAT 3
Laundry room	MTArb[27] 3
Pickup and delivery service	MTArb 2
Sewing shop	MTArb 4
Collection	MTArb 3

Table 5-15: Tariff allocation of the employees

In addition, the personnel costs for 12 months as determined by the required number of employees and the tariff allocation (see table 5-15) were calculated for every hospital and related to the total annual laundry consumption in kg. The third cost factor is the dirty laundry price per kg that each house has to pay.

Figure 5-74 shows the total costs for the laundry price of each hospital. The data are arranged according to the cost factors; one or two space cost pools (beige for HNF3, purple for HNF4), one personnel cost block (blue) and the dirty laundry price (green) per kg laundry.

In summary it can be concluded that a high potential exists in the optimisation of the cost factors space and in-house personnel. All of the laundries in the OPIK Iran project have an annual budget with which mostly cloth for the sewing shop (that mostly produces bed sheet) is bought. The cost of the personnel is calculated separately. The monthly salary of an employee (depending on his contract) is around 150.000 to 200.000 Tooman (around 110 to 180 €).

---

[26] The BAT (Bundesangestellten Tarif -German civil service pay scale) depends to different factors like age, family status, number of children etc.)

[27] The MTArb (Manteltarif der Arbeiter – Collective agreement of the worker)

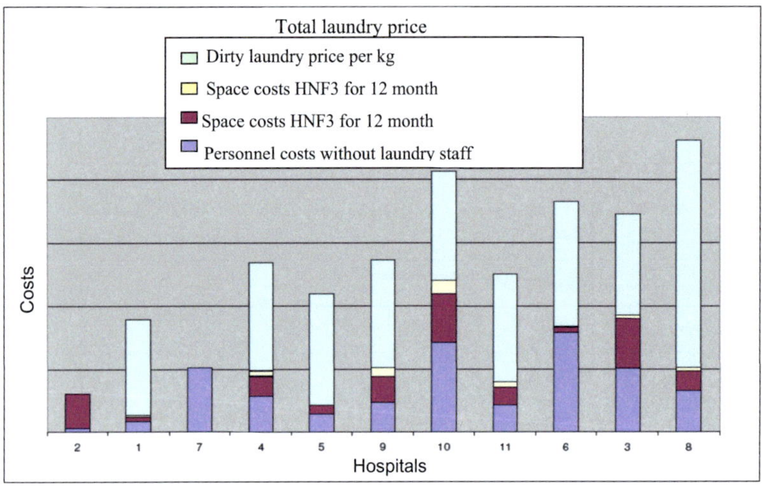

Figure 5-74: Total costs for the laundry (Germany) [LENNERTS2001-2007d]

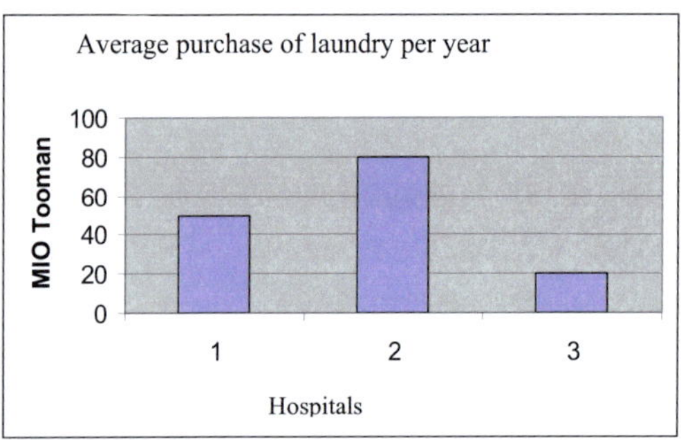

Figure 5-75: Average annual purchase of laundry (Iran) [AUTHOR2007]

It is interesting that no budget is available for maintenance or repair. If devices do not function properly or are damaged, the laundry staff itself does the repair or the department of physical management is asked for assistance.

If a major investment is needed, the budget must first be submitted to the hospital management or the university. That process is time-consuming and linked with a high administrative effort.

The wear and tear due to the use of space is not included as a cost factor. The property management is mainly performed by the technical office of the university, which is responsible for all university buildings and hence only pays attention to the laundry in emergency cases.

None of the hospitals was able to specify a price per kg of laundry. This again demonstrates that low priority is given to the laundry in the organisation of the hospital.

5.4.5 Summary laundry management

Laundries exist in both countries. While in Iran only in-house laundries are running the German hospitals use more and more specialised suppliers that assume the whole life cycle. On modern factory floors the partially automated processes are conducted. The trend in Germany points towards renting of laundry (instead of giving the own laundry to a supplier). So the hospitals declare their needs and the supplier adopt the whole process. Interesting is that even this suppliers buy -because of high production cost (contrary to Iran) - most of the laundry abroad (mostly India, Taiwan or China) while in Iran rolls of fabrics are ordered at the bazaars and needled in the in-house sewing shops. Another noticeable difference is that the German hospitals assume in general the whole laundry while in the Iranian hospitals only provide bed linen and only in exceptional cases staff clothes. These must be bought, washed and repaired by the single staff members.

By using barcodes every single piece of cloth is marked and can be allocated and controlled from the first beginning till it is worn out and disposed. So interesting data, how often the single cloth has been washed, to what hospital / department it belongs etc. can be gathered what enables an informative basis for planning and management of laundries in German hospitals. In Iran the belonging of the single pieces of one hospital is marked by an identification stamp; the departments get different colours so that they could not get confound. The planning of needs is based only on experiences.

Conspicuous was the waste of space in the Iranian laundries. Huge areas were not used while in other departments shortage of space prevailed, disregarded operating costs or the land price of around 2.0 mil. Toomans (around 1500,-€).

Compared with the medical equipment and the technical facilities the laundry management has a mean position in the management hierarchy especially in the Iranian hospitals. This accordingly applies for the social hierarchy of the employees that are working in this department. But in recent years their position is focused, so that they get controlled by hygiene departments and also get considered by the audits of the Institute of Public Health.

5.5 Resume

Contrary to the opinions expressed in the first statements in the kick-off meeting in Tehran, which claimed that the facility management processes of Germany and Iran "could not be compared with each other", it proved possible to analyse and present these processes.

After evaluating this comparison, (see chapter 5.1-5.4) it can be stated that the main features of the process flow and the basic aims are the same in the two countries. Differences arise primarily in the implementation phase and in the external basic conditions, which are dealt with below.

The **legal foundation** in Germany is very extensive, worked out in detail and available for all processes. This scenario does not apply for Iran. Laws and regulations do not exist in all fields. In lot of cases the existing laws and regulations are not or only partly implemented and often the control organs are too weak. The Ministry of Health and Medical Education does, however, attempt to fill the existing gaps with professional committees and technical boards. In addition, only a few professional federations and unions –public as well as private institutions – have been founded in the country. The existing institutions do not have an influence comparable to that of their German counterparts. For example, the trade unions and employees associations have a weaker position. It is evident that this affects the terms of employment and the working standards.

In Iran the **organisation**, the **management** and thereby also the **hierarchical structures** are not subjected to regulations and vary strongly between the hospitals. Plans for annual-, strategy-, work-, budget-, or time plans for every single department usually do not exist, so that the distribution of the resources (monetary as well as equipment or manpower) is primarily determined by the hospital management. Because of the lack of standards and documentation view statistics can be made and used what consequently affects to the planning that is mostly done by experience.

The Iranian hospital directors (Reis Bimarestan) und hospital managers (Modir e Bimarestan) have more scope for decisions and flexibility than their German colleagues. They are, however, consulted and controlled by the regulatory authority and the control units. But the difficulty is the high fluctuation of the top managers whose term of offices is around 2-3 years. Combined with the lack of strategic long term plans this affects the whole processes in the hospital. It is interesting that most of the directors and managers in Germany are physicians and not economists or managers like their colleagues in that position in Iran.

The position of the physicians in the hospital is different in the two countries. In Iran, well known physicians have particularly great influence with considerable effects, e.g. for the purchase of biotechnical equipment.

In general the Iranian system is much more dependent on individuals, which has both positive and negative consequences. One of the positive consequences is that decisions can be made more quickly; one of the negative consequences is that the processes of approval are time consuming (for instance an order for the purchase of pills must be submitted for approval, passes through 3 to 6 levels depending on the hospital and requires the signature of the hospital manager).

The **historical** and **cultural background** of course needs to be considered carefully. The impacts can be seen on the one hand in **education** and **training** as well as the **state of education and research** in the country. On the other hand, the **values, traditions** and **attitudes** play a significant role in the daily association and customs of the people and their attitude towards devices. Closely connected to these issues are the working morale and the sense of responsibility that have direct influence on the manner of work. In this connection, the **religion** and the **philosophy of life** are important factors.

Contrary to the situation in Germany, a patient in Iran is rarely seen as a customer; the supply of services is distinctive. In addition, other perceptions of value exist with respect to real estate as an investment, regular maintenance or the sense of time.

A great advantage for Germany is that it has developed for decades into a modern industrial nation where, for instance, the machine and device industry grew parallel to practically orientated research. In addition, the training facilities especially for technical careers are very broadly diversified and on a high level by international standards.

The lack of training in the practical, particularly in the technical area causes difficulties because the staffs in technical departments mostly draw on experience and are often dismissed because of their low level of education.

In this connection **policy** plays a key role - the policy for the country as a whole, but also the policy for the universities and hospitals.

The sanctions which are imposed on Iran with respect to certain machines and devices as well as the tense political situation hinder the exchange especially in the economic field. This particularly affects the situation in the technical departments of the hospitals. For instance, maintenance contracts for technical and medical equipment cannot be concluded because few companies in the country can afford these services and the German quality standards.

The **economic** and **monetary possibilities** play indeed a most important role. The availability of an ample budget allows the purchase or the repair of modern, power-saving machines, buildings and investments in education and advanced training to secure the satisfaction of the internal as well as the external costumers.

6 Concept development and solution process

The juxtaposition of the FM processes in chapter 5 and the presentation of background knowledge in the preceding parts of the thesis are now followed by an evaluation of the collected insights. A concept, strategy or model (*FM-transferability-tool*) will be developed which can be applied to support the transfer of FM-processes and FM-systems. In order to demonstrate and to test the developed *FM-transferability-tool*, the transfer process is implemented for medical equipment.

In order to be able to investigate the transferability of FM systems between hospitals in different countries, it is first necessary to analyse the concept of "system transferability" precisely. This analysis is followed by a specific definition and interpretation of the concepts "FM-system" and "transferability" in the context of this dissertation.

6.1 System transferability: General definitions

System / system theory

The word system is descended from the Greek „systema (syn + histanai)" which means *collection, orderly entirety*. According to the Langenscheidt dictionary of borrowed words, system is defined as follows: "Entity, functional entity consisting of several component parts that serves for the execution of a specific task or of a sequence of tasks".

The foundation of system theory was laid in 1969 by Ludwig von Bertalanffy who introduced the concept of "system science". His research results together with cybernetics and information theory constitute the basis of the scientific approach. Niklas Luhmann has dealt with systems especially from the point of view of sociology. He uses the *Difference between System and Environment* as starting point for the formation of concepts. "In this context, a differentiated system no longer simply consists of a certain number of components and relationships between components". [ALLEFELD1999, S.22] He does not define a social system as a collection of people but as an "operatively closed process of communication" [LUHMANN1985].

Systems are treated scientifically in the discipline called "system theory". The natural sciences (physics, mathematics, chemistry, biology, and geography), the technical disciplines (informatics, electrical engineering) and the humanities (educational theory, logic, physiology, sociology, psychology, ethnology, semiotics, literature and philosophy) investigate this topic in interdisciplinary fashion, each discipline treating it from its own point of view. System science is decomposed as shown in Fig. 6-1:

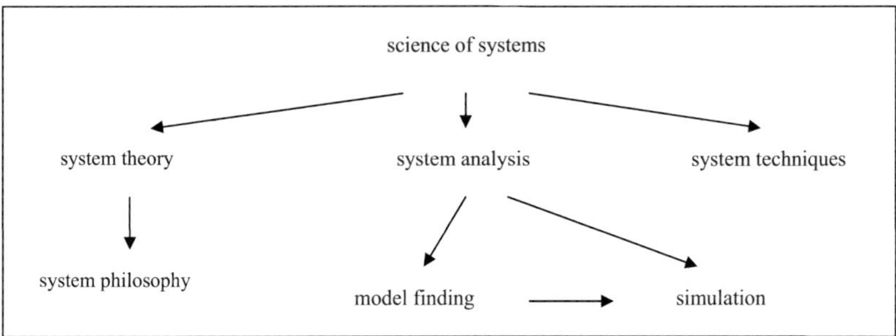

Figure 6-1: Branches of system science [cf. MATTHIS2002, S.4]

Systems are collections of elements that interact, are related to each other and develop coherency. A system is a combination of entities for specific tasks, semantics and purposes. The **system boundaries** separate the system from its environment. The boundary makes it possible to decide what belongs to the system. In general the type of the demarcation at the boundary is determined by the purpose of the system. Systems react to changes of their environment; this reaction is called the **behaviour** of the system. "A system can be described on the one hand through its structure or on the other hand by its behaviour. "[MATTHIS2002, S.12]. The **structure** of a system is comprehended as the arrangement or pattern (shape) of the elements of the system and the interactions that occur between them and explain the regularities. The numbers of influences which can act on a system as well as the number of possible system responses depend on the structure of the system. The quantities which completely describe the state of the system at every point in time are called the **state variables**. The behavioural quantities which describe the influence of the system on its environment can be derived from the state variables. State variables are independent of each other. They also serve as the memory of the system. A state variable consists of the complete set of changes in state during the observed period of time; in this way the state variables also represent the history of the system. The state variables must be distinguished clearly from the **changes in state** and the parameters (flow variables) which describe them [MATTHIS2002,

S.11]. Systems are called **closed** if they exchange only energy (and no mass) with their surroundings. A system is called **open** if it exchanges mass or mass and energy[28] with its surroundings. In general the behaviour of real systems of the technical, social or ecological environment is very complicated and therefore difficult to understand. A considerable reduction of complexity can be achieved by decomposition of the system into subsystems. [MATTHIS2002, S.13]. Systems can be **static** or **dynamic**. They can also be assigned to a **macro level** or a **micro level**. In this connection the system is treated as a whole on the macro level while the elements of the system are considered at the micro level. The structure as well as the attributes and the interactions at the micro level determine the properties of the total system at the macro level. The properties which are determined in this manner form a general framework which acts on the elements at the micro level. The relations between elements of the micro level are the effects of exchange processes (material, energy, information flows). Furthermore, systems can be evaluated **qualitatively** (according to the nature of their influence) or by **quantitative** aspects (according to the strength of their influence).

The analysis of a system starts with an **effect analysis**. In this analysis, the interconnectedness of a system is decomposed into links between the system elements. The impact of a system element on another system element is called an **influence**. An essential tool for the effect analysis is the **effect graph** that shows the links in graphical form. The elements of the system are mapped to the nodes, the effects to the edges of the graph. According to graph theory, every graph can be represented by a **matrix.** The effect graph can be converted into an effect matrix and in this manner subjected to a qualitative effect analysis [MATTHIS2002, S.4]

A given system is itself a component of an ensemble of systems and together with these influences the attributes of a higher system. In order to be able to analyse a part of reality that is of interest, a simplified mapping is generally required. The image of the map is called a **model**; the specific form of the model is determined by the definition of the task and the purpose of the model. According to system theoreticians a system can be neither correct nor incorrect as a model of reality, but is more or less suitable. The identification of a system always depends on the viewer; it is therefore subjective and must be adjusted to the context of the current investigation. According to Bertalanffy, models "...can be both substantial,

---

[28] Alive systems are open systems that need energy and mass exchange for their survival.

technical or organic entities and semantic (theoretical) entities, i.e. verbal or mathematical descriptions. The only decisive point is that they have a specific relation with the mapped original. The process of the construction and testing of a model is called modelling. The mapping of the dynamic behaviour of systems to models is called **simulation**. The purpose of systems analysis, model construction and simulation is to gain a reliable representative description of the behaviour of a real system."

The results of systems analysis are often called **scenarios**; more precisely these are descriptions of system behaviour for specific values of a set of parameters. It is possible to derive predictions from scenarios, which represent the most likely behaviour of the system. These predictions are always subject to uncertainty. Typical applications of systems analysis lie in the area of scientific research, system development in the technical area, system or project management and development planning.

Transferability

"The potential or authority of acquisition or use of an object or an attribute". [Brockhaus2006]

6.2 Development of a "FM- transferability model"

As described in chapter 6.1 a special system definition with a corresponding map of the system is required for the development of a „FM-transferability model". This definition depends on the definition of the task and on the purpose of the model, which must be specified exactly because they determine the "choice of the system boundaries and of the perspective of the view" [MATTHIS2002, S.4]. Consequently a precise description of the goals must take place first and the purpose of the model must be elucidated.

6.2.1   Specific definition and modelling process for this thesis

This thesis deals with the **management of hospital processes** in different countries. The research is conducted on the basis of special processes in Germany and Iran.
It was possible in the analysis to identify similarities, differences, regularities and common characteristics and to compare the processes under the prevailing general conditions.
This knowledge will now be used to develop first a "FM-System Transferability Model" and then a "Tool" which is based on this model. This tool should assist facility managers to transfer a FM system into another environment or to implement a "foreign" FM system in

their own organisation (in our case a hospital). Figure 6-2 shows the definitions which are used in this context, together with the chapters in which they were treated.

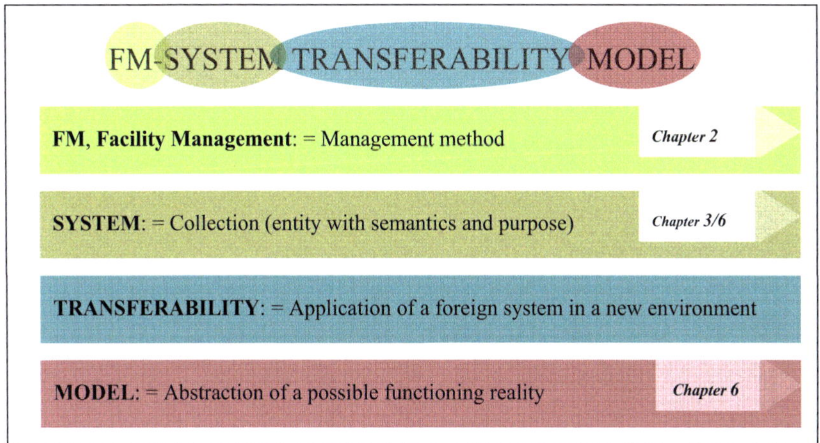

Figure 6-2: Definition FM-system transferability model [AUTHOR2007]

FM is comprehended as the management of a hospital according to EN 15221/1 *"The integration of processes within an organisation to maintain and develop the agreed services which support and improve the effectiveness of its primary activities."* (cf. chapter 2). The system, which is treated in this manner, comprises the services which are provided in a hospital. In the model shown in figure 6-3 these are symbolised by the Red Cross (symbol of the German FM in hospitals) and the Red Crescent (standing for the FM services in Iranian hospitals).

Two **levels of analysis** are distinguished, the micro level and the macro level. At the **micro level** or hospital level all processes, structures, functions and influence factors are investigated that exist in facility management and are decided or performed at the hospital level. The parameters influencing the FM processes at the country or state level are, however, analysed at the **macro level**.

As can be seen in figure 6-3, numerous parameters of influence act on the system, both from the micro and at the macro level. These parameters are grouped according to subject and collected under generic names in order to make it possible to analyse their effects on the system and their interactions with each other.

The parameter groups are both quantitatively and qualitatively measurable and rateable; they are assessed according to their significance and influence[29]. This signifies, while some of the parameters have fixed values that can be determined reliably, e.g. Gross domestic product (GDP) of the counties (see Table 6-1 see page 145) , other parameters must be estimated or are very difficult to assess, for instance working morale, sense of responsibility, religiousness, etc.

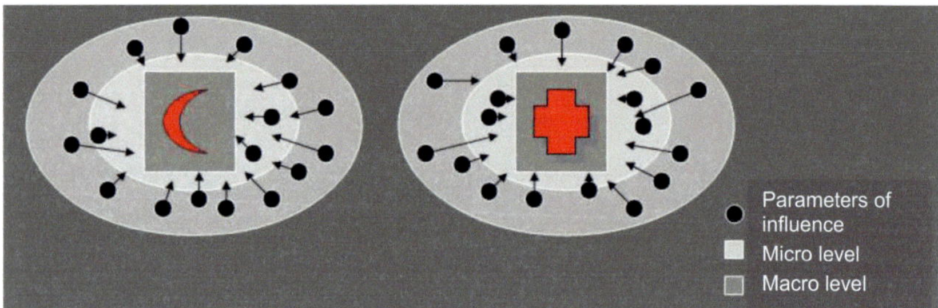

Figure 6-3: Parameters of influence [AUTHOR2007]

It is necessary to point out, as Goethe has already observed, that "nothing happens in live nature that is not connected with the whole". This means that concepts or subjects that are closely related and partially directly connected are drawn apart and treated in different subject groups. To conclude, the significance of a holistic approach is emphasized, which will be treated in greater detail in later sections of this thesis. At first, small "parts of the puzzle" must be defined which will then be used to create an image that corresponds to reality.

The experiences of the OPIK project (workshop, questionnaires…), as well as numerous discussions with experts and interviews, conducted in Iran as well as in Germany, provided important impulses for the determination of the generic terms that are used for the parameters of influence of the system.

The numerous parameters were divided in nine groups:

Economy/ financial strength / monetary potential opportunities

All areas that relate to the economy and therefore financial aspects are collected under this heading. The factors describe aspects both of national and business economics, but mainly are

---

[29] The assessment depends on the viewer und can fluctuate with country and perception

parameters of health economy, for example health expenditure per gross domestic product, annual national budget for health and education, earnings of the hospital, etc…

Politics

According to Niklas Luhmann [HORSTER2005 p.159 and following] politics are the "complex of social processes that serve especially to guarantee administrative (material) decisions … Politics should assume responsibility, legitimise and provide the power base required for the implementation of objective administrative decisions." For the purposes of this thesis, the term politics is defined to include the determination of objectives as well as goal-oriented actions, activities and strategies to reach the specified objectives. A policy includes tactics, which is defined as the path that is followed and the -short and long term- measures that are taken on this path, for example cooperate management or strategy as well as economic, regulatory, structural and process policy.

Management, organisation, administration

The terms leadership, organisation of groups and hierarchy are collected in this group of influence factors. The term management (in Latin *manum agere* "leading the hand") refers to "a process whose goal is to achieve desired or planed results by performing actions and using resources". [BROCKHAUS2006]

The comprehension, guidance and supervision of the transition of a set of tasks from a stochastic system to a stable deterministic system, including the assumption of leadership and responsibility, are called administration. The term organisation (in Greek *organon* = tool) describes the manner of planning and implementation of a project. Organisation is comprehended as an institution, a function or an instrument.

State of education and research

The state of education and research in a country describes the abilities of its people, based on their natural gifts and guided by their education. "The term education refers to both the process (to form oneself) and the state (to be lettered)". It stands for the life-long development process of human beings that extends their spiritual, cultural and practical abilities as well as their personal and social competences.[30]

---

[30] Expert discussion with Prof. Dr. Farsin Banki, in Teheran (September 2006) [see Appendix - A21]

Research is the methodical and systematic search for new knowledge. Research, which is usually conducted in a scientific environment, according Taylor "contributes to the extension of human knowledge and relies on established knowledge. Research also attempts to disprove the existing systems, rules or theories in order to achieve a new understanding of the phenomena in our surroundings. "

Judicative, legal regulations

The judicative provides the legal framework for institutions, fields of activity and persons. This framework is created by the legislators and promulgated in the form of laws which are arranged at different levels of hierarchy. [BROCKHAUS2006]

Culture and Social Structure

The term social structure is defined [according to SCHÄFERS2003] in this thesis not only as a set of possible subdivisions of societies according to social attributes, as is done in sociology, but also as the social structure of groups and organisations (in our case hospitals), which in sociology is treated as part of small group or organisation sociology. In this context, social structures are comprehended as **social classes** (social status, income, education, training, professional position).

Demography and demographical development also are significant aspects of social structure (age, nationality, ethics origin, environment, social environment, life style, consumer behaviour, linguistic groups and mobility of groups or social change which is defined as the change of social structure with time).

The term culture denotes the historical and cultural background. It comprises historical events that have influenced the culture of the country and the hospital directly or indirectly. In this context the term culture is comprehended as the entire set of knowledge, customs, traditions and skills which a member of a society acquires. Another definition of culture is given by Dave Wilson [WILSON2006, S.99ff] in his paper "Getting cultured". In his opinion "culture is ... the way life is organised from societal norms and expectations through to personal behaviour". Behaviour is comprehended as language, customs and traditions, where the social norms embody the standards and values that are approved by society. Other essential factors influencing social structure are **religion, ideology and philosophy**. These terms describe different cultural phenomena which influence human behaviour, way of thinking and concepts of values.

Public and private institutions

According to the Brockhaus lexicon an institution is „a social, state or religious facility where certain tasks are performed, usually in a manner that is regulated by law" [BROCKHAUS2006]. This term therefore includes all public and private institutions as well as all NGOs[31] of a country. In the context of this dissertation the health- and socio-political institutions like the medical association, health insurances and support groups are addressed.

Infrastructure

This term descends from the Latin „*infra*" (under, below). It denotes all „long-term basic facilities of personnel, material and institutional type that guarantee the functioning of a national economy based on division of labour " and includes administration, traffic, trade and production. [BROCKHAUS2006]

Geography, climate, environmental conditions

The term geography denotes spatial structure and events, as well as their influences on human beings. Climatology is subdivided into geographic and meteorological climatology. Geographic climatology is defined by Blüthgen [BLÜTHGEN, WEISCHET1980, p.1] as follows: "The geographic climate is the summary of the atmospheric conditions and weather-related events which occur near to or influence the surface of the earth during a longer period of time and are typical for a location, a landscape or a larger region.".

Hendel [RISTO2007, p.5] defines the meteorological climate as follows: "Climate is the typical local frequency distribution of atmospheric states and processes during an adequate reference period which is to be chosen in such a way that the frequency distribution of the atmospheric states and processes does justice to the typical circumstances at the reference location". In this thesis, the focus will be on environmental conditions which directly or indirectly influence the increasing spread of illnesses, the conditions for the construction of hospitals, etc..

---

[31] NGO (Non-Governmental Organization)

6.2.2 Model investigation

After the development of the „cognition model" and the definition of the parameters of influence *FM System Transferability Tools* should be developed. This tools should enable to make predictions concerning to FM System transfer and should identify issues which may cause problems if a transfer is attempted. To approach such tools different ways and methods are executed to approach to a solution. It must be noticed that such tools are only models or methods that depict trends and tendencies.

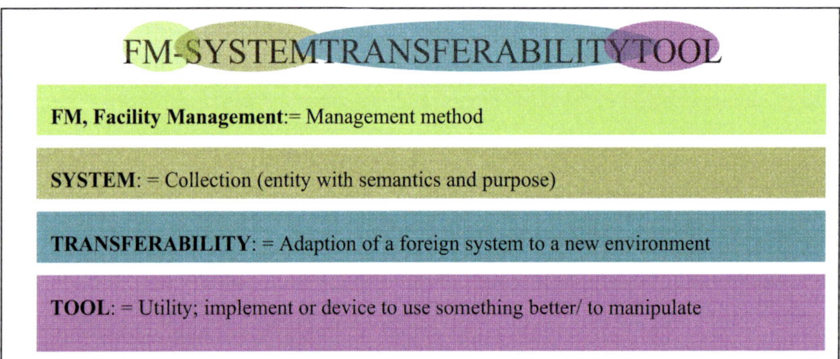

Figure 6-4: Definition of a FM System Transferability Tool [AUTHOR2007]

Development of tools with the help of models:

6.2.2.1 Parameters of Influence Model

The effect analysis is initiated by considering the relations between the nine parameters of influence (see figure 6-5) that were emerged for this transferability model. One approach to differ this nine main influence parameters could be to separate them by their impact. Therefore a distinction is made between direct and indirect impact of the influence parameters among themselves to define a trend or tendency; the direct effect graphs have greater significance in the subsequent evaluation because their effects are noticed directly. For example the legal conditions of a country or an organisation like a hospital have direct impact to the management or the strategy planning while the impact of the climate is not high. By using the experiences made during the OPIK-Iran project the following relations between the single influence parameters were determined by the author for Iran (see figure 6-5).

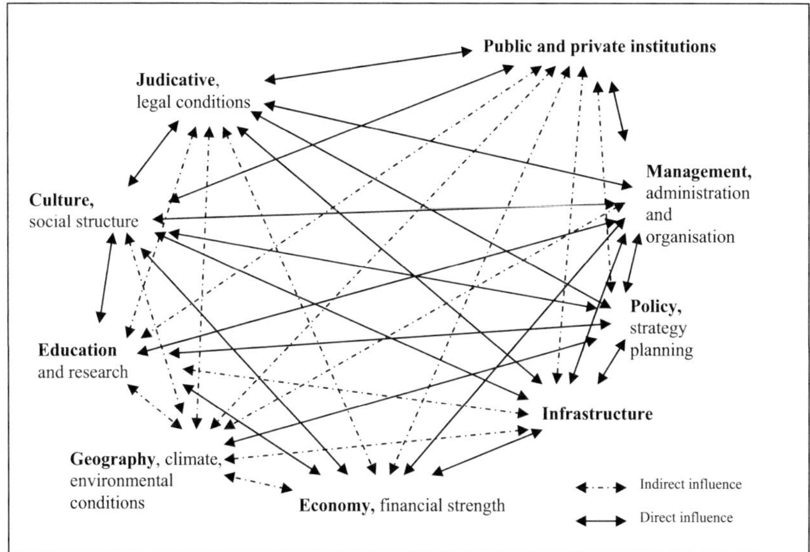

Figure 6-5: Relations between the parameters of influence [AUTHOR2007]

After the determination of the relation of the parameters the numbers of the direct or indirect graphs of influence were separated in Figure 6-6 to get a tendency of the position of the single parameters. It is remarkable that management, policy and culture affect nearly all other areas directly, followed by the judicative, economy and infrastructure. In contrast herewith, the direct influence of public and private institutions and geography on the other influence parameters is small.

This figure draws attention to the important position of the management in a system. It is evident that nearly all parameters of this group have direct influence on the system, which emphasises once again the high rank of facility management in hospitals.

It must, however, be remembered that these results reflect only the direct and indirect influences and can only refer to tendencies and trends and must be analysed and evaluated for every single country anew.

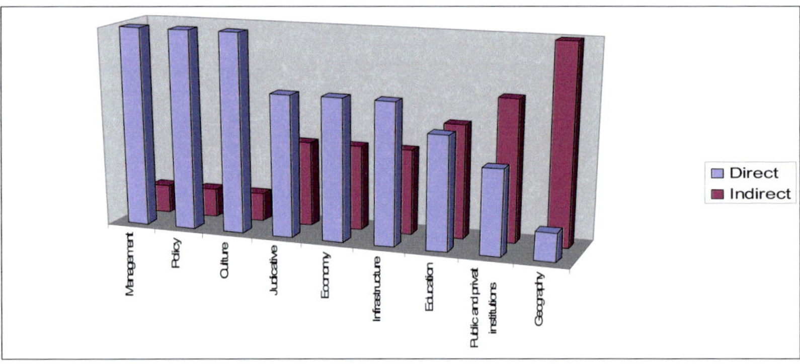

Figure 6-6: Direct and indirect influence of the graphs [AUTHOR2007]

6.2.2.2 The Matrix Model

Since every graph according to graph theory can be represented by a matrix, another idea was to use a matrix model for this thesis. (See figure 6-7).

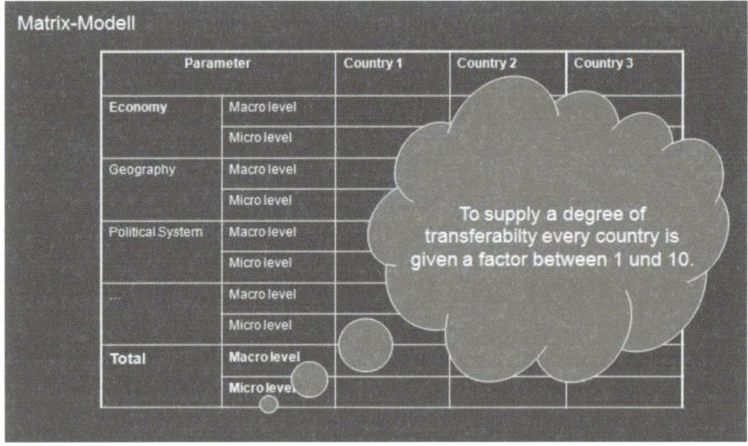

Figure 6-7: Matrix Model [AUTHOR2007]

In this model the parameters of influence are spitted into the macro and micro level. The countries which are investigated are arranged on the horizontal axis. Next a degree between 1 and 10 is determined for every parameter that must be given by an appraiser. This degree specifies the grade of transferability; the degree 1 indicates an easy transfer, the degree 10 a complicated transfer. The advantage of this model is that a quick overview and a direct comparison of the countries are gained. The straightforward presentation with constant values allows the calculation of a final sum. This model can be a good solution to compare multiple countires.

The big challenge in this model is the determination of the degree. Depending to the appraiser (expert or organisation) a high fluctuation can be obtained. This model as well can only give - dependent to the preciseness of the appraiser- trend and tendencies for the transferability of a system.

6.2.2.3 The Share Model

The "Share Model", which is presented in figure 6-8, also contains the parameters of influence. In addition, the weights of the various parameters which are assigned to the macro or micro levels, can be read from this model. In this figure all parts have the same size, because they're not weighted yet.

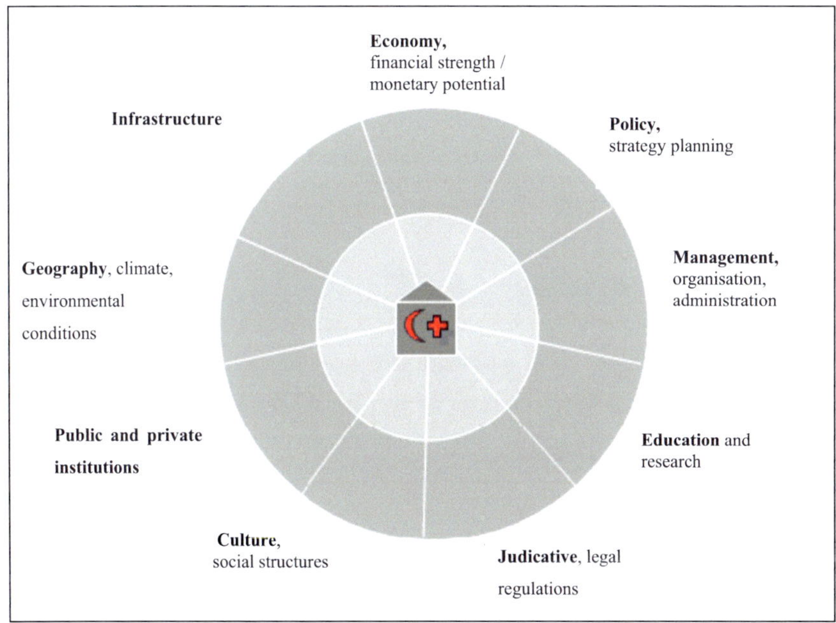

Figure 6-8: FM-Share-Model [AUTHOR2007]

This model offers the advantage that the weighting is readily observed visually, so that the critical transfer areas become evident "with one look ".

6.2.2.4 Comparison of the models

While the Parameter of Influence Model indicates the reciprocal relation of the single influence parameters the Matrix and the Share Model focus on the grade of transferability using the macro and micro level of the several influence parameters. As distinguished from the Parameter of Influence Model this models regard the macro and the micro level what provides an insight into the different levels of our system and hence a detailed review.

But in all models the question of the evaluation arises. How can the dependency, the shares or the degree of transferability be determined? According to which criteria should - for example- a value of 5 been given for the infrastructure of a country? How does this emerge and how can it be proved that this value should not be 4 or 3? This problem has been one of the main topics in the discussions with the experts[32] and has turned out to be rather complicated.

At the beginning some participants expressed the opinion that especially the presentation in the matrix model is not suitable. It turns out that it is indeed an impossible undertaking to determine an exact degree of transferability, since there are simply too many factors of influence, many of which cannot be captured with numbers and it always depends to the view and perception of the appraiser.

6.2.3   Analysis

It is not denial that the exact weighting of the single parameters of influence is not possible but there are analyses by using **qualitative** and **quantitative indicators** of the project countries that provide getting an exact comparison. Two different analyses are introduced to approach a comparison.

6.2.3.1 Expert Analysis

One method to get an overview is the inquiring of experts. This analyse allows to get

---

[32] Beneath the Iranian experts (listed in AppendixA-21) this issue has been discussed in detail with Prof. *Prof. Dr. R. G. Heinze,* chairman of the Institute of Work and Economic Sociology at the Ruhr-University Bochum, *Prof. Dr. rer. pol. Dipl.-Soz. G. Wachtler,* Institute of Sociology at the Bergische-University Wuppertal and *Prof. Dr. H. Lietzmann,* Institute of Political Science at the Bergische-University Wuppertal.

expeditious information by specialists with experience in the specific analysed field. This method was the first analyse that was made in the OPIK-Project in Iran, too. The experts of the project (see Appendix-A21) were assigned to weighting the nine influence parameters by an ABC analyse, while A represents a high influence, C marks a low influence of the parameter.

The advantages of this method are the short time that is needed and the expert opinion that mostly comes along with many years of experience in the special analysed topic. The objectivity and the universally valid must be regarded as critical points of that analyse.

Figure 6-9 shows the result of the expert analyses carried out by 17 experts in Iran (named in Appendix A-20).

Management has the top ranking, followed by politics and culture in second place. While politics have a stronger influence at the macro level, culture plays a bigger role at the micro level. The judicative follow in the fourth place, followed by economy. Also in this context, the ranking of geography is lowest.

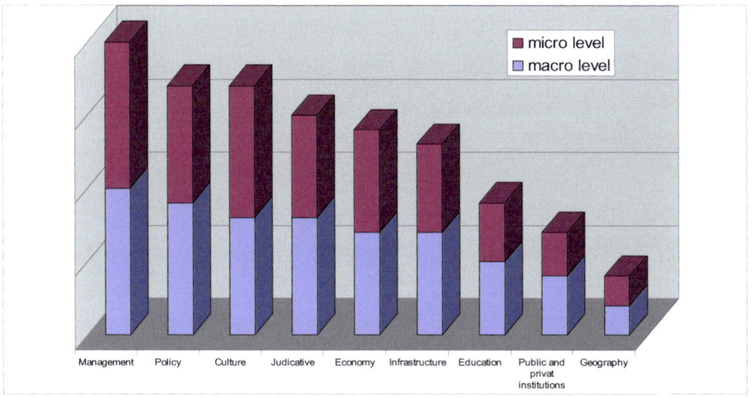

Figure 6-9: Ranking of the parameters of influence by expert opinions [AUTHOR2007]

6.2.3.2 Indicator Analysis

While the Expert analyse is a highly quantitative analyse of the parameters of influence the indicator analyse mainly uses quality indicators to define the parameters that are mostly based on OECD and WHO indicators. The determination of these influence indicators will assist greatly in the determination of the degree of transferability. It is evident that only the most important indicators can be taken into consideration[33] in this thesis, since not all effects can be considered and this would go beyond the scope of the thesis. The aim is to obtain an estimate that permits a rough assessment.

The indicators are determined for the micro as well as for the macro level. The macro level reflects the country or the health system, whereas the hospital- and process domains are treated at the micro level.

Economy/ financial strength / monetary potential

Indicators of the macro level:	
Country	*Health sector*
• GDP (per capita) in US $	• Portion of the expenditures for health of GPD in %
• Economic growth to sectors (real, %)	• Per capita government expenditure on health at average exchange rate US $
• GDP – emergence (%)	• Expenditures for medicine (per capita) US $
• Unemployment rate (%)	• Cost structure of ambulant health facilities
• Inflation rate (%)	• Expenditure for rehabilitation
• Foreign depts. (Bil. US $) • Export/ import • Currency, monetary value	• Total expenditure for health in US $ per capita
Indicators of the micro level:	
Hospital	*Department (e.g. medical equipment)*
• Costs of the hospitals according to cost type and hospital size range	• Expenditure of material, personnel, training
• DRG-case-based lump sum, revenues and benefits • Income	• Budget of the department • Budget for maintenance/ repair • Budget of purchase

Table 6-1: Economy [AUTHOR2007 based on the OECD and WHO indicators]

---

[33] The "Action program of the community of the EU, division of public health,, works with more than 400 health indicators in the field of health alone, excluding economy and politics. See
(http://ec.europa.eu/health/ph_information/indicators/indic_data_de.htm) or
http://www.gesis.org/en/social_monitoring/social_indicators/Data/Eusi/index.htm

Politics

Indicators of the macro level:	
Country	*Health sector*
• Number and quality of the reforms	• Number and quality of health reforms
• Voter Turnout	• Degree of unionisation
• Number of party members	
• Memberships in international organisations	
Indicators of the micro level:	
Hospital	*Department (e.g. medical equipment)*
• Degree of unionisation	• Degree of unionisation
• Strategy planning (five year plan, …)	• Strategy planning (five year plan, …)

Table 6-2: Politics [AUTHOR2007 based on the OECD and WHO indicators]

Management, organisation, administration

Indicators of the macro level:	
Country	*Health sector*
• Governmental system	• Hospital planning
• Administration system	
Indicators of the micro level:	
Hospital	*Department (e.g. medical equipment)*
• Hospital management	• Working processes
• Organisation chart of the hospitals	

Table 6-3: Management [AUTHOR2007 based on the OECD and WHO indicators]

Education and research

Indicators of the macro level:	
Country	*Health sector*
• Number and quality of the universities, research centres	• Number and quality of the universities, research centres, university hospitals
• Illiterate ratio	• Number of physicians

• Number and quality of education and training facilities	• Physicians per 1000 population
• Number and quality of congresses and exhibitions	• Number and quality of the training facilities for medical staff
Indicators of the micro level:	
Hospital	*Department (e.g. medical equipment)*
• Number and quality of total staff/ technicians	• Number and quality of technicians
• Number and quality of the medical staff	• Number and quality of trainings
• Number and quality of trainings	

Table 6-4: Education [AUTHOR2007 based on the OECD and WHO indicators]

Judicative, legal regulations

Indicators of the macro level:	
Country	*Health sector*
Number, quality and control of the accomplishment of laws and regulations	Number, quality and control of the accomplishment of laws and regulations
Indicators of the micro level:	
Hospital	*Department (e.g. medical equipment)*
Number, quality and control of the accomplishment of laws and regulations	Number, quality and control of the accomplishment of laws and regulations

Table 6-5:Judicative [AUTHOR2007 based on the OECD and WHO indicators]

Culture, Social Structures

Indicators of the macro level:	
Country	*Health sector*
• Official language	• Life expectancy (m/w)
• Total population	• Infant mortality rate (per 1000 Life birth)
• Population density	• Percentage of Population aged 65+ years
• Population growth rate (%)	• Percentage of Population aged 14 -
• Population portion of the city dwellers	• Age ratio (Relation of the 20+ to 60-)
• Mobility, agglomeration, migration	• Population with access to local health services (%)

• prosperity, differential social stratum	
• self- contentment, luck, anxiety index	
Indicators of the micro level:	
Hospital	*Department (e.g. medical equipment)*
• Number of admissions per year	• Age distribution of the staff
• Average period spent in the hospital (days)	• Fluctuation of the staff
• Disability days per staff member	

Table 6-6: Culture [AUTHOR2007 based on the OECD and WHO indicators]

Public and private institutions

Indicators of the macro level:	
Country	*Health sector*
Number and influence of the institutions, offices and associations	Number and influence of the institutions, offices and associations
Indicators of the micro level:	
Hospital	*Department (e.g. medical equipment)*
Number and influence of the institutions, offices and associations	Number and influence of the institutions, offices and associations

Table 6-7: Public and private institutions [AUTHOR2007]

Infrastructure

Indicators of the macro level:	
Country	*Health sector*
• Streets (km), highways (km)	• Number of beds (per 10 000 capita)
• Railway line (km)	• Population with access to sanitary facilities (%)
• Airports	• Number of hospitals
• Transport service	• Population with access to clean drinking water (%)
• Extent and options for use of available traffic equipments	

Indicators of the micro level:	
Hospital	*Department (e.g. medical equipment)*
• Number of beds (total)	• Route per day (km)
• Number of elevators	

Table 6-8: Infrastructure [AUTHOR2007 based on the OECD and WHO indicators]

Geography, climate, environmental conditions

Indicators of the macro level:	
Country	*Health sector*
• Wood (wood- and settlement area)	• Quality of drinking water, pollution of waters
• Concern about the protection of the environment (lack of green space, destruction of the landscape….)	• Atmospheric pollution
• Expenditure for environmental protection on GDP	
• Emissionen per inhabitant	
Indicators of the micro level:	
Hospital	*Department (e.g. medical equipment)*
• Expenditure for environmental protection on the income of the hospitals	• Amount of waste (t)
• Amount of waste (t)	
• Usage of water	

Table 6-9: Geography [AUTHOR2007 based on the OECD and WHO indicators]

6.3   Solution approach by using the process medical equipment

After some methods and tools to get a values and tendencies for the transferability of Facility Management in hospital a solution approach should be realized by using a concrete example; the process of medical equipment. This process is chosen because the medical equipment market is worldwide increasing[34] [BRÄUNINGER2008, Page 5] especially in the emerging economies (like Iran (see figure 6-10)).

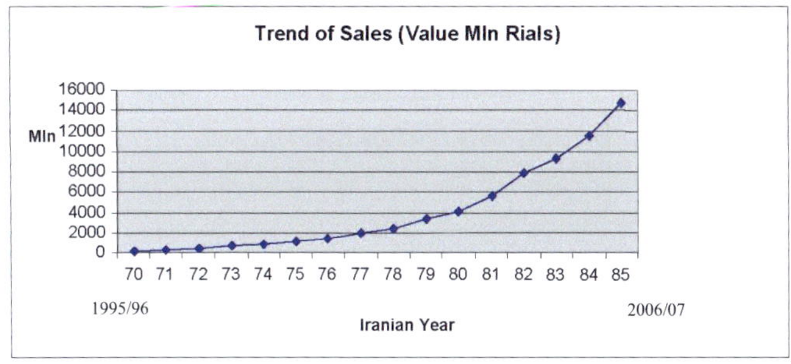

Figure 6-10: Trend of sales in the Iranian medical equipment market [MOHME2008]

This tendency is predicted because of the demographic development from the shift of the morbidity spectrum to chronically diseases [compare Chapter 3]. Behind the USA (31%)[35], Germany (15%)[35] is belonging to the leading medical equipment exporting countries, while Japan releases rank three to Great Britain (6,5 %)[35]. Additionally must be noticed that more that 50% of the medical equipments produced in Germany are exported. [BMBF2005]

So it is obvious to analyse the transferability of the Facility Management process medical equipment in hospitals from Germany to Iran. The scenario of this system is mapped with the aim to obtain a prognosis and to determine a probable degree of transferability. Because the relations between the parameters of influence have been already evaluated in Chapter 6.2.2.1 the next step, model finding is chosen. In the last chapter two models are introduced; the matrix and the share model. Here the share model is elected because we only compare two countries with each other; and not specific values but tendencies could be evaluated. Hence in the first step the indicator analyse is carried into execution; the indicators of the parameters of

---

[34] Between 1995 and 2002 the market of medical equipment increases 7,5% per year , twice as much as the global trading with industrial products (3,4% per year)

[35] According to VDE2005: Deutschland in der Medizintechnik international führend

influence of the project countries compared and interpreted. The assessment follows in the second step. Afterwards all methods and analysis (Parameters of Influence; the share model, and the expert analyse) are faced and compared and a final solution is evaluated.

6.3.1 Analyse of the parameters of influence

The project countries Iran und Germany and their health systems are considered at the macro level. At the micro level, the hospitals and the department of medical equipment (the FM-processes) are analysed.

Because some of the topics (especially the process analysis of medical equipment) have been discussed in detail in the previous chapters, only brief comparisons and references to the respective chapter are presented in this chapter.

The project countries

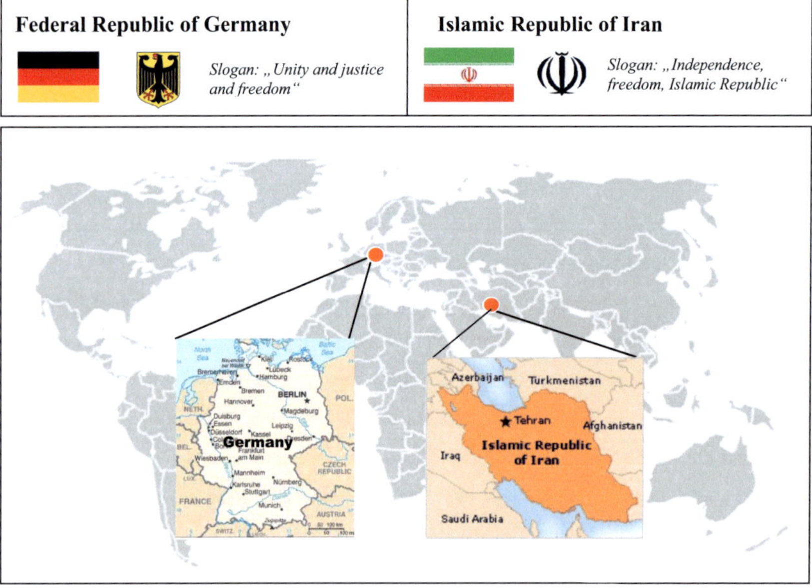

Figure 6-11: The project countries [AUTHOR2007]

a. Economy/ financial strength / monetary potential

Macro level: Country	
D[36]	IR[37]
Germany is relatively poor in raw material; its economy is concentrated mainly on the industrial and service providing sector. Large surfaces of the county are used agriculturally, while only 2-3% of the employees are active in agriculture. With a GDP of 2200 bio € (2004) Germany is the third-richest national economy of the world. The export with 894 bio € (2006) makes Germany the "export world champion". The automotive, utility vehicle, electro-technical, machine manufacture and chemical industries are regarded as the most competitive global branches of German industry. The population capable of work (between 15-64 years of age) is about 55.7 mil.. The unemployment rate is 9.1% (May, 2007).	The most important economic sector in Iran is the rich oil and natural gas resources. Other important economic sectors are the textile industry, agriculture as well as cement and building material production. With approx. 200 mil. t of raised oil Iran ranks fourth and with approx. 80 bio. m³ of raised natural gas seventh in the worldwide production of oil and natural gas. It controls approx.18 bio. t of known oil reserves, the third largest oil reserve in the world. With a natural gas reserve of about 27000 billion m³ Iran ranks second worldwide. The population capable of work in Iran is 23.68 mil. people. The unemployment rate is about 11.2%. The service providing sector offers 45%, agriculture 30% and industry 25% of the workplaces. The government maintains a very big administrative apparatus.

Economy	D[36]	IR[37]
Inflation ratio (%)	2,0 (2005)	13,0 (2005/06)[38]
Economic growth (%)	0,9 (2006)	6,10 (2005/06)
Unemployment (%)	12,6 (2006)	15,7 (2005/06)
Foreign debt (Mrd. US$) in bn. €	2.689 (2004); 3.023 (2005)	23,07 (2004/05)

---

[36] Based on the OECD and the Statistisches Bundesamt (2006)

[37] Based on the OECD and the Iranian Statistic Center (ISC) (2005/2006)

[38] 2006 equates to the Iranian year 1383 (from 21th of March 2005 to 20th of March 2006)

GDP (per habitants) in US$	25 608 (2006)	1 996 (2005-06)
GDP- Emergence (%)	Industry 25,8; Trade, Tourist and Traffic 18,0; Construction Industry 3,9; Agriculture and Forest Industry 0,9; others 22,3(2005)	Oil sector 25; Agriculture 11,2; Industry, Mining Industry 11,9; Construction Industry 4,4; Services 47,5 (2004-05)
Economic growth by sectors (real, %)	Industry +2,6; Construction Industy3,8; Agriculture and Forest Industry -4,9 Trade, Tourist and Traffic +1,4; Financing, Letting, Service +1,7 (2005)	Oil sector +2,6; Non- Oil sector +5,1; Agricultural Industry +2,2; Manufacturing Industry; Mining Industry +11,9; Construction Industry - 4,1; Service +4,6 (2004-05)
Import in Mio US $	585 000 (2006)	25 260 (2005/06)
Export in Mio US$	696 900 (2006)	29 880(2005/06)
Currency	1 Euro abbr. 1.35 US $ (2006)	1 Iranian Rial abbr. 0,000108 US $ (2005/06)

Table 6-10: Economy indicators-country [AUTHOR2007]

Macro level: Health system			
D (see chapter 3.1.1.2)		IR (see chapter 3.1.2 .2)	
The German health system is derived from the Bismarck model. It is based on governmental and non-governmental institutions and divided into ambulant and stationary care. The system is primarily financed by insurances. The "Family doctor" (Hausarzt) model obtains. About every ninth person is employed in the health economy.		The Iranian health system can rather be assigned to the Beveridge model. The health care system is guided mainly by the state and primarily finances itself from state subsidies or medical costs (Out of Pocket). There is no health insurance obligation although the rural population has a free state health insurance.	
Health Sector		D[39]	IR[40]
Portion of the expenditures for health of GPD in %		10,9 (2004)	5,7 (2005/06)
Total expenditures for health in $ per capita		3204 (2003)	158 (2005/06)

---

[39] Based on the Statistisches Bundesamt (2006)

[40] Based on the Statistical Centre of Iran (SCI) and MOHME (2006)

Per capita government expenditure on health at average exchange rate US $	2506 (2003)	75 (2005/06)
Expenditures for medicine (per capita) US $	318 (2006)	157.8 (2004/05)

Table 6-11: Economy indicators- health system   [AUTHOR2007]

Micro level: Hospital		
D (see chapter 3.2.1.2)	**IR (see chapter 3.2.2 .2)**	
Germany has a dual financing system; while the investment costs are carried by the states, the health insurances takes over the running operating expenses and medical costs. Since 2003 the account system follows to case-based lump sum in the form of DRG's. 70% of all hospitals costs are personnel expenditure.	The Iranian hospitals are financed by the state (Ministry of Health and Medical Education, universities) or by private investors. Hospital construction is primarily financed through the budget of the hospital, through subsidies of the ministry or by donations. Expensive are not so much the personnel costs but in particular foreign machines, devices and their spare parts.	
Hospital	**D[41]**	**IR[42]**
Total hospital personnel	1 079 832 (2004)	31394[43] (2004/05)
Medical personnel in the hospitals	132 100 (2001)	509 165 (2005/06)
Non medical personnel in the hospitals	899 300 (2001)	183 299 (2005/06)
Personnel expenditure in hospitals in %	65,3 (2004)	45 (2004/05)

Table 6-12: Economy indicators-hospital [AUTHOR2007]

---

[41] Based on the Statistisches Bundesamt (2006)

[42] Based on the Statistical Centre of Iran (SCI) and MOHME (2006)

[43] Estimated based on average number of medical and non-medical personnel per bed

Micro level: Department medical equipment	
D (see chapter 5.1)	IR (see chapter 4.2.2.2)
An annual budget is assigned to the department of medical equipment by the hospital. The budget is administered by the supervisor of this department and used for maintenance and repair.	This department exists only in around 10% of the hospitals. The budget is administered by the management (Modiriat). Any bills must be handed over to the management which refers them to the financial department (Bakhshe Mali).

b. Politics

Macro level: Country	
D[44]	IR[44]
The political system is federal and organised as a parliamentary democracy. The federation exists of 16 partially-sovereign federal states. The state is regulated by the basic law. Head of state is the Federal President (Bundespräsident) with representative tasks. By protocol the president of the German Bundestag, the Federal Chancellor and the acting President of the Bundesrat follow. Head of the government of Germany is the Federal Chancellor. He (or she) has the guideline competence for the policy of the Federal Government (Kanzlerdemokratie). Germany is organised federally, i.e. there are two levels in the political system: the national level which represents the whole country of	Since the revolution of 1979 the uppermost jurist ("revolutionary leader") is either the *Rahbar* (i.e. "leader") or in his absence a council of religious office-bearers. The revolutionary leader has unlimited power, he appoints the highest judges (all ecclesiastics) and is also supreme commander of the armed forces. He is chosen by the expert's council for life. This council is chosen every eight years by the people. The guard's council must approve the candidates. The head of state, the religion leader who is also the head of the government is determined in a general election for a term of 4-years. The president appoints the members of the cabinet and also presides over the cabinet. He coordinates the governmental work and presents the government bills to parliament. The power of the president, the government and the parliament is however, severely limited, because all candidates and all laws must be confirmed by the guard's council. Besides, the revolutionary leader takes the last decision in all matters.

---

[44] Based on the OECD and the Brockhaus (2006)

<table>
<tr>
<td>

Germany to the outside and the state level (Bundesland), which exists in every federal state individually.

Every level has its own state organs of the executive (executive power), legislative (legislative power) and judicative (judicial power). Decisions on the federal laws are taken jointly by the Bundestag and the Bundesrat. They have the competence to change the basic law, the constitution of Germany, with a two-thirds majority in both houses.
In the federal states the state parliaments decide on the laws.

Although basic law determines that the representatives in the parliaments are not subject to directives, preliminary decisions in the parties dominate the legislation.

</td>
<td colspan="2">

The guard's council consists of 6 ecclesiastical and 6 secular jurists. The ecclesiastics are appointed by the revolutionary leader. It is their task to check every law for its conformity with the Islamic principles. The lawyers are appointed by the chief justice, the head of the judiciary. Their task is to check the conformity of legislative acts with the constitution. The chief justice on his part is appointed by the revolutionary leader. The guard's council is authorised to reject every law or to declare it invalid a posteriori, and to refuse the nomination to candidates for parliament and the presidential office The guard's council can decide by simple majority. In case of a split vote the revolutionary leader has the decisive vote.
In the Iranian constitution the state power, consisting of the legislative, executive and judicative power, is subordinated to the religious leadership (welayat-e faghi). Therefore all three powers are not autonomous in their decisions, but dependent on the leader (Rahbar).
The Iranian unicameral parliament (Majles e-Shura ye-Eslami consists of 290 representatives that are chosen directly by ballot in general elections for a 4-year term of office.

</td>
</tr>
<tr>
<td>Politics</td>
<td>D[45]</td>
<td>IR[45]</td>
</tr>
<tr>
<td>Establishment</td>
<td>23. Mai 1949</td>
<td>11. April 1979</td>
</tr>
<tr>
<td>System of government</td>
<td>Republic</td>
<td>Islamic Republic</td>
</tr>
</table>

Table 6-13: Policy indicators-country [AUTHOR2007]

---

[45] According to [Brockhaus2006]

Macro level: Health system	
D	IR
Health policy is the task of the Federal Ministry of Health dealing with the planning, organisation, financing and control of the health system. This consists mainly of negotiations with the federations of the health insurance companies, the hospital owners and operators, physicians, chemists and the pharmaceutical industry.	The health policy of Iran is influenced very strongly by the parliament (Majles). According to the constitution, every inhabitant has the right of "access to public health services". A determined effort is made to offer at least a minimum of health care particularly for the rural population.

Micro level: Hospital	
D	IR
The hospital policy or strategy planning is determined mainly by the general management of the institution (consisting of medical, nursing and administrative managers). Most houses adopt long-term plans, which state their goals and the required measures in 5 and 10 year plans.	Iranian hospital policy is determined strongly by the management (Raise Bimarestan) and the policy of the university. There are rudimental strategy plans, but these are seldom executed and redefined after every change of the leadership.

Micro level: Department Medical equipment	
D (see chapter 5.1)	IR (see chapter 5.1)
The policy or strategy of the hospital is continued in detail in the department.	

c. Management, organisation, administration

Macro level: Country	
D[46]	IR[44]
Germany is subdivided into 16 federal states which are organised - except for free cities and counties - in local authorities associations.	Iran is administratively divided into 28 *Ostans* (provinces), each run by an *Ostandar* (Governor General) appointed by the Ministry of Interior. Each

---

[46] According to [Brockhaus2006]

There are 14 cities with more than 500,000 inhabitants. The specific feature of Germany is the decentralised distribution of the cities which lie near to the borders. The five biggest cities in Germany are: Berlin (3.4 Mil.), Hamburg (1.75 Mil.), Munich (1.3 Mil.), Cologne (0.89 Mil.), Frankfurt am Main (0.65 Mil.).	province is in turn divided into a number of *Shahrestan*s (districts) administered by a Farmandar (Governor) appointed by the Minister of Interior. Currently there are 278 *Shahrestan*s (districts). Each district includes a number of urban centers (cities/towns) and villages. The biggest cities are the capital of Teheran with officially 7.1 Mil. (12 Mil. in the conurbation), Maschhad (2.1 Mil.), Isfahan (1.5 Mil.), Karaj (1.4 Mil.), Täbris (1.4 Mil.), Schiraz (1.2 Mil.), Qom (1.0 Mil.) inhabitants.

Macro level: Health care system	
D (see chapter 3.1.1.1)	IR (see chapter 3.1.2.1)
Power in the health care system is subdivided between the federation (Bund) and the federal states (Länder). The self-government of the associations and federations plays a big role in the execution of the legal framework. The Federal states are responsible for the hospital infrastructure.	The PHC (Primary Health Care) network is active on national (health ministry), provincial (health centres) and district level (district health centres, urban health centres, urban health posts, rural health centres and health houses). Besides, the country is subdivided into 30 provinces, 300 districts, 676 cities and 66000 villages.

Micro level: Hospital	
D (see chapter 3.2.1.2)	IR (see chapter 3.2.2.2)
Reorientation of the concept from pure management to a hospital seen as an enterprise. Establishment of facility management departments which are fully responsible for all secondary services. Process optimisation and transparency enhanced by the use of IT.	Hospitals are subordinate to the medical universities. The very hierarchically structured hospital is administered by the hospital manager. (Only recently proposals have been made by the health ministry to establish supervisory boards.)

<table>
<tr><th colspan="2">Micro level: Medical equipment</th></tr>
<tr><td>D (see chapter 5.1)</td><td>IR (see chapter 4.2.2.2 und chapter 5.1)</td></tr>
<tr><td colspan="2">The administration and the management of the department follow the policy of the hospital management.</td></tr>
</table>

d. Education and research

<table>
<tr><th colspan="2">Macro level: Country</th></tr>
<tr><td>D[47]</td><td>IR[48]</td></tr>
<tr><td>Education is the responsibility of the federal states. It is co-ordinated by conferences of the ministers of culture. Depending on the federal state, compulsory education extends over 9 to 13 years. The educational system is mostly three-tiered. After the elementary school there are the main, secondary and high school (Grund- (1-4), Haupt- (5-9), Realschule (5-10), and Gymnasium (5-13)). After graduation vocational schools and occupational companies are visited by trainees, while students can choose between universities and advanced technical colleges. 4.6% of the GDP are spent on education.</td><td>The education has very high significance in Iran. Parents often work very hard and accept privations to give their children a good education. The compulsory education lasts 12 years. The educational system is four-tiered (Dabestan (1-5), Rahnamai (6-8), Dabirestan (9-11), Pishdeneshgahi (2). The selection procedures at the universities are very strict, only the best pupils of a year are accepted. The state promotes the education of its citizens, about 15% of the budget flows into the education system.</td></tr>
</table>

Country	D[47]	IR[48]
Students	1 800 000 (2006)	700 000[49] (2005/06) (45% male; 55% female)
Students per 1000 capita	21,84 (2006)	10, 14 (2005/06)
Illiterate ratio (%)	1,0 (2006)	20,6 (2005/06)

Table 6-14: Education indicators-country [AUTHOR2007]

---

[47] According to [Brockhaus2006] and the Statistisches Bundesamt

[48] According to the Statistical Centre of Iran (SCI) and Ministry of Culture and Islamic Guidance (MCIG)

[49] Universities and Islamic Azad Universities

Macro level: Health system	
D (see chapter 3.1.1.3)	IR (see chapter 3.1.2.4)
Physicians study at medical faculties of the universities which are under the care of the Federal Ministry of education and research while the nursing staff completes a three-year training at health and nursing schools.	The medical universities in Iran are subject to the health ministry. Physicians, as well as nurses and midwives must conclude 6years of study. Training exists for the staff of the health centres and health houses.

Health Care	D[50]	IR[51]
Medical University	35 (2006)	41 (2005/06)
Physicians (number)	306 000 (2006)	63 000 (2005/06)
Physicians per 1000 population	3,7 (2006)	0,9 (2005/06)

Table 6-15: Education indicators – health system [AUTHOR2007]

Micro level: Hospital	
D (see chapter 5.1)	IR (see chapter 5.1)
Particularly continued training is of great significance for the entire staff of the hospital. Training courses are offered in great number and can be attended; in some cases they must be attended.	There are almost no continued training courses offered at the hospitals.

Micro level: Department medical equipment	
D (see chapter 5.1)	IR (see chapter 5.1)
The continued training of the employees, in particular of those in the technical areas, is indispensable. A series of training courses is offered, in particular by manufacturers.	Because the branch is poorly represented, continued training is rare. English language user instructions, if still available, are the only source of information. During the last years the interest in the use of medical equipment as a profession has grown very strongly (especially among the female students). The low level of practical experience hampers the entrance into professional life very much.

---

[50] According to the Statistisches Bundesamt

[51] According to MOHME (2006)

e. Judicative, legal regulations

Macro level: Country	
D[52]	IR [53]
The federal republic regards itself as a constitutional state. That means state activity can only be justified by law and is limited by law. The judges are independent and are not subject to directives. The jurisdiction is exercised essentially by courts of the federal states.	Islamic right, the sharia, has been reintroduced as law by the Islamic revolution. Because the sharia has never been codified, the administration of justice and development of the Islamic jurisprudence are implemented in a kind of Case Law system. There is no division of power in Iran. The highest ecclesiastical leader has far-reaching competence.

Macro level: Health system	
D	IR
The health system in Germany is subject to numerous laws, which are made by committees of the Bundesrat or Bundestag	The laws for the health system in Iran are also made by committees of the parliament (Majles).

Micro level: Hospital	
D	IR (see chapter 3.2.2.2)
The hospitals are subject to a large number of laws and regulations (hospital financing law, hygiene) which are controlled by experts (health authority).	The laws and regulations are controlled by consultants of the ministry of health and medical education or by representatives of the universities (Department of assessment Arzeshyabi).

---

[52] According to [Brockhaus2006]

[53] According to Iranian Constitution and to [Brockhaus2006]

<table>
<tr><td colspan="2">Micro level: Department medical equipment</td></tr>
<tr><td>D (see chapter 5.1)</td><td>IR (see chapter 5.1)</td></tr>
<tr><td>

The following laws must be considered:

- law on medical products (Medizinproduktegesetz (MPG))
- regulation for medical product operation (MPBetreibV)
- Radiation control regulation (Röntgenverordnung (RöV))
- Regulation on medical products (Medizinprodukteverordnung (MPV))
- Radiation Protection Ordinance (Strahlenschutzverordnung (Str.SchV))
- Regulation on detection, assessment and prevention of risks due to medical products (Verordnung über die Erfassung, Bewertung und Abwehr von Risiken bei Medizinprodukten)
- Requirements on hygiene during the processing of medicine products (Anforderungen an die Hygiene bei der Aufbereitung von Medizinprodukten)

</td><td>

There are hardly any laws that apply to medical equipment; however, the health ministry has constituted a committee which deals with laws and regulations for

- import and installation of the devices
- use
- repair measures and
- maintenance measures.

</td></tr>
</table>

f.   Culture, Social Structures

<table>
<tr><td colspan="2">Macro level: Country</td></tr>
<tr><td>D[54]</td><td>IR[54]</td></tr>
<tr><td>

Demography: With more than 82 million inhabitants Germany has the largest population in the European Union. It belongs to the countries with the densest populations in the world. The birth rate is one of the lowest worldwide. The city of Chemnitz even has the lowest birth rate in the world. The average age in 2004 was 42.1 years, whereas the average age of the world population in 2004 was 27.6 years. About 75 million inhabitants (91%) are German citizens. Seven million of these citizens have a migration background.

</td><td>

Demography: The population of Iran (approximately 70 million persons) is comprised of Persians, Azerbaijanis, Kurds, Gilaki and Mazandarani, Arabs, Turkomanen, Luren, Belutshs and some smaller minorities, like Christian Armenians, Assyrians, Georgians and Jews. Many refugees from the neighbouring countries have been accommodated during the last years: over 2 Mio. Afghans and Iraqis. Approximately 62% of the people live in the cities. Iran has a very young population, 36% of the

</td></tr>
</table>

---

[54] According to the OECD and [Brockhaus2006]

<table>
<tr><td>

Religion: The majority of the German citizens are Christian: Roman-Catholic churches 31.5%, Protestant church 31.1%, about 31 % of the population do not belong to any confession. (In the former GDR around 70 % did to belong to any confession). Islamic municipalities count approximately 3.2 Mio. members (3.9 % of the inhabitants). Of these, 732,000 are German citizens (0.9% of the German citizens); the witnesses of Jehovah count approximately 164,000 or 0.2% of the population. 106,000 persons belong to Jewish communities. After France and the United Kingdom, Germany has the third-biggest Jewish community in Europe.

</td><td>

Iranians are less than 15 years old.

Religion: 98% of the population are Muslims, of these 90% Shiites and 8% Sunnites. In addition there is a Christian population of around 280,000 persons, of which 90% are followers of the Armenian-apostolic church. The rest are around 20,000 to 40,000 Assyrian Christians, about 3,000 Chaldeans and a few Protestants. The Jewish community is around 11,000 to 30,000 persons, besides 33,000 Parse, some thousand Medes and approximately.

</td></tr>
</table>

Country	D[55]	IR[56]
Official language	German	Persian /Farsi
Capital	Berlin	Teheran (Tehran)
Total population	82.424.700 (2006)	70.049.826 (2005/06)
Area km²	610.357 (2006)	1.648.195 (2005/06)
Population density inhabitants per km²	231 (2006)	42 (2005/06)
Population growth rate (%)	0, 02 (2006)	1,07 (2005/06)
Urban population (%)	87,27 (2006)	64,54 (2005/06)
Average age	41,70 (2006)	23,5 (2005/06)

Table 6-16: Cultural indicators-country [AUTHOR2007]

The social structures (historical background, ethics, and sense of values …) were discussed in detail in chapter 3.2, hence only some of the most important indicators are presented here as examples.

---

[55] According to the Statistisches Bundesamt

[56] According to the Statistical Centre of Iran (SCI)

Macro level: Health system		
Health Care	**D[57]** (see chapter 3.2.1.1)	**IR [58]** (see chapter 3.2.2.1)
Life expectancy (w)	82 (2004)	73,17 (2003)
Infant mortality rate (per 1000 Life birth)	4,0 (2004)	26,0 (2004)
Percentage of Population aged 14 - years	14,7 (2006)	28 (2005/06)
Percentage of Population aged 65+ years	18,3 (2006)	4,8 (2005/06)

Table 6-17: Cultural indicators – health system [AUTHOR2007]

Micro level: Hospital		
Country	**D[59]** (see chapter 3.2.1.1)	**IR[60]** (see chapter 3.2.2.1)
Beds in hospital (per 10 000 population)	64,4 (2004)	30 (2005/06)
Number of admissions per year	16.801.649 (2004)	10 324 734[61] (2005)
average period spent in the hospital (days)	9,8 (2001)	3,6 (2005)

Table 6-18: Cultural indicators –hospital [AUTHOR2007]

Micro level: Department medical equipment	
D (see chapter 5.1)	IR (see chapter 4.2.2.2, 5.1)
The biomedical technology in Germany, also in the hospitals, looks back on long, historically grown development and experience.	The department of medical equipment is a very young institution and exists in only approximately 10% of the Iranian hospitals. This becomes evident in the handling of the equipment.

---

[57] According to the Statistisches Bundesamt

[58] According to the Statistical Centre of Iran (SCI)

[59] According to the WHO

[60] According to MOHME

[61] Estimate based on mean length of stay and occupancy rate of hospital beds

g. Public and private institutions

Macro level: Country	
D	IR
Germany is a federal state and a member of the United Nations, NATO and the G8 as well as foundation member of the EU.	Iran is a member in the United Nations, the OPEC (Organisation of Petroleum Exporting Countries) and the ECO (Economic Cooperation Organisation), an economic alliance of the central Asian states.

Macro level: Health system	
D (see chapter 3.1.)	IR (see chapter 3.1.2.4)
There are numerous public and private institutions in Germany which represent interests in unions, federations and representations.	The presentation of public and private health institutions in Iran is very weak and, hence, plays a subordinate role.

Micro level: Hospital	
D (see chapter 3.1.)	IR (see chapter 3.1.2.4)
At hospital level there are hospital societies, union of hospital managers, employer, employee representations and professional federations.	See 6.3.7.2

Micro level: Department Medical equipment	
D (see chapter 3.1)	IR (see chapter 3.1.2.4)
The situation in the hospital is continued in detail in the department.	

h. Infrastructure

Macro level: Country	
D[62]	IR[63]
Due to the central position of Germany in Europe, the traffic volume in Germany is very high. In particular, it is an important transit country for the freight traffic that in the past decades	Urbanization and migration from the rural areas makes it necessary to improve the urban infrastructure in Iran constantly. The construction and extension of the water supply and sewage disposal requires high investments in the cities as

---

[62] According to the Brockhaus 2006

[63] According to the Ministry of Road and Transportation (MRT)

<table>
<tr><td>has shifted steadily from the rail to the road.

In addition, the individual traffic has increased strongly, so that the amount of traffic on the roads is very high by international standards. For the future a strong increase of the traffic is expected.</td><td>well as the rural areas. The rail beds must be electrified and high-speed trains must be provided. In the transport sector, roads, metros, railways, airports and harbours are being developed. The plans provide for 1,300 km of motorways, 1,000 km of streets, 3,400 km of railroad connections as well as the construction and extension of metros.</td></tr>
</table>

Country	D (chapter 3.2.1.1)[64]	IR (chapter 3.2.2.1)[65]
Streets (km)	650 000 (2006)	94 109 (2005/06)
Highways (km)	11 515 (2006)	890 (2005/06)
Passenger car per 1000 capita	523,24 (2006)	35,79 (2005/06)
Railway line (km)	46 039 (2006)	7 203 (2005/06)
Airports	32 (2006)	72 (2004)

Table 6-19: Infrastructural indicators-country [AUTHOR2007]

Macro level: Health system		
Health Care	D (chapter 3.2.1.1)[66]	IR (chapter 3.2.2.1)[67]
Population with access to local health services (%)	100 (2006)	94 (2000)
Population with access to clean drinking water (%)	100 (2002)	93 (2002)
Population with access to sanitary facilities (%)	100 (2006)	84 (2002)
Number of Hospitals	2 166 (2004)	750 (2005)
Number of beds (total)	531.333 (2004)	203 666 (2004)[68]

Table 6-20: Infrastructural indicators-health system [AUTHOR2007]

---

[64] According to the Statistisches Bundesamt

[65] According to the Ministry of Road and Transportation (MRT)

[66] According to the WHO

[67] According to MOHME

[68] According to the Statistical Centre of Iran (SCI)

Micro level: Hospital

The infrastructure depends on the specific hospital or the authority which is responsible for it.

Micro level: Department Medical equipment

The infrastructure of the department depends on the possibilities of the hospital.

i. Geography, climate, environmental conditions

Macro level: Country	
D [69]	IR [69]
Geography, geology, climate and vegetation	**Geography, geology, climate and vegetation**
Germany lies in Central Europe and has common borders with Denmark, Poland, Czech Republic, Austria, Switzerland, France, Luxembourg, Belgium and the Netherlands. In the north the natural borders of the country are formed by the North Sea and the Baltic Sea. The Alps are the only high mountain range with the highest mountain Zugspitze (2962 m). Germany is geologically diverse. The crystalline mountains (e.g., the Black Forest) have originated in the early age of the earth. On the northern edge formations from the carbon age are found in which immense coal deposits are embedded (Ruhr area). Germany belongs completely to the temperate climate zone of Central Europe in the west wind zone and lies in the transition area between the maritime climate in Western Europe and the continental climate in Eastern Europe. The climate is influenced by the Gulf Stream (amongst other factors) which is responsible for the unusually mild climate for the degree	Iran borders on seven states and two seas: in the west and northwest on Iraq, Turkey and Azerbaijan, on Armenia and the Caspian Sea; in the east and southeast on Turkmenistan, Afghanistan and Pakistan as well as the gulf of Oman and the Persian Gulf. It is a distinct mountain country, with the exception of the deserts. Geologically the country is divided by several almost parallel mountain chains sweeping to the southeast and reaching a height of more than 4000 m in the Zagros- and Kuhrud mountains. The eastern part consists mostly of desert basins like the saline Dasht-e Kavir, Iran's largest desert, in the north-central part of the country, and the Dasht-e Lut, in the east, as well as some salt lakes. The highest mountain of Iran is the 5.671-m-high Damavand in the Elburs Mountains. The difference in height to the Caspian Sea, which is only 60 km distant, is nearly 6000 m. Such a steep incline is not even found in the Chilean Andes. In the south and south-west the country has an

---

[69] According to the Brockhaus

of latitude. The composition and quality of the soils is very different in different regions. All together 53.5 % of the surface area of Germany is used agriculturally; woods cover another 29.5 %. On account of high population density and rising mobility 12.3 % of the area is used for settlements and public thoroughfares, water surfaces amount to 1.8 percent, the remaining 2.4 percent are spread over other types of surfaces, mostly wasteland.	approximately 2,000 km long coast on the gulf of Oman and on the Persian Gulf, which are separated by the Strait of Hormuz, a narrow stretch of sea that is important for the oil routes. The climate in Iran is very diverse. Arid climatic conditions are predominant, but annual precipitations of more than 2.000 mm occur regionally - in particular in the north and at some mountain edges. The salt lakes are driest in the inland. 53% of the land surface of Iran is desert area, 27% pastureland, 9% arable land and 11% woods.

The climatic and geologic environmental conditions play a big role in the emergence and propagation of pathogenic agents.

Micro level: Hospital

The climatic conditions mainly play a role for hospital construction and must be considered. Besides the climatic conditions, environmental protection is becoming more and more significant.

Micro level: Department medical equipment

Phenomena such as increased humidity must be taken into account by modification of the maintenance measures.

6.3.2 Implementation of the transferability model using the share model

After the macro (country, health system) and micro level (hospital, department) are analysed an interpretation of the quality and quantity occurs to get values for the share model.

Economy

Germany belongs to the emerged economies and the leading (second place compare chapter 6.3 [BMBF2005]) countries particularly in the area of technology and machine construction (especially in the field of medical devices). Iran is part of the middle income countries and earns most of its income from the oil and natural gas business. The service sector has established itself in Germany, while it is still emerging in Iran.

The big difference in the GDP of the countries directly affects the health expenses. Thus the total expenditures for health in Germany are about 20 times, the governmental expenses even 33 times as large as the Iranian expenses. This implies that fewer funds are available for the hospitals in Iran, and even less for the area of medical equipment.

It is remarkable in the comparison that the number of co-workers in Iranian hospitals is considerably larger than in German hospitals. One reason may be attributed mainly to the low wages in Iran.

Politics

Politics provide the legal framework and, hence, play a fundamental role for the overall strategy of the country. While health policy is the task of the Federal Ministry in Germany the health policy of Iran is influences very strongly by the parliament (Majles). Because of the influence of politics in the universities in Iran this tendency is noticeable in the hospitals, too. That affects to the strategic plans that mostly vary with a governmental or personnel change while in Germany changes have hardly that strong influence.

Management, organisation, administration

The administration of Iran is divided in provinces and districts, but nevertheless the main governmental power is going out from the capital. Germany has a federal structure with federal states (Länder) that are to a great extent self-sufficient.

The Iranian state operates a huge administrative apparatus that extends up to the state hospitals. This trend is continued in the hospitals. The Iranian hospital management has considerably more power than its German counter-part, which is under continuous control of the supervisory board.

Education

Approx. 700,000 students (around 55% of these female)[70] annually complete their studies at the Iranian State (Daneshgahe Sarasari) and Free (Daneshgahe Azad) Universities. Hardly any apprenticeships are available; these also do not enjoy a high reputation in society.

---

[70] According to the Statistical Centre of Iran (SCI) [2005]

While theoretical lectures are emphasized in Iranian universities their German counterparts offer a very practically oriented course of study. There are many apprenticeships which are very practically and socially oriented. In addition, a wide spectrum of measures for qualification and continued education is offered by the chamber of industry and commerce, affected associations or additional institutions.

Judicative

Germany is known as "the country of law and order"; these exist in all areas and are continuously controlled. In Iran there are some areas (e.g. medical equipment) where no laws are available. Often when there are laws, these are not applied due to lack of knowledge or lack of control. The right for health care for all citizens is guaranteed by the Iranian constitution; since the Islamic Revolution the situation for the rural population has improved (free insurance, etc.)

Culture

Differences can also be found in values related to culture, such as working morale, sense of responsibility, sense of time. These phenomena cannot be assessed. It must, however, be pointed out that big differences exist which directly affect the transferability of the FM-systems.

While the average age is 24 in Iran this indicator is almost double in Germany (43 years) what causes different situations for the macro and micro level specially in the health system and the hospitals that are confronted with a corresponding other clients.

Public and private institutions

Germany has numerous public and private institutions which represent interests in unions, federations and representations, while the presentation of public and private health institutions in Iran is very weak and, hence, plays a subordinate role.

Infrastructure

A comparison of the population densities (231 in Germany and 42 inhabitants per km² in Iran) and the transportation infrastructure points to the difficulties which Iran encounters in its effort to guarantee the health care of the entire population. In spite of the PHC system high technical health supply cannot be provided in the whole country. In contrary Germany with the high density and a tight infrastructural net can warranty almost country wide a high standard medical care.

Geography, climate

The climatic and geologic environmental conditions are different. While Germany has a balanced climate in Iran can daily be find temperature differences of around 30-40 grade Celsius. From snow-capped mountains (still in summer) to hot and dry deserts to humid wood can be found in the different regions of the land. Besides the climatic conditions, environmental protection is becoming more and more significant. Phenomena such as increased humidity must be taken into account by modification of the maintenance measures in the medical equipment departments.

Evaluation

After the comparison and interpretation of the parameters of influence, a ranking by means of an indicator analysis is undertaken. This method includes all indicators (parameters of influence) that imply the system.

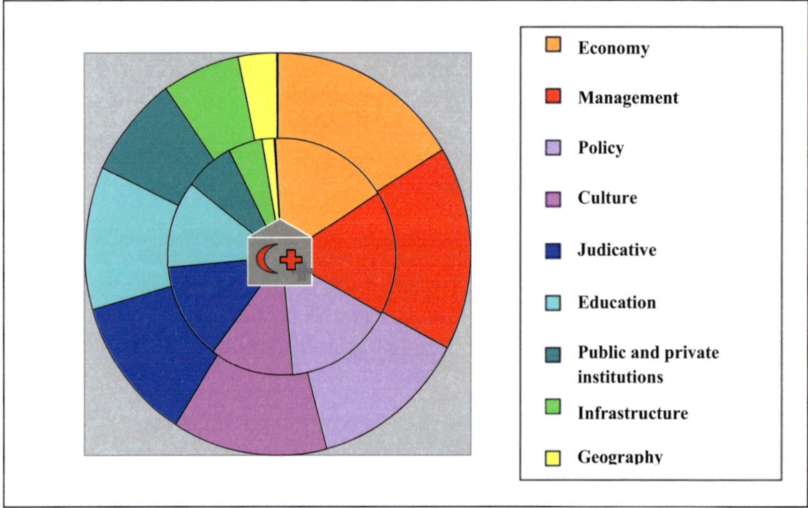

Figure 6-12: Ranking of the macro-and micro parameters by the indicator analysis [AUTHOR2007]

Figure 6-12 shows the weighting of the parameters of the interpretation of the indicator analysis.

Next to the management, the economy plays the most important role. If financial opportunities exist, higher-value medicine products and machines as well as staff with higher qualification can be employed. The management is, however, closely involved with the administration, preservation and sustenance of the available resources and opportunities.

The third place is taken by politics and strategy planning. In this connection, more weight is given to the micro level because good leadership and tactics at the hospital level directly influence a successful FM system.

The situation is reversed for the social structures, where the macro level is dominant, because the working behaviour in the hospital can be influenced to some extent by different measures (continued training, organizational measures …).

Education follows after the judicative in sixth place. At the macro level its weight equals that of the judicative, whereas at the micro level the compliance with legal requirements is to be assessed higher, because this can entail direct consequences. The public and private institutions, as well as the infrastructure and geography, are weighted lighter because they do not have significant influence on the transferability of the FM system.

6.3.3 Conclusion

After the analyses and the single ranking have been performed to assess the diverse influence factors, the last step is now taken to obtain an overall assessment.

All three rankings are compared directly in Figure 6-13. The ranking of the direct and indirect parameters of influence (compare Figure 6-6), the ranking by the experts (compare Figure 6-9) and the result of the indicator analysis of the main indicators (compare Figure 6-12) placed in the "Share Model".

Figure 6-13 provides an overview of the three assessments. It must be taken into account that while the evaluation of the parameters of influence are based on a model and can only declare a general view the indicator analyse and the expert opinion differ between macro and micro level and are based on qualitative and quantitative data, respectively on expert experiences.

Due to this difference, their results are of greater significance for the system transferability than the results of assessment the parameters of influence, which only indicates whether the parameter of influence causes a direct effect or not.

Ranking	Parameters of Influence Model	Indicator Analysis	Expert Opinion
1	Management	Economy	Management
2	Politics	Management	Politics
3	Culture	Politics	Culture
4	Judicative	Culture	Judicative
5	Economy	Judicative	Economy
6	Infrastructure	Education	Infrastructure
7	Education	Public and private institutions	Education
8	Public and private institutions	Infrastructure	Public and private institutions
9	Geography	Geography	Geography

Figure 6-13: Comparison of the rankings [AUTHOR2007]

Because the assignment of reliable numbers or values is very difficult and depends on the situation at a particular point in time, the ranking that leads to the identification of trends in the assessment of the parameters of influence will be retained. Figure 6-14 shows the rough result of the comparison. To show the tendencies the parameters are separated in high, middle and low ranked parameters.

Rank	Transferability parameters		
high	Management	Economy	Politics
middle	Culture	Judicative	Education
low	Infrastructure	Public and Private institutions	Geography

Figure 6-14: Result of the comparison [AUTHOR2007]

This ranking shows that monetary opportunities, management and politics including strategy planning are the most important parameters of influence. Culture, education and the legal framework that applies to the FM system occupy the middle third of the ranking. The parameters public and private institutions, infrastructure and geography are of less importance.

During the investigation of the solution approach the parameters of influence and their indicators have been decomposed into small "puzzle parts" in order to gain precise insight into their nature and properties. Now these insights must be evaluated and assembled into an overall picture, so that a conclusion can be drawn for the system transferability of FM.

To bring the outcomes of the abstract theory into a usable implementation following steps are recommended. As the results in figure 6-14 demonstrate the high ranked parameters are the management, the economy and the policy. These are the main parameters where the approach should start.

To stay in the micro level, in the hospital, a Facility Management Department should be established similar to the German one, with the exception that the Iranian one is directly connected to the University FM-Department, too.

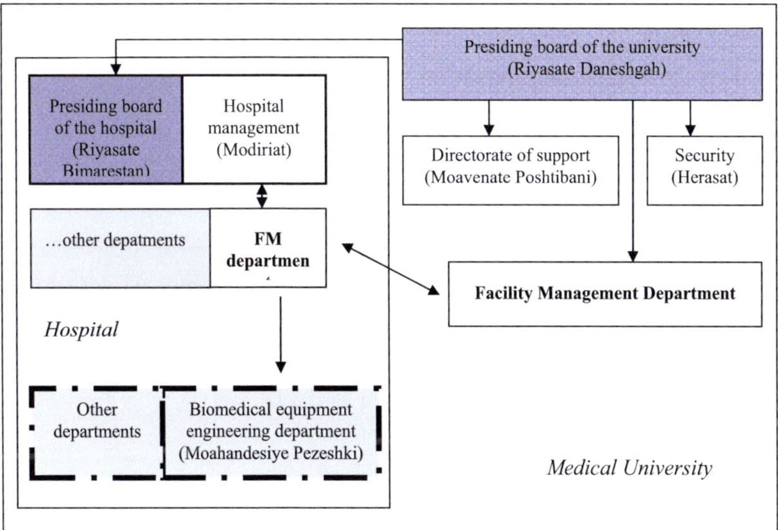

Figure 6-15: Suggestion of a FM department [AUTHOR2007]

The application of the FM department as a coordinating and controlling body of the single services (medical equipment, laundry...) would release the tasks of the hospital management that could concentrate on other areas. This presumes a good documentation work and transparency (what now is failing in Iranian hospitals) so that informative reports can be provided and plans could be designed.

Important for the success of a FM department is a certain authority to decide. The department, subjected to the hospital management, is obliged to meet the requirements and should stay in continuous exchange. But it is reasonable that they have scope of action and an appropriate budget to react quickly. This was a point that was noticeable in the system running now in Iranian hospitals, that for every little step (e.g. repair of a bulb) a long administrative way must be passed.

An appropriate budget includes the second parameter "economy". The finance of a hospital and also of a FM department is a basic for a successful operating. Important is to make the rank and importance of FM aware and to be all set to invest in that field. This philosophy must apply to the university with its own FM department, too. Whereas facility management is specially an effective tool for real estate and resource management. Recommendable would be a close cooperation between the FM departments of the university and the hospitals; in e.g. monthly meetings experiences, challenges and news could be exchanged (what could as well been organized for the single departments).

The third high ranked parameter is politics. Here especially the existence of the philosophy (including values, ethic and goals) and strategic plans aimed for the hospital are opined. It is of great importance that long term strategic plans are introduced and realized, even if there is a change in the management board. This people–orientation and the high fluctuation in hospitals evocate an instability and is especially problematical for long term investment and projects. This phenomenon could be seen e.g. by the procurement of giant equipment. Very expensive equipment was bought, but because of the lack of planning no maintenance plan was made what not only caused a high maintenance backlog but also the earlier destruction as the durability.

Another factor for an early destruction is the nescience of the users. Insufficient instructions and the lack of advanced training cause a wrong handling of the equipment what leads to a breakdown and to high costs for repair. The parameter of education is not only necessary for trainings. The fill of vacancy with fully-fledged personal is basic requirement for the success of a FM department and a hospital. A further support to delegate a successful hospital is the introduction of laws and regulations and their control. The implementation of laws is certainly task of the country or ministry; but in the micro level regulations could be decreed for a hospital or university, too. Important is the application and realization that must be regularly controlled by independent organs.

The characteristics and conditions that result from the history and the culture must be considered as well. Beginning with the comprehension and exposure to illness and sick person[71] the influence of the culture ranges from the social position of labor and jobs (e.g. huge difference between a physician and a worker in the laundry - from salary and social prestige) to the perception of time.

Likewise the cultural and the infrastructural mobility play a major role. The good infrastructure and accessibility are important for the hospitals as well as for the single departments (supply of spare parts). This also must be considered in the hospital (state of the departments, availability of operation room) with a deliberated space and resource management.

Public and private institutions must be established and organized so that specialized divisions can interchange, transfer knowledge and build working groups. This also must happen in the macro (e.g. Iranian Facility Management Association) and micro level (e.g. hospital associations) to advance single fields and to strengthen the positions.

These all are first suggestions for the transfer of facility management and is not exhaustive; but should be a first approach and give a jump start to implement this helpful and profitable system.

---

[71] While a German physician is obligated to tell the patient the diseases, in Iran normally diagnostic findings are told to the family who normally doesn't tell the patient the truth (especially if it is a terminal illness).

6.4   Summary and outlook

During the presentation of the kick-off meeting "OPIK- Iran project" in Tehran in spring 2006 the idea of developing a transferability method of FM appeared to be an impossible task to some participants.

But after numerous data acquisition, workshops, expert questioning, site survey and research studies possibilities to achieve that goal were found.

First of all the definitions and the status quo of facility management in hospitals for both project countries were established. Afterwards the general requirements of the health system with a special view to the hospital were compared. In this respect not only the administrative, financial and education system, but also the historic development and impact was analysed.

In order to be able to compare the single FM processes in Germany and Iran first the project hospitals in Iran and the three analysed processes (maintenance of medical equipment, maintenance and repair process of technical facilities and laundry management) were introduced. The processes, including process steps and responsibilities, characteristic variables (cost and quality factors) and interfaces as well as the data analysis were considered.

The dissection of the single processes and their environment made influence parameters observable that were used to develop a transferability concept. These influence parameters are: management, economy, politics, culture, judicative, education, public and private institutions, infrastructure and geography. Before models and methods could be designed the boundary conditions and the definitions of the system were established.

The model investigation was started with the "Parameter of Influence Model" that analyses the internal relationship between the parameters of influence; this model cannot differentiate between micro and macro level, but gives a rough estimation about the trend of the importance of the parameters. Another model is the "Matrix Model" that can be a good solution if more than two countries are compared with each other. Here a degree between 1 and 10 is given by an appraiser for the transferability of the single micro and micro level of the influence parameter. The third model presented is the "Share model" that is described by a circle. This model offers the advantage that the weighting is readily observed visually, so that the critical transfer areas become evident "with one look". The challenge in these models is the determination of the degree that is depending to the appraiser. They can only give trends and tendencies for the transferability of a system. To substantiate the findings of the models two analysing methods were implemented; one, the expert opinion or "Expert Analysis" and the "Indicator Analysis". The "Expert Analyse" is more subjective; but depending on the experience and the technical knowledge the existing situation can be reflected very well. The "Indicator Analysis" compares the qualitative and quantitative indicators of the project countries and is more objective. It must be once again emphasised that most of the indictors have a dynamic behaviour. Therefore each analysis and conclusion should be based on updated and current data.

To achieve a general conclusion the analysis included in a chosen model are ranked. This assessment serves as a guide or a "trend barometer". The level of the ranking indicates the degree of importance. It gives a direction which parameters should be considered for the successful transferability of a FM-system.

Hence, facility management is transferable; the system must however be matched to the boundary conditions and be adapted to the specific situation in the country or even in the hospital to which they are applied.

The successful implementation demands to establish the philosophy of facility management that is based on three basic principles. These are: transparency, integrity and the consideration of the life cycle of a system.

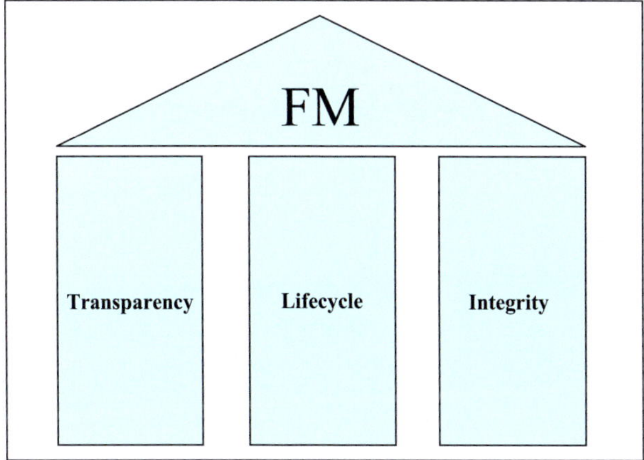

Figure 6-15: Basic columns of FM [NÄVY2003, p.2]

The first column means the transparency in an organisation. "The goal is to have a pellucid organisation where all information about resources and their operation, organisation and administration are available. The information is up to date and anytime recallable". [NÄVY2003, p.6]

The life cycle considers every single stage of life of the resource e.g. for a building from the first idea to the planning, construction and utilisation phase to the point of demolition and recycling. This enables to think and act in long ranges; so already in the planning phase it is thought about the demolition and recycling what executes a completely different philosophy now ruling in Iran "besaz va befrush" (build and sell).

The third column, integrity contains the holistic view of the resources. Hereby the three P's of facility management **People**, **Place** and **Process** are considered in the same time.

This new philosophy that demands a new way of thinking must be implemented and lived in the organisation.

To establish this mind change first the idea and knowledge of facility management must be transferred and introduced. A good opportunity beside seminars in hospitals and ministries is the education. In addition to training courses and further professional training the development of a facility management study course at the university would establish the existence of facility management and would raise this philosophy. The introduction could also get supported by laws and regulations. Here ministries or universities can enact laws or regulations which lead to the implementation of a facility management department. Also private or public associations (e.g. the foundation of the "Iranian Facility Management Association") that put exhibitions and conferences and the inception of a Facility Management Competence Center could expedite the distribution of facility management in Iran.

Certainly it requires time, patience, work, research and courage to establish facility management in Iran. But incipiencies are done and a market with high potentials is waiting to be discovered.

LITERATURE

[ABOLHASSANI2006] Abolhassani, Farid, 2006 Extract Primary Healthcare in Islamic Republic of Iran, 2006

[AHK2006] [Internetseite]Deutsch- Iranische Industrie- und Handelskammer Teheran, Landesinfo, 2006, http://www.iran.ahk.de

[ALFAGHED2004] Alfaghed Aidin, 2004, Untersuchung der Instandhaltung der Medizintechnik, Diplomarbeit der Medizinischen Universität Iran, Institut für Management und Information

[AIG1996] Arbeitsgemeinschaft Instandhaltung Gebäudetechnik der Fachgemeinschaft AIG, Allgemeine Lufttechnik im VDMA (Hrsg.) (1996) Instandhaltungsinformation Nr. 12: Gebäudemanagement, Definition, Untergliederung. Frankfurt/Main, S. 2

[ALLEFELD1999, S.22] Allefeld Carsten, Erkenntnistheoretische Konsequenzen der Systemtheorie: Die Theorie selbstreferentieller Systeme und der Konstruktivismus; Magisterarbeit vorgelegt am Fachbereich Philosophie und Sozialwissenschaften I der Freien Universität Berlin im Sommersemester 1999, S.22

[AMINI1983] Amini and M.A. Barzgar The Iranian Experiment in Primary Health Care: The West Azerbaijan Project (Principal investigations), School of Public Health, Ministry of Health and Social Welfare, Teheran, Oxford University Press, Oxford, UK., King Maurice (Ed), 1983

[BARKMANN2004] Nataly Barkmann, Gesundheitssysteme im europäischen Vergleich, Studienarbeit, GRIN Verlag, 2004, p.13, ISBN: 978-3-638-89070-0

[BLÜTHGEN,WEISCHET1980] Joachim Blühtgen, Wolfgang Weischet, Allgemeine Klimageographie, Lehrbuch der Allgemeinen Geographie, published by Walter de Gruyter, 1980, p.1, ISBN 3110065614

[BMG2006] Bundesministerium für Gesundheit [Internetseite], Abgerufen am 5.12.2006 www.bmg.bund.de

[BMBF2005] Bundesministerium für Bildung und Forschung (2005): Studie zur Situation der Medizintechnik in Deutschland im internationalen Vergleich – Zusammenfassung,

Bonn, Berlin 2005

[BRÄUNINGER2008] Dr. Michael Bräuninger, HWWI; Dr. Eckardt Wohlers, HWWI, Medizintechnik in Deutschland, Zukunftsbranche Medizintechnik-Auch im Norden ein Wachstumsmotor; Studie im Auftrag der HSH Nordbank AG, 2008

[BROCKHAUS2006] der Brockhaus 12.aktualisierte Auflage ed. Leipzig; Mannheim: Brockhaus Enzyklopädie ISBN: 3-7653-1682-1

[BUNDSAGENTUR FÜR ARBEIT2008] Bundesagentur für Arbeit [Internetseite], Aufgerufen am 28.08.2008, verfügbar unter http://infobub.arbeitsagentur.de

[BUNDESMINISTERIUM FÜR GESUNDHEIT 2005] Bundesministerium für Gesundheit Statistisches Taschenbuch Gesundheit 2005 Herausgeber: Referat Öffentlichkeitsarbeit 11055 Berlin, Stand: Oktober 2005

[CEN2006] CEN. Norm DIN EN 15221-1: Facility Management- Teil 1: Begriffe. Berlin: Beuth Verlag GmbH; 2006

[CHRISTEN2006] Christen, Marc, Convenor INB/TK 196 - ETH-Rat Immobilien Facility Management, Vortrag Oktober 2006 auf der Expo Real in München

[COTTS1999] David G. Cotts, The Facility Management Handbook, Second Edition, AMACOM Div American Mgmt Assn, 1999, ISBN 0814403808, Page 3

[DKG2006] [Internetseite] Deutsche Krankenhaus Gesellschaft, Daten von 2006; Aufgerufen am 20.01.2007 verfügbar unter http://www.dkgev.de

[FEDERAL MINISTRY OF HEALTH2007] Gesundheitsministerium 2007, [Internetseite] Aufgerufen am 02.06.2007, verfügbar unter http://www.bmg.bund.de/

[FROSCH2001] Frosch, Eduard: Outsourcing und Facility Management im Krankenhaus: Strategien- Entscheidungstechniken- Vorgehensweisen; mit Fallbeispielen aus der Praxis/ Eduard Frosch/ Gerd Hartinger/ Gerhard Renner- Wien/ Frankfurt; Wirtschaftsverlag Uberreuter, 2001, S.20ff

[GEFMA2004] GEFMA 100 Bonn: Deutscher Verband für Facility Management e.V.: 2004

[GUDAT2005] Horst Gudat, Optimierung der Einkaufs- und Logistikprozesse eines 1.400-Betten- Krankenhauses in Deutschland, 1. Europäische Konferenz über Krankenhaustechnik, Baden-Baden 5.-7. Sept. 2005, Tagungsband S.2

[GOETHE UNIERSITÄT FRANKFURT 2007] Johann Wolfgang Goethe Universität Frankfurt 2007 [Internetseite], Daten von 2006; Aufgerufen am 26.04.2007] verfügbar unter http://www.kgu.de/kgu/content.asp?bereich=ueberuns&FolderID

[GRUNDDATEN DER KRANKENHÄUSER 2006] [Internetseite], "Grunddaten der Krankenhäuser 2006" - Fachserie 12 Reihe 6.1.1, Statistisches Bundesamts www.ec.destatis.de

[HELBING2000] Helbing Management Consulting GmbH, Facility Management in Deutschland –Status und Perspektiven, Marktstruktur 2000, München, Februar 2000, P. 5

[HELLERFORTH2006] Michaela Hellerforth, Handbuch Facility Management für Immobilienunternehmen, Springer Verlag, 2006, Page 11, ISBN: 978-3-540-32196-5

[HORSTER2005] Detlef Horster, Niklas Luhmann, Published by C.H. Beckm 2005, p.159 ISBN 3406528120

[IFM2008] IFM 2008 [Internetseite], Daten von 2008 Aufgerufen am 06.05.2008 verfügbar unter http://www.imf.org/external/pubs/ft/weo/2008/01/weodata/groups.htm

[IFMA2007] IFMA2007 [Internetseite], Daten von 2006; Aufgerufen am 08.05.2007 verfügbar unter http://www.ifma.org

[IMMOBILIENZEITUNG2005] Ergebnis einer Umfrage unter 625 Immobilienprofis; Immobilien Zeitung Ausgabe21/2005, Wiesbaden, 2005

[KBV2007] Kassenärztlicher Bundesverband, [Internetseite], Daten von 2005; Aufgerufen am 08.05.2007 verfügbar unter http://daris.kbv.de

[KERN2002] A.O.Kern, Bericht zum Kolloquium über „Gesundheitssysteme im internationalem Vergleich" am Lehrstuhl für Finanzwissenschaft und Sozialpolitik, Universität Augsburg, 2002

[KRANKENKASSEN2008] Krankenkassen Deutschland [Internetsite], aufgerufen am 28.09.2008, verfügbar unter http://www.krankenkassen.de/private-krankenversicherung/

[LENNERTS2006] Lennerts, Kunibert, Prof., Vorlesungsskript, Grundlagen des strategischen Facility Management, 2006

[LENNERTS2002a] Lennerts, K., Abel, J., Pfründer, U: Optimisation and Analysis of processes in hospitals (OPIK) – a research project IFHE, Bergen, Norway May 2002

[LENNERTS2002b] Lennerts, K., Abel, J., Pfründer, U: Optimierung und Analyse von Prozessen in Krankenhäusern – das Forschungsprojekt OPIK Tagungsband/Proceedings Facility Management Messe und Kongress Düsseldorf, Veranstalter: MESAGO Messe und Kongress GmbH, Stuttgart, Juni 2002

[LENNERTS2002c] Lennerts, K., Abel, J., Pfründer, U.: Hospital Facilities Management – the OPIK research project IFMA, World Work Place Congress, Paris, France, July 2002

[LENNERTS2001-2007d] Lennerts, K., Diez, K., Abel, J. Pfründer, U., Script Krankenhaus OPIK2001.2007, Universität Karlsruhe]

 [LOCHMANN1998] Lochmann, Hans- Dieter, Köllgen Rainer, Andersen Consulting, (1998), Facility Management - Strategisches Immobilienmanagement in der Praxis, Betriebswirtschaftlicher Verlag Dr. Th. Gabler/ GWV Fachverlage GmbH, Wiesbaden 1998, S. 44 ff

[LUHMANN,1985] Luhmann N., Soziale Systeme. Grundriss einer allgemeinen Theorie. Suhrkamp, Frankfurt/M. (1985)

[LÜNENDONK2004] Lünendonk 2004, [Internetseite]; Aufgerufen am 28.11.2007 verfügbar unter http://www.luenendonk.de/download/press/LUE_FM_2005_f080605.pdf

[Lyne1995] Lyne, J. IDRC's real estate revolution: occupancy costs plummet, productivity crests. *Site Selection Magazine.* 1995, April.

[MAY1996] May, Alexander/ Breitenstein, Oliver/ Lennerts, Kunibert, The latest development in facility management research , Leipzig annual civil engineering report, Band 1, 1996, Leipzig S. 375

[MATTHIES2002, S.2] Matthies Michael Prof. Dr., Einführung in die Systemwissenschaft

WS 2002 / 2003, Studiengang Angewandte, Systemwissenschaft an der Universität Osnabrück S.2

[MUNRKEN1988] Murken, Axel Hinrich: Vom Armenhospital zum Großklinikum: die Geschichte des Krankenhauses vom 18. Jahrhundert bis zur Gegenwart, Köln: DuMont, 1988

[NAGHAWI2005] Naghawi, Mohsen, Todesursachen in 23 Provinzen des Landes des Jahres 1382 (entspricht 2003/2004), 1384

[NÄVY2003] Jens Nävy, Wolfgang Löwen, Facility Management: Grundlagen, Computerunterstützung, Einführungsstrategie, Praxisbeispiele, Springer Verlag, 2003, p.2, ISBN 3540441670

[NORD/LB 2002] NORD/LB, Global Markets Krankenhausmarkt im Umbruch, November 2002, p. 10, www.nordlb.com

[OECD2007] [Internetseite]; aufgerufen am 13.04.2007, Seite verfügbar unter: http://www.oecd.org/

[OVERSBY2007] Overby, Stephanie, 2007, ABC: An Introduction to Outsourcing verfügbar unter www. CIO.com.

[PEIRAVI2000] Peiravi Khamene Mohammad, Philosophie der Entstehung der Gesundheitsnetzwerke des Landes und Position der Lehre in ihren Anfängen, Diplomarbeit, Universität Teheran 1370/ 71, p. 33-35

[PREUSS 2006] Preuß, N., Schöne, L.B. (2006), Real Estate und Facility Management, Aus Sicht der Consultingpraxis, 2.erw. und akt. Auflage, Springer Verlag 2006, p. 48

[RISTO2007] Peter Risto, Erhebungsbericht, Klima, Luft und Wasser, Fachbereich Landschaftsplanung und Gartenkunst, Department für Raumentwicklung, Infrastruktur- und Umweltplanung Technische Universität Wien, 2007, p.5

[SCI2005] Statistical Centre of Iran, [Internetseite]; aufgerufen am 24.07.2007, verfügbar unter: http://www.sci.org.ir/portal/faces/public/sci_en/

[SCÄFERS2003] Schäfers, Bernhard,: Grundbegriffe der Soziologie. 8. Aufl. Opladen: Leske+Budrich, 2003 (UTB-Taschenbuch)

[SCHÖNIG2001] Werner Schönig, Perspektiven institutionalistischer Ökonomik (Reihe : Soziale Marktwirtschaft und europäische Perspektive) Bd. 1, 2001, p. 202, ISBN 3-8258-5349-7

[SHADPOUR1994] Shadpour, K., The PHC Experience in Iran, The Council for Expansion of PHC Networks, Ministry of Health and Medical Education, UNICEF, Teheran (1994), p. 11 ff

[SGB] Sozialgesetzbuch (SGB) Fünftes Buch (V) - Gesetzliche Krankenversicherung - (Artikel 1 des Gesetzes v. 20. Dezember 1988, BGBl. I S. 2477), §12 (1)

[STAPF-FINE2003] Stapf-Fine/ Schölkopf, Die Krankenhausversorgung im internationalen Vergleich- Zahlen, Fakten, Trends- Deutsche Krankenhaus Verlagsgesellschaft mbH Düsseldorf, 2003

[STATISTICAL CENTER OF IRAN2006] SCI [Internetseite], Aufgerufen am 1.12.2006 verfügbar unter http://www.sci.org.ir

[STATISTISCHES BUNDESAMT2006] [Internetseite], Daten von 2006; Aufgerufen am 13.11.2006 verfügbar unter http://www.destatis.de (2006)

[TUMS2005] Tehran University of Medical Science, [Internetseite], Daten von 2005; Aufgerufen am 24.12.2006; verfügbar unter http://www.medicine.tums.ac.ir

[WILSON2006, S.99ff] Wilson Dave S. 99ff. , PG Dip Mgt, Mace Macro, Director, Getting cultured: delivering FM across Europe, Limited, Exhibition Europe - Frankfurt am 7-9 March 2006 - Tagungsband /Proceedings (2006), VDE-Verlag Berlin, ISBN 978-3-8007-2938-8

[UN2005] United Nations. Standard country or Area Codes for Statistical Use. Series M, No. 49, Rev. 4 (United Nations publication, Sales No. M.98.XVII.9). Available in part at http://unstats.un.org/unsd/methods/m49/m49regin.htm

[UNDEP2006] [Internetseite], Report, UNDP Country Programme for the Islamic Republic of Iran (2005-2009); Aufgerufen am 13.11.2006 verfügbar unter http://www.undp.org.ir/reports (2006)

[VDE2005] VDE: Deutschland in der Medizintechnik international führend, Presseveröffentlichung 2005; [Internetseite] aufgerufen am 13.03.2009; verfügbar unter http://www.invent-a-chip.de/VDE/Presse/2005-Oeffentlich/2005-56.htm

APPENDIX

Appendix- A1: General Questionnaire

Appendix- A2: Questionnaire Data

Appendix- A3: Questionnaire Management

Appendix- A4: Questionnaire Medical equipment –Personal

Appendix- A5: Questionnaire Medical equipment –Management

Appendix- A6: Questionnaire Medical equipment – Maintenance

Appendix- A7: Questionnaire Medical equipment – Repair

Appendix- A8: Process Comparison- Medical equipment

Appendix- A9: Interfaces of the Process- Medical equipment

Appendix- A10: Identity Values –Medical equipment

Appendix- A11: Questionnaire Technical Facilities – Maintenance

Appendix- A12: Questionnaire Technical Facilities - Repair

Appendix- A13: Process Comparison- Technical Facilities

Appendix- A14: Interfaces of the Process - Technical Facilities

Appendix- A15: Identity Values – Technical Facilities

Appendix- A16: Questionnaire Laundry

Appendix- A17: Process Comparison - Laundry Management

Appendix- A18: Interfaces of the Process - Laundry Management

Appendix- A19: Identity Values – Laundry Management

Appendix- A20: Expert Analyse

Appendix- A21: Expert list

APPENDIX - A1

General Questionnaire

Universität Karlsruhe (TH)
Prof. Dr.-Ing. Kunibert Lennerts
Facility Management

Krankenhaus

Trägerschaft

Versorgungsstufe

Bettenzahl

Anzahl stationäre Patienten 2000

Belegtage 2000

Anzahl Mitarbeiter 2000

Anzahl Mitarbeiter (vollzeit) 2000

Anzahl Fachabteilungen

Anzahl OP's

BGF   m²   NNF   m²

NGF   m²   VF   m²

HNF   m²   FF   m²

Gesamtetat in 2000 in Euro

APPENDIX – A2

Questionnaire Data

Universität Karlsruhe (TH)
Prof. Dr.-Ing. Kunibert Lennerts
Facility Management

Krankenhaus

Bettenzahl   HNF   m²

Anzahl PC's   Anzahl TV Geräte:

Anzahl Telefonanschlüsse:

Wäscheleistung 2000:   Tonnen

Anzahl Hygieneuntersuchungen:

Anzahl Sterilguteinheiten 2000:

Anzahl aufbereitete Betten 2000

Anzahl Essen 2000:

Frühstück: Mittags   Abend:

Wärmeverbrauch 2000:   kWh

Kälteverbrauch 2000:   kWh

Stromverbrauch 2000:   kWh

Wasserverbrauch 2000:   m³

Wiederbeschaffungswert Medizintechnik: Euro

Gesamtmenge Abfall 2000:   Tonnen

Anzahl Patiententransporte 2000

Anzahl Fahrzeuge gesamter Fuhrpark:

OPIK بهینه سازی و تجزیه و تحلیل فرآیندهای بیمارستانی

پرسشنامه 2:

1. نام بیمارستان: ــــــــــــــــــــــــــــ

2. تعداد کامپیوترها : ــــــــــــ

3. تعداد تلویزیونها : ــــــــــــ

4. تعداد خط : ــــــــــــ

5. تعداد رخت شسته در سال 1384: ــــــــــــ تن

6. تعداد آزمایشهای تشخیصی در سال 1384: ــــــــــــ

7. تعداد واحدهیه استریل در سال 1384: ــــــــــــ

8. تعداد تختهای اشغل در سال 1384: ــــــــــــ

9. تعداد غذاهای آماده شده در سال 1384: صبحانه: ــــــ ناهار: ــــــ شام: ــــــ

10. مصرف گرما در سال 1384 ــــــــــــ kWh

11. مصرف سرما در سال 1384 ــــــــــــ kWh

12. مصرف برق در سال 1384 ــــــــــــ kWh

13. مصرف آب در سال 1384 ــــــــــــ kWh

14. ارزش تجهیزات پزشکی ــــــــــــ ریال

15. مقدار زباله : ــــــــــــ تن

16. تعداد بیمار اعزامی در سال 1384 : ــــــــــــ

17. تعداد ماشین در بخش حمل و نقل: ــــــــــــ

194

APPENDIX – A3

Questionnaire Management

FRAGEBOGEN

ORGANISATION UND VERWALTUNG

Finanzielle Fragen

1. Können Sie Angaben zur derzeitigen Kostenverteilung machen?

    a. Personalkosten (gesamt)   ______________

        i. Ärztlicher Dienst   ______________

        ii. Pflegedienst   ______________

        iii. Medizinisch technischer Dienst   ______________

        iv. Funktionsdienst   ______________

    b. Sachkosten (gesamt)   ______________

        i. Medizinischer Dienst   ______________

2. Wie ist bei Ihnen die Kosten/Leistungsrechnung organisiert?

    a. Patientenabrechung (Umlageverfahren)

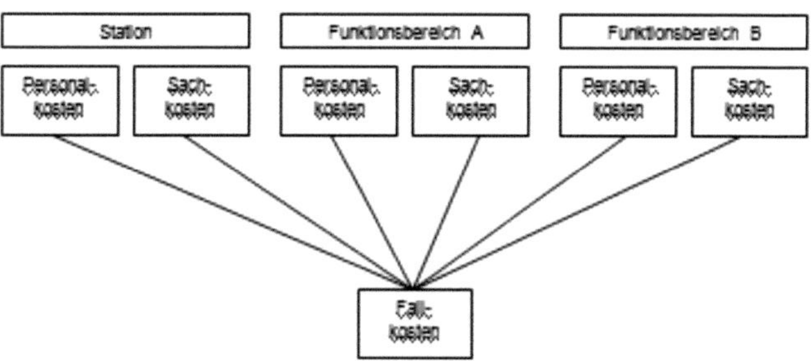

Fragebogen

Termin am 25./26.04.2002

b. facilitäre Leistungen

Werden die FM-Kosten auf
Stationen/Abteilungen/Fälle umgelegt?   ☐ ja ☐ nein

Wie werden die Kosten umgelegt?________________________________

3. Wo sehen Sie im Bereich des Facility Management im Krankenhaus die
   größten Einsparpotentiale?
   (Zutreffendes der Wertung nach mit ◎ ◎ ◎ ◎   )

Technik		
	Energiemanagement	
	Betreiben	
		Instandhaltung
		Beheben von Störungen
		Sanieren
Bewirtschaftung		
	Reinigung	
	Hausmeisterdienste	
	Parkraum	
	Umzug	
	Sicherheit	
	Verpflegungsdienste	
	Winterdienste	
	IT/Telekommunikation	
Bauen im Bestand		
	Modernisieren	
	Sanieren	
	Umbauen	
Medizintechnik		
	Betreiben	
		Instandhaltung
		Beheben von Störungen
		Sanieren
Einkauf/Logistik		
	Entsorgung	
	Interne Postdienste	
	Transportdienste	
	technischer Einkauf	
	Einkauf Medizintechnik	
sonstiges		

Tabelle 1

197

OPIK ...ینه سازی و تجزیه و تحلیل فرآیندهای بیمارستانی

پرسشنامه II:

بیمارستان: ــــــــــــــــــــــــ

مدیریت و اداره

سوال های بودجه:

1. هزینه ها را لطفا با % یا ریالی بفرمایید

   1.1. بودجه پرسنل (کل)   ــــــــــــــــــــ

      1.1.1. پزشکان   ــــــــــــــــــــ

      1.1.2. پرستاران   ــــــــــــــــــــ

      1.1.3. خدمات پزشکی و فنی   ــــــــــــــــــــ

   1.2. بودجه کالا (کل)   ــــــــــــــــــــ

      1.2.1. خدمات پزشکی   ــــــــــــــــــــ

2. هزینه های جانبی چگونه حسب می شوند؟ بخش / ایستگاه / هزینه مریض

   ــ

3. بیشترین صرفه جویی مدیریت امکلات و تجهیزات را در چه بخشهایی میبینید؟

◎	◎	◎	◎		
				Maintenance of technical facilities	نگهداشت تجهیزات فنی
				Maintenance of biomedical equipment	نگهداشت تجهیزات بهداشتی
				Logistic of pharmaceuticals	پشتیبانی دارو
				Logistic of medical products	پشتیبانی محصولات پزشکی
				Cleaning Management	...اخداری -گروه مدیریت نظافت
				Sterilization	گروه استریلیزه
				Waste Management	گروه مدیریت پسماندهای جامد
				Waste Water Management	گروه مدیریت پسماندهای مایع
				Organization	گروه مدیریت ساختار سازمانی
				Service Management	گروه آماده باش
				Energy Management	گروه مدیریت انرژی
				Repair Management	گروه تعمیرات ساختمانی
				Catering	گروه مدیریت خوراک
				Laundry Management	گروه مدیریت رخت شویی
				Telecommunication Management	گروه مدیریت مخابرات
				IT- Management	گروه مدیریت اطلاعات

سوال های مدیریتی

1. آیا در بیمارستان فرایندها را آنالیز کرده اید؟

1.1. برای بخش پزشکی

☐ بلی   ☐ خیر

اگر بلی برای کدام فرایند ....

1.2. برای بخش پرستاری

☐ بلی   ☐ خیر

اگر بلی برای کدام فرایند ....

1.3. برای بخش اداری

☐ بلی   ☐ خیر

اگر بلی برای کدام فرایند ....

1.4. برای بخش مدیریت و تجهیزات-پشتیبانی

☐ بلی   ☐ خیر

اگر بلی برای کدام فرایند ....

Ablauftechnische Fragen

1. Gibt es vorangegangene Prozessanalysen bzw. –beschreibungen
   a. Für den ärztlichen Dienst ☐ ja ☐ nein

   wenn ja, welche..

   b. Für den Pflegedienst ☐ ja ☐ nein

   wenn ja, welche..

   c. Für die Verwaltung ☐ ja ☐ nein

   wenn ja, welche..

   i. Für das Facility Management ☐ ja ☐ nein

   wenn ja, welche..

2. Gibt es Aufbau- und Ablaufpläne?

3. Gibt es Stellenbeschreibungen?

Organisationsfragen

1. Was verstehen Sie unter Facility Management?
   - ☐ Gebäudetechnik
   - ☐ Management der sekundären Dienstleistungen
   - ☐ Gebäude Lebenszyklus
   - ☐ Outsourcing
   - ☐ Sonstiges_________________________________

2. Wo sehen Sie im Bereich des Facility Management im Krankenhaus die größten Optimierungspotentiale? (Organisation/Ablauf/Schnittstellen)

siehe Tabelle 1 (Zutreffendes mit **O** kennzeichnen)

3. Welches der aufgeführten Organigramme entspricht im wesentlichen Ihrer Organisation?

4. Welches der Organigramme halten Sie für eine sinnvolle Organisationsform?

5. Welche Erwartungen haben Sie an qualifizierte Mitarbeiter im Facility
    Management? (1 hoch; 5 gering)

    ❶❷❸❹❺

    a. Personalkompetenz   ☐☐☐☐☐
    b. Technische Kenntnisse   ☐☐☐☐☐
    c. Teamfähigkeit   ☐☐☐☐☐
    d. Kaufmännische Kenntnisse   ☐☐☐☐☐
    e. Sonstiges   ☐☐☐☐☐

6. Wie sieht Ihrer Meinung nach die begriffliche Trennung von Primär- und
    Sekundärprozessen im Krankenhaus aus?

7. Hat aus Ihrer Sicht die Einführung der DRG's Einfluss auf das Facility
    Management bezüglich der

    a. Abrechnungsmodalitäten   ☐ ja ☐ nein
    b. Organisationsstruktur   ☐ ja ☐ nein
    c. Prozessabläufe   ☐ ja ☐ nein

ینه سازی و تجزیه و تحلیل فرآیندهای بیمارستانی

سوال های سازمانی

1. برای شما Facility Management چیست؟

☐ تاسیسات ساختمان
☐ مدیریت خدمات جانبی
☐ چرخ زندگی ساختمان
☐ Outsourcing
☐ غیره _______________________________

2. بیشترین پتانسیل پیشرفت مدیریت امکانات و تجهیزات را در چه بخشهایی میبینید؟

◎	◎	◎	◎		
				Maintenance of technical facilities	نگهداشت تجهیزات فنی
				Maintenance of biomedical equipment	نگهداشت تجهیزات بهداشتی
				Logistic of pharmaceuticals	پشتیبانی دارو
				Logistic of medical products	پشتیبانی محصولات پزشکی
				Cleaning Management	گانخداری - گروه مدیریت نظافت
				Sterilization	گروه استریلیزه
				Waste Management	گروه مدیریت پسماندهای جامد
				Waste Water Management	گروه مدیریت پسماندهای مایع
				Organization	گروه مدیریت ساختار سازمانی
				Service Management	گروه آماده باش
				Energy Management	گروه مدیریت انرژی
				Repair Management	گروه تعمیرات ساختمانی
				Catering	گروه مدیریت خوراک
				Laundry Management	گروه مدیریت رخت شویی
				Telecommunication Management	گروه مدیریت مخابرات
				IT- Management	گروه مدیریت اطلاعات

3. چه انتظاراتی از همکاران بخش Facility Management پشتیبانی دارید

زیاد 1- کم5

                                        ❶❷❸❹❺
a. قابلیت کاری   ☐☐☐☐☐
b. دانش فنی   ☐☐☐☐☐
c. کار در تیم   ☐☐☐☐☐
d. دانش بازرگانی   ☐☐☐☐☐
e. غیره   ☐☐☐☐☐

APPENDIX – A4

Questionnaire Medical equipment –Personal

Medizintechnik

Fragen an Prozessbeteiligte aus dem Bereich ‚Heilen und Pflegen'

Datum: ___________

Bemerkung	Fragestellung	Wertung der Fragestellung	
		1 bis 6	1 = sehr gut / sehr wichtig 6 = ungenügend / unwichtig
	Wie beurteilen Sie aus Ihrer Sicht die Gesamtqualität der Instandhaltung der medizintechnischen Geräte?		☐ 1 ☐ 2 ☐ 3 ☐ 4 ☐ 5 ☐ 6
	Besteht Ihrer Ansicht nach mehr Schulungsbedarf?		☐ 1 ☐ 2 ☐ 3 ☐ 4 ☐ 5 ☐ 6
	Wie beurteilen Sie die Dauer der jeweiligen Instandsetzung von medizintechnischen Geräten?		☐ 1 ☐ 2 ☐ 3 ☐ 4 ☐ 5 ☐ 6

Welcher der aufgeführten Gruppen sind Sie zugehörig?   Pflegedienst (leitende Fkt. / nicht leitende Fkt.), ärztlicher Dienst (leitende Fkt. / nicht leitende Fkt.), sonstige ______________________________

(zutreffendes bitte unterstreichen, bzw. ausführen)

بهینه سازی و تجزیه و تحلیل فرایندهای بیمارستانی   **CPIK**

تجهیزات پزشکی

تاریخ: ____________

نظر	سوال	ارزش سوال	
		1 تا 6	1= خیلی خوب / خیلی مهم 6 =بد/ بی اهمیت
	به نظر شما سیستم نگهداشت تجهیزات پزشکی از نظر شما چگونه است؟	□ □ □ □ □ □ 6 5 4 3 2 1	
	آیا به نظر شما نیاز به آموزش بیشتری وجود دارد؟	□ □ □ □ □ □ 6 5 4 3 2 1	
	نظر شما در باره زمان تعمیرات تجهیزات پزشکی چیست؟	□ □ □ □ □ □ 6 5 4 3 2 1	

شما جز کدام گروه هستید ؟

□ پزشک   □ پرستار   □ غیره

دانشکده کارسروده		تجهیزات پزشکی / پرسشنامه 1
1/1	تاریخ	بیمارستان

APPENDIX – A5

Questionnaire Medical equipment–Management

<table>
<tr><td>**OPIK**</td><td>Fragebogen</td></tr>
</table>

MEDIZINTECHNIK

Finanzielle Fragen
1. Welcher Anteil der Leistungen (in %)
   wird durch Fremddienstleister erbracht?   ______ %

2. Wie hoch ist der Wert des derzeitigen
   Gerätebestandes? (Wiederbeschaffung)   __________ €

3. Auf welche Höhe belaufen sich
   die Anschaffungskosten in den letzten 5 Jahren?   __________ €

4. Wie hoch sind die jährlichen Kosten für die
   Instandhaltung der Medizintechnik?   __________ €

Ablauftechnische Fragen
1. Welche Instandhaltungsstrategie verfolgt Ihre Klinik?
   a. führen Sie betriebsabhängige Instandhaltungen durch?

2. wer trifft die Anschaffungsentscheidung für medizintechnische Gerätschaften

3. existiert ein Schulungsplan   ☐ ja ☐ nein
   a. werden durchgeführte Schulungen dokumentiert   ☐ ja ☐ nein

Organisationsfragen
1. Wie viel Mitarbeiter beschäftigen Sie in der Medizintechnik

   _____________ Mitarbeiter

2. Personalkosten Medizintechnik

   _____________ €

3. Wie viele Aufträge werden im Jahr abgewickelt?

   _____________ Aufträge

Die Eintragungen entsprechen den im Interview gemachten Angaben.

_______________________   _______________________
(Ort / Datum)   (Unterschrift)

تجهیزات پزشکی

بهینه سازی و تجزیه و تحلیل فرآیندهای بیمارستانی **OPIK**

پرسشهای مالی

1. چند در صد کارائی بوسیله پیمانکار انجام می شود؟ _________ تومان
2. قیمت دستگاه های موجود یا تهیه دوباره آن چه قدر است؟ _________ تومان
3. مبلغ خرید تجهیزات پزشکی در 5 سال گذشته چه قدر بوده است؟ _________ تومان
4. هزینه سالانه نگهداشت تجهیزات پزشکی چه قدر است؟ _________ تومان

پرسشهای فرایندهای فنی

1. برنامه ریزی استراتژیک بیمارستان شما چیست؟   ☐بلی   ☐خیر

a. آیا نگهداری توسط همکاران بیمارستان براساس برنامه ریزی سازمان انجام می گردد؟
☐بلی   ☐خیر

2. تصمیم گیری خرید تجهیزات پزشکی با کیست؟

3. برنامه ریزی برای آموزش وجود دارد؟

a. آیا آموزشهای انجام شده مستندسازی می شوند؟

پرسشهای سازمانی

1.تعداد نیروی انسانی بخش تجهیزات پزشکی چند نفر می باشند؟ __________نفر

2.هزینه نیروی انسانی بخش تجهیزات پزشکی چند تومان می باشد؟ _______تومان

3.چند درخواست در سال انجام می شود؟ _________در خواست

تجهیزات پزشکی / پرسشنامه 1

APPENDIX – A6

Questionnaire Medical equipment – Maintenance

Optimierung und Analyse
von Prozessen in Krankenhäusern

Wartung/Inspektion Medizintechnik

Gerätehersteller:		Gerätetyp:	
Ansprechpartner:		Station/Funktionsbereich:	

Auftragskurzbeschreibung:	

Zeiten (in Minuten)

Nr.	Bezeichnung	Wegezeit	Information	Arbeitszeit
1.	Arbeitsplanerstellung			
2.	Terminplanung			
3.	Durchführung			
4.	Kostenübermittlung			
5.	Auftragsabweichungsanalyse			
6.	Inspektionsanalyse			

Material

Nr.	Anzahl	Bezeichnung	Preis in €
1.			
2.			
3.			
4.			
5.			
6.			
7.			

Qualität

(Durch den Erbringer der Leistungen auszufüllen)

1. War dieser Wartungseinsatz terminiert? ☐ ja ☐ nein
2. Wenn ja, wurde der Auftrag termingerecht durchgeführt? ☐ ja ☐ nein
3. Wie viele Anläufe wurden benötigt, um den Auftrag erfolgreich abzuschließen? ____ Anläufe
4. Ist die Erzeugung eines Instandsetzungsauftrages notwendig? ☐ ja ☐ nein

بهینه سازی و تجزیه و تحلیل فرایندهای بیمارستانی — CPIK

| مراقبت و نگهداشت تجهیزات پزشکی

تولید کننده دستگاه:

اسم \ نام دستگاه:

مکان:

دوره نگهداشت:

وضعیت فعالیتها به طور مختصر/ مشکل دستگاه چه بوده است؟

توضیح عملکرد/ تعمیر دستگاه چگونه انجام شده است؟

☐ خارج از بیمارستان   ☐ داخل بیمارستان   تاریخ:

اگر از خارج از بیمارستان نگهداشت انجام می گیرد لطفا توضیح دهید

لوازم مصرف

شماره	نام کالا	تعداد	هزینه (تومان)
1.			
2.			
3.			
4.			
5.			
6.			
7.			

زمان (دقیقه)

شماره	فعالیت	زمان اطلاع رسانی	زمان اطلاع رسانی تا اقدام	زمان کار
1.	برنامه ریزی نگهداشت			
2.	تهیه برنامه ریزی کاری			
3.	برنامه ریزی زمانی			
4.	انجام نگهداشت			
5.	مستندسازی			
6.	تجزیه و تحلیل فعالیت			
7.	نظارت و بازرسی			

کیفیت

| 1. | آیا سفارش در زمان در نظر گرفته انجام گردیده؟ | خبر ☐   بلی ☐ | 2. | تا انجام نهایی کار چندبار مراجعه لازم بوده؟ ___ مراجعه |
| 3. | | | | آیا نیاز به تعمیرات است؟   خبر ☐   بلی ☐ |

Universität Karlsruhe (TH)	دانشگاه علوم پزشکی تهران	تجهیزات پزشکی / پرسشنامه 3
Facility Management (TMB)	تاریخ	۲۱

APPENDIX – A7

Questionnaire Medical equipment – Repair

Optimierung und Analyse
von Prozessen in Krankenhäusern

Instandsetzung Medizintechnik

Gerätedaten

Gerätehersteller:		Gerätetyp:	
Ansprechpartner:		Station/Funktionsbereich:	

Auftragskurzbeschreibung:	

Zeiten (in Minuten)

Nr.	Bezeichnung	Wegezeit	Information	Arbeitszeit
1.	Schadensprüfung			
2.	Berichte erstellen			
3.	Ursachenanalyse			
4.	Ursachenauswahl			
5.	Ursachenbeseitigung			
6.	Instandsetzung			
7.	Wiederinbetriebnahme			

Material

Nr.	Anzahl	Bezeichnung	Preis in €
1.			
2.			
3.			
4.			
5.			
6.			

Qualität

(Durch den Erbringer der Leistungen auszufüllen)

1. Ist ein Fehler gleicher Ursache bereits aufgetreten □ ja □ nein

2. Wie viele Anläufe wurden benötigt, um den Auftrag erfolgreich abzuschließen? ____ Anläufe

(Durch den Kunden auszufüllen)

3. Wie beurteilen Sie die Qualität der Leistung?

   1 2 3 4 5 6
  □ □ □ □ □ □

4. Wie beurteilen Sie die Qualität der Information über den Auftrag

  □ □ □ □ □ □

5. Verbesserungsvorschläge:

بهينه سازى و تجزيه و تحليل فرآيندهاى بيمارستانى OPIK

تعميرات و بازرسى تجهيزات بهداشتى

توليد كننده دستگاه:
سال توليد:
اسم \ نام دستگاه:
مكان :
بخش كارى :
توضيح فعاليت بطور مختصر:
توضيح عملكرد :

زمان (دقيقه)

لوازم مصرف

زمان كار	زمان اطلاع رسانى	زمان راه	مشتس سازى	شماره		هزينه (تومان)	توضيحات	تعداد	شماره
			تعيين اشكل	1					1
			تهيه گزارش	2					2
			تجزيه و تحليل علت	3					3
			تعمير	4					4
			كار گيرى دوباره	5					5

كيفيت

1	آيا تا به حل اشتباهى با همچين حالتى پيش آمده است؟ خير ☐ بلى ☐ 2	تا انجام نهايى كارچندبار مراجعه لازم بود؟ مراجعه___
3	كيفيت اطلاع رسانى اين درخواست را ارزشيابى فرماييد؟	1 2 3 4 6 8 ☐☐☐☐☐☐ 1 خيلى خوب 6 نارسا
4	پيشنهادهاى اصلاحى	

Gl Universität Karlsruhe (TH)	دانشكده علوم پزشكى تهران	تجهيزات پزشكى / پرسشنامه 2
Facility Management (TMB)	تاريخ:	V1

APPENDIX – A8

Process Comparison- Medical equipment

Deutschland

Lebenszyklus Medizintechnik

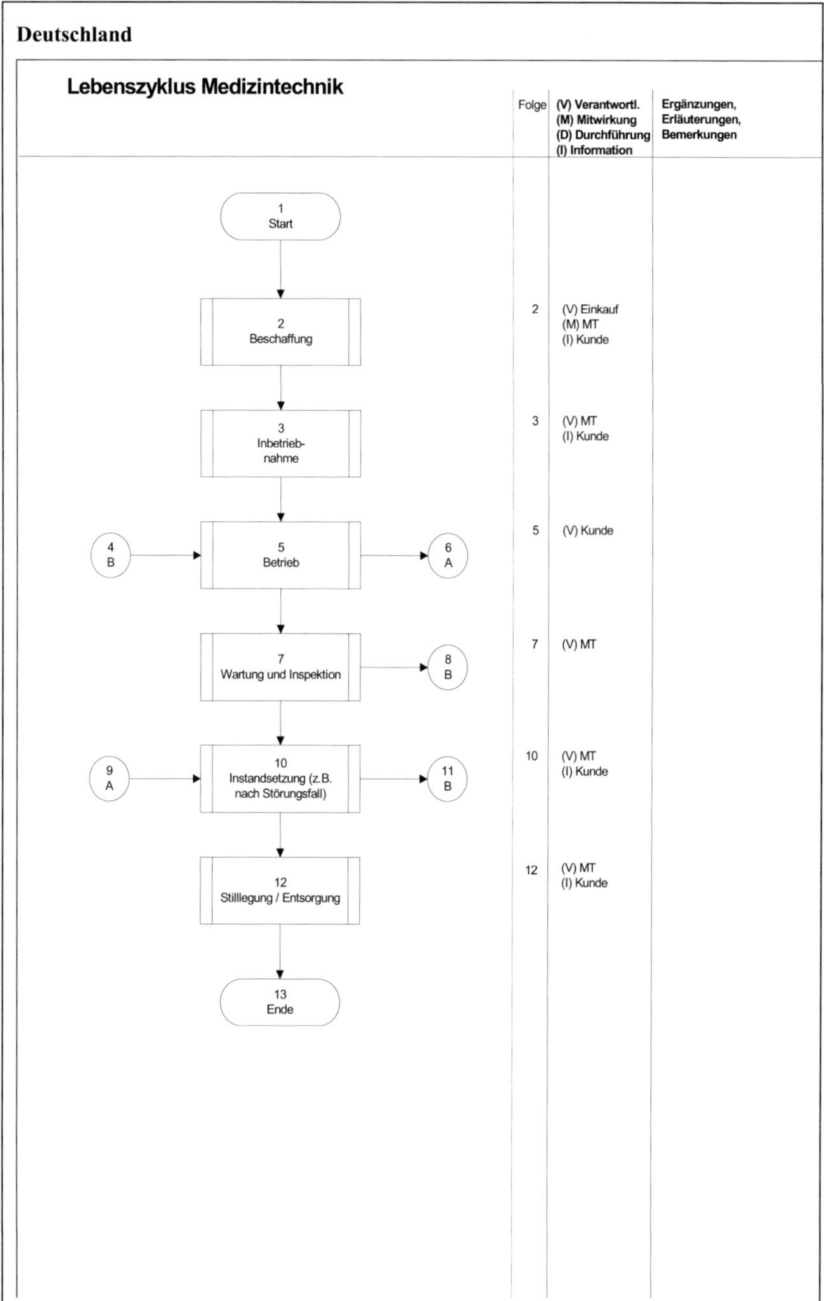

Iran

Lebenszyklus Medizintechnik

	Folge	(V) Verantwortl. (M) Mitwirkung (D) Durchführung (I) Information	Ergänzungen, Erläuterungen, Bemerkungen
1 Start			
2 Beschaffung	2	(V,D) Einkauf MT,Einkaufs-kommitee Management, (M) Finanzabteilung (I) Kunde, Krankenhausleit ung, Unterstützungs-einheit der Universität	Es muss insesondere Unterschieden werden zwischen: Ausländische-Inländische Währung Einweg- Mehrwegprodukte
3 Inbetrieb-nahme	3	(D) MT, Subunternehmer , Niederlassung, (I) Management, Fiananzabteilung ,(M) Abteilungsleiter, Einkauf, Abteilung Technische Abteilung	
4 Betrieb	4	(D) Abteilung/ Station	
5 Wartung und Inspektion	5	(D) MT, Subunternehmer , Niederlassung, Nutzer/ Abteilung (I) Management, Finenzabteilung, Abteilugsleiter	
6 Instandsetzung (z.B. nach Störungsfall)	6	(V, D) MT, Subunternehmer, Niederlassung, Einkauf (I) Management, Finanzabteilung, Abteilungsleiter	
7 Stilllegung / Entsorgung	7	(V,D) MT, Subunternehme r, Niederlassung (M) Abteilung/ Station, (I)Management, Krankenhausleit ung, Finanzabteilung der Universität	
8 Ende			

Deutschland

Beschaffung Medizintechnik

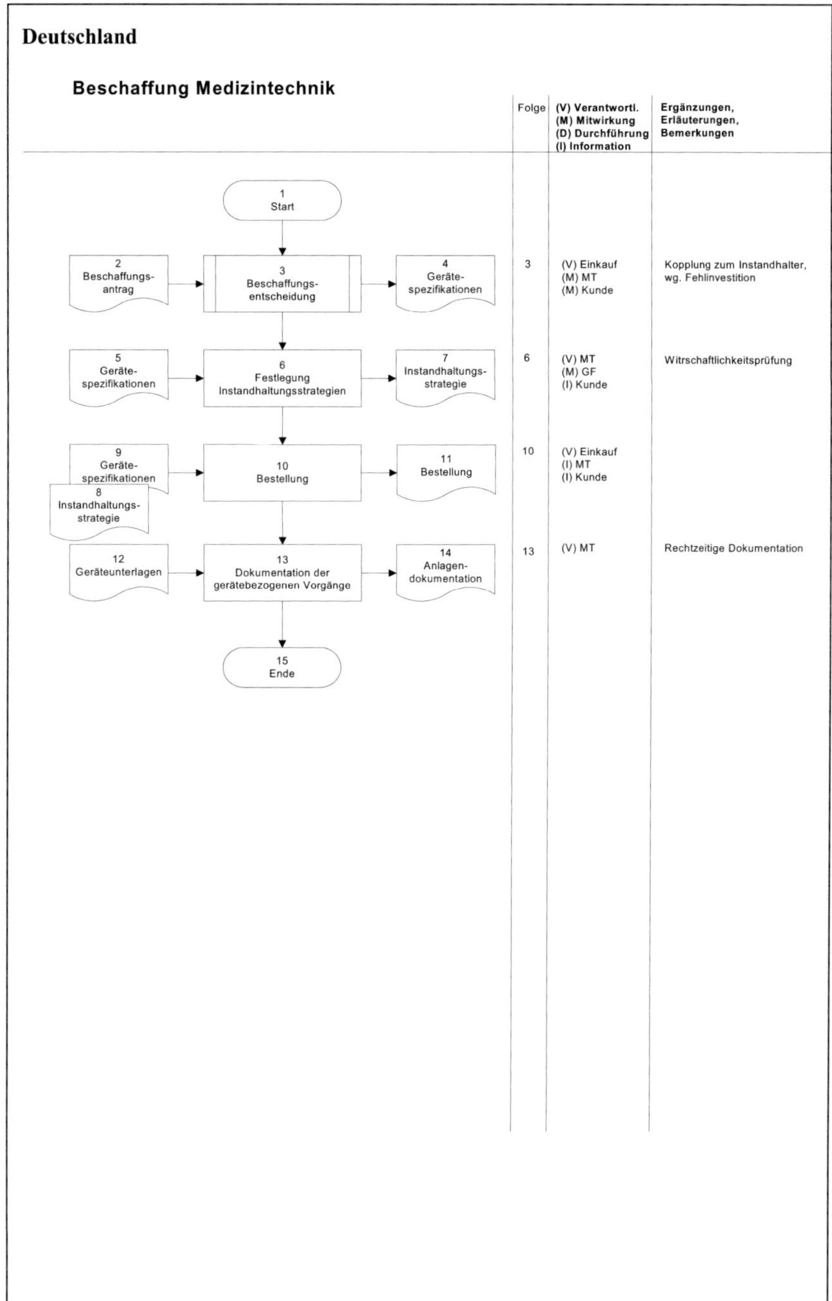

Folge	(V) Verantwortl. (M) Mitwirkung (D) Durchführung (I) Information	Ergänzungen, Erläuterungen, Bemerkungen
3	(V) Einkauf (M) MT (M) Kunde	Kopplung zum Instandhalter, wg. Fehlinvestition
6	(V) MT (M) GF (I) Kunde	Witrschaftlichkeitsprüfung
10	(V) Einkauf (I) MT (I) Kunde	
13	(V) MT	Rechtzeitige Dokumentation

Iran

Beschaffung Medizintechnik
Folge
(V) Verantwortl.
(M) Mitwirkung
(D) Durchführung
(I) Information
Ergänzungen,
Erläuterungen,
Bemerkungen
1
Start
2
Beschaffungs-
antrag
3
Beschaffungs-
entscheidung
4
Geräte-
spezifikationen
3
(V,D) Abteilung/
Stationen, MT,
(M) Management,
Finanzabteilung,
Einkaufskommitee
(I) Gesundheits
Abteilung ,
Unterstützungs-
einheit der
Universität
90% der des Bedarfs wird von
den Abteilungen/ Stationen
beantragt
4
Geräte-
spezifikationen
5
Festlegung
Instandhaltungsstrategien**
6
Expertenmeinung*
5
(V,D) MT,
Einkaufskommitee
(M) Management,
Finanzabteilung,
(I) Gesundheits
Abteilung ,
Unterstützungs-
einheit der
Universität
*Indirekt und unregelmäßig
6
Expertenmeinung
7
Bericht und
Strategie
8
Entscheidung
9
Bestellung
8
(V,D) Einkauf,
Einkaufskommitee
(M) Management,
MT,
Finanzabteilung,
(I) Gesundheits
Abteilung ,
Unterstützungs-
einheit der
Universität
**In einigen Krankenhäusern (
je nach Politik des Hauses)
existiert ein Einkaufskommitee
bestehend aus:
- Abteilungsleiter
- Krankenhausleitung,
Management,
- Abteilungsleiter
Medizintechnik
- 2 Mitglieder des
wissenschatlichen Beirats
10
Geräteunterlagen
11
Dokumentation der
gerätebezogenen Vorgänge
12
Anlagen-
dokumentation
11
(V) MT,
Abteilung/
Station,
Finanzabteilung
13
Ende

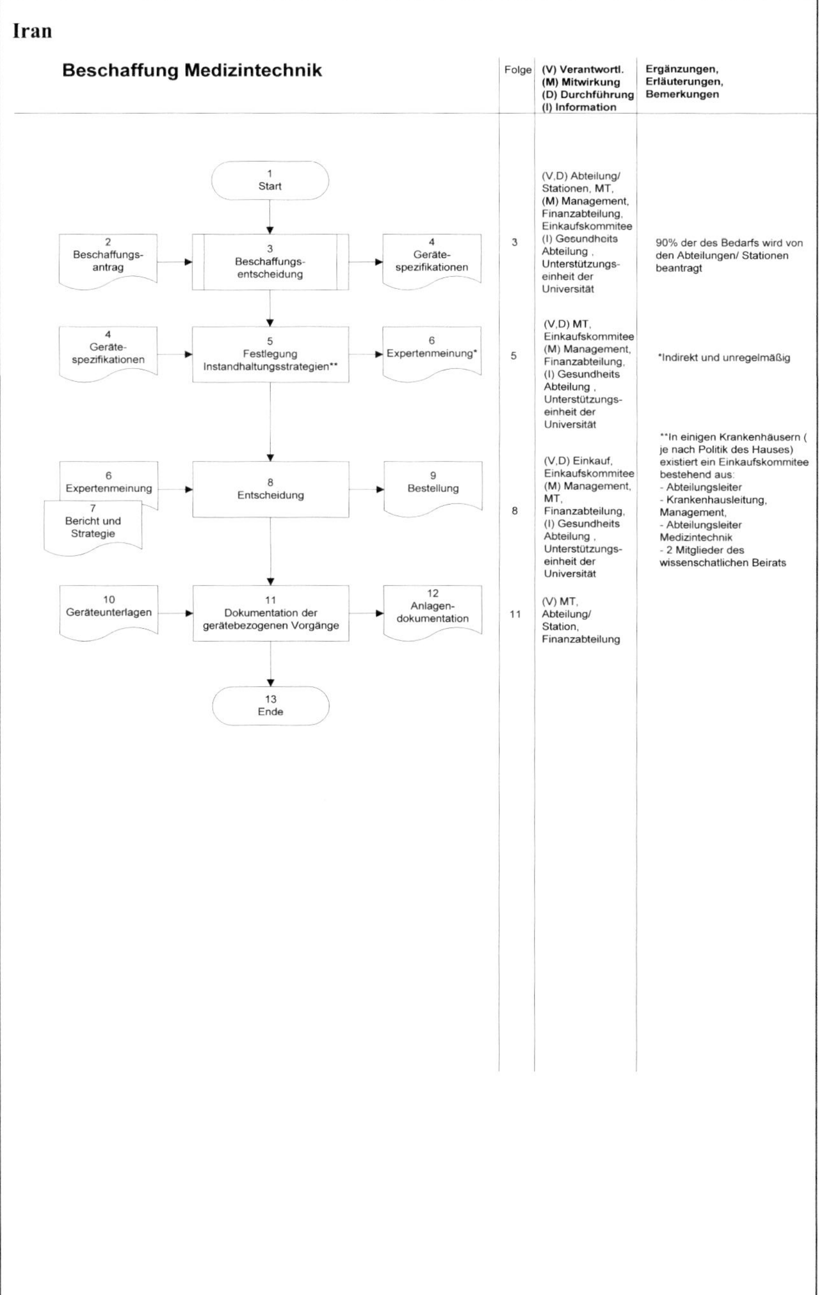

Deutschland

Inbetriebnahme
Medizintechnik

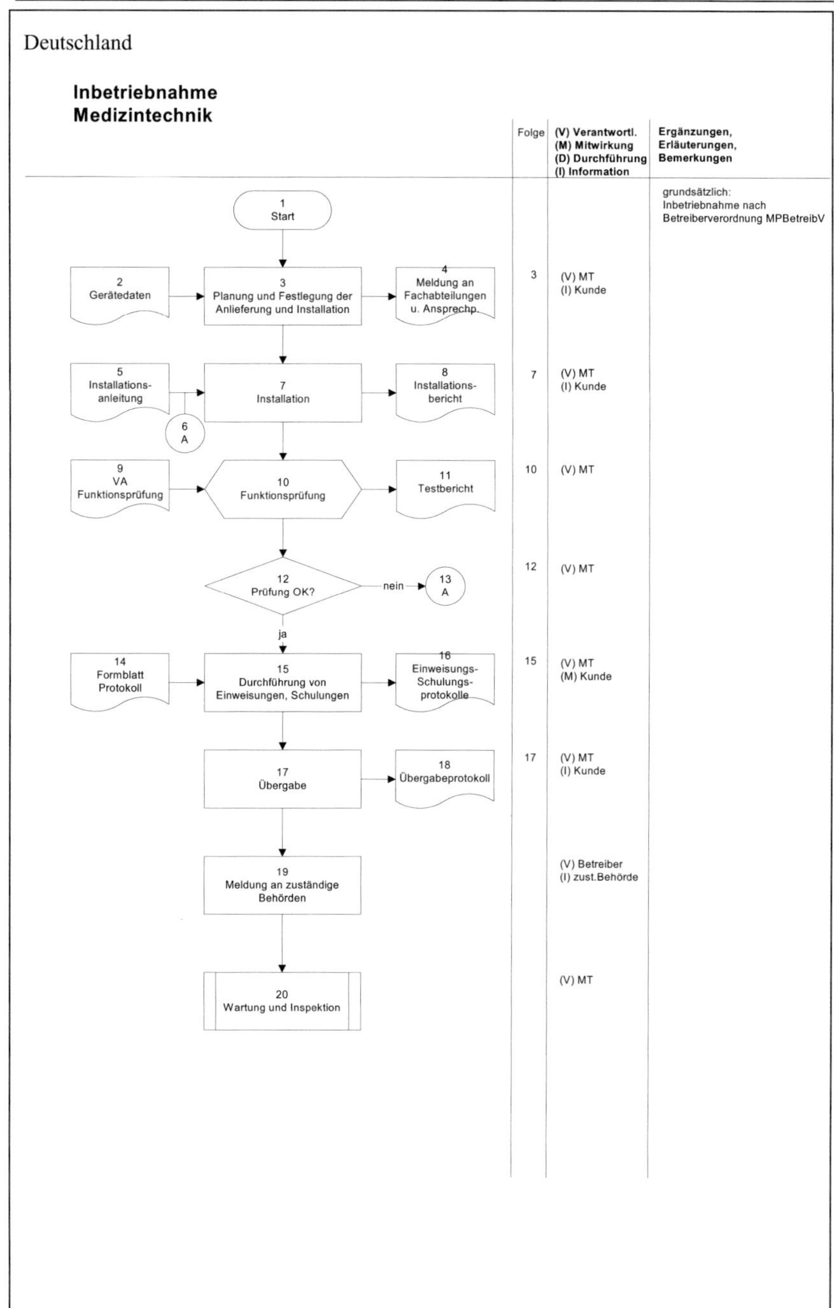

Folge	(V) Verantwortl. (M) Mitwirkung (D) Durchführung (I) Information	Ergänzungen, Erläuterungen, Bemerkungen
		grundsätzlich: Inbetriebnahme nach Betreiberverordnung MPBetreibV
3	(V) MT (I) Kunde	
7	(V) MT (I) Kunde	
10	(V) MT	
12	(V) MT	
15	(V) MT (M) Kunde	
17	(V) MT (I) Kunde	
	(V) Betreiber (I) zust.Behörde	
	(V) MT	

Iran

Inbetriebnahme Medizintechnik

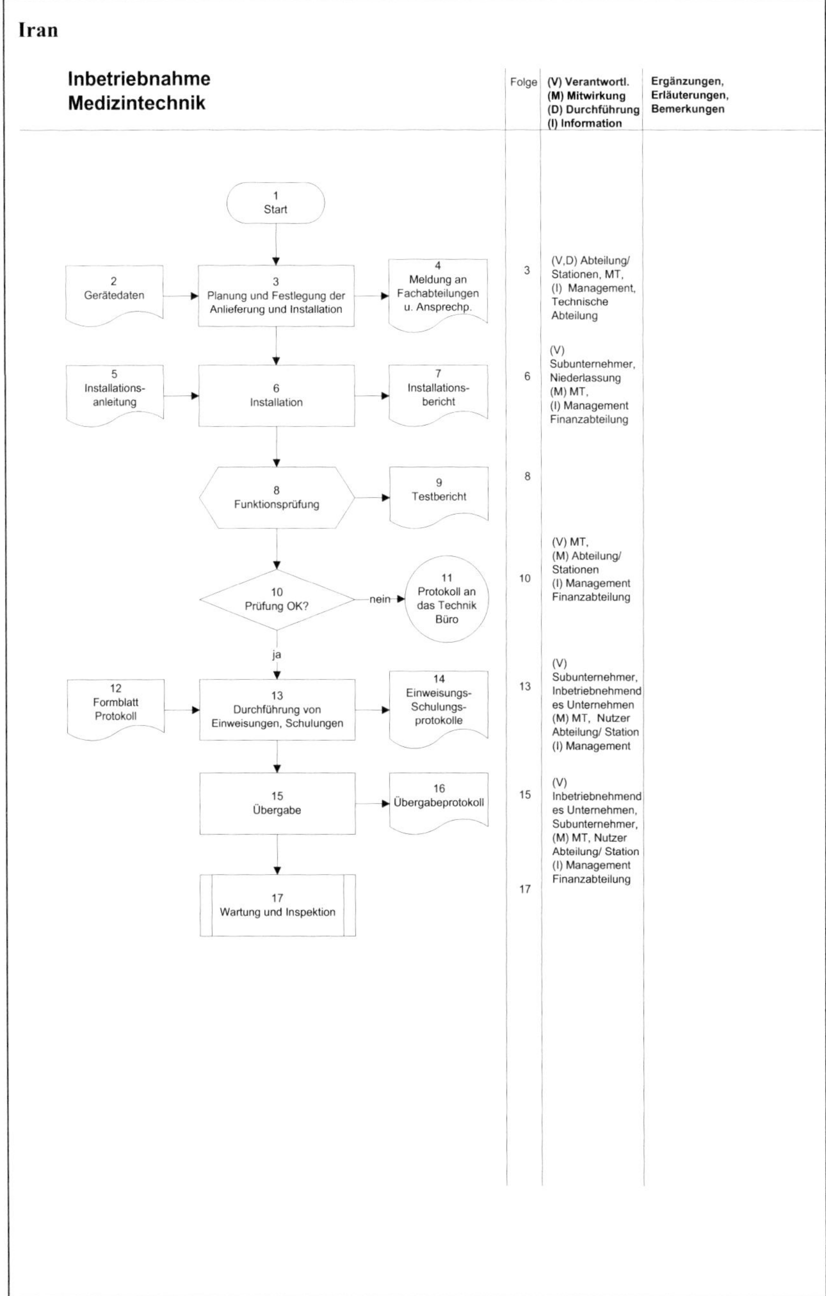

Deutschland

Wartung und Inspektion

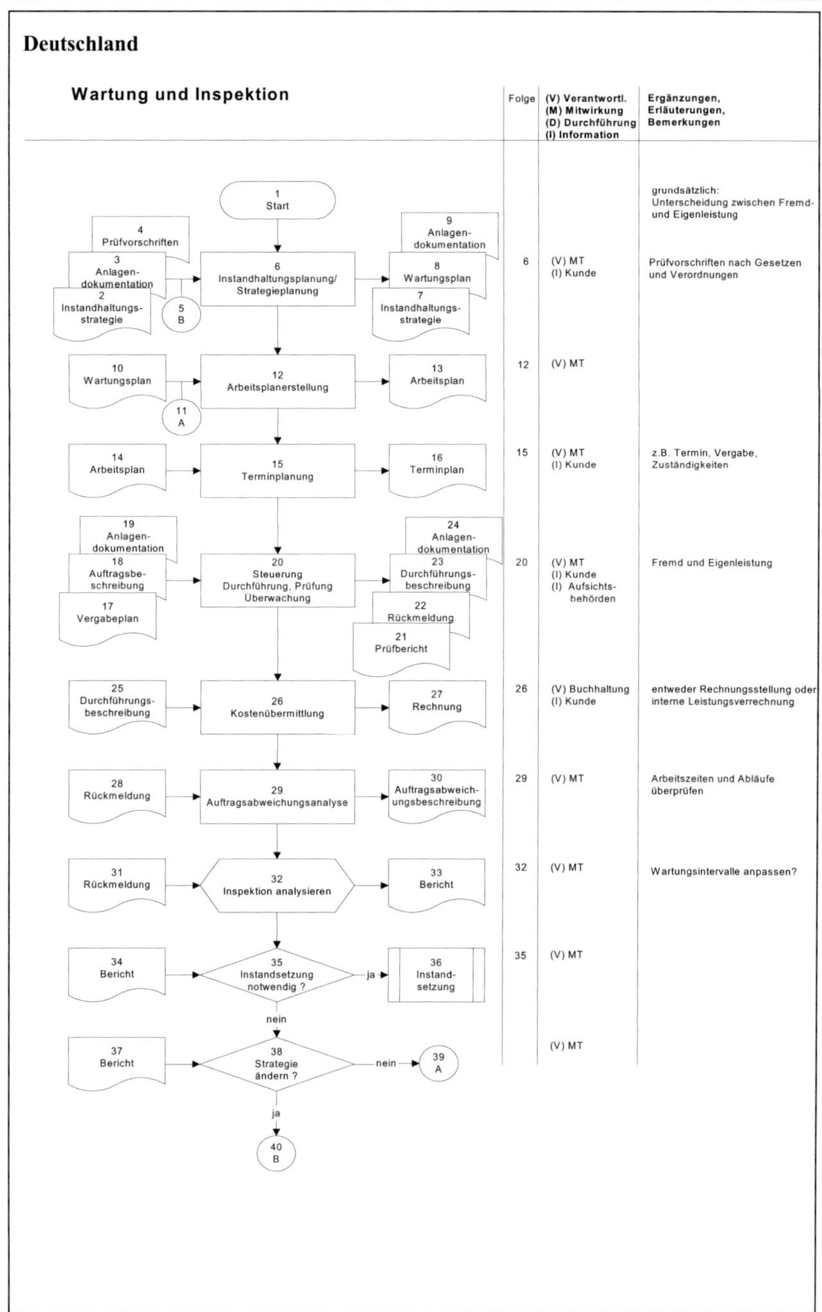

Iran

Wartung und Inspektion

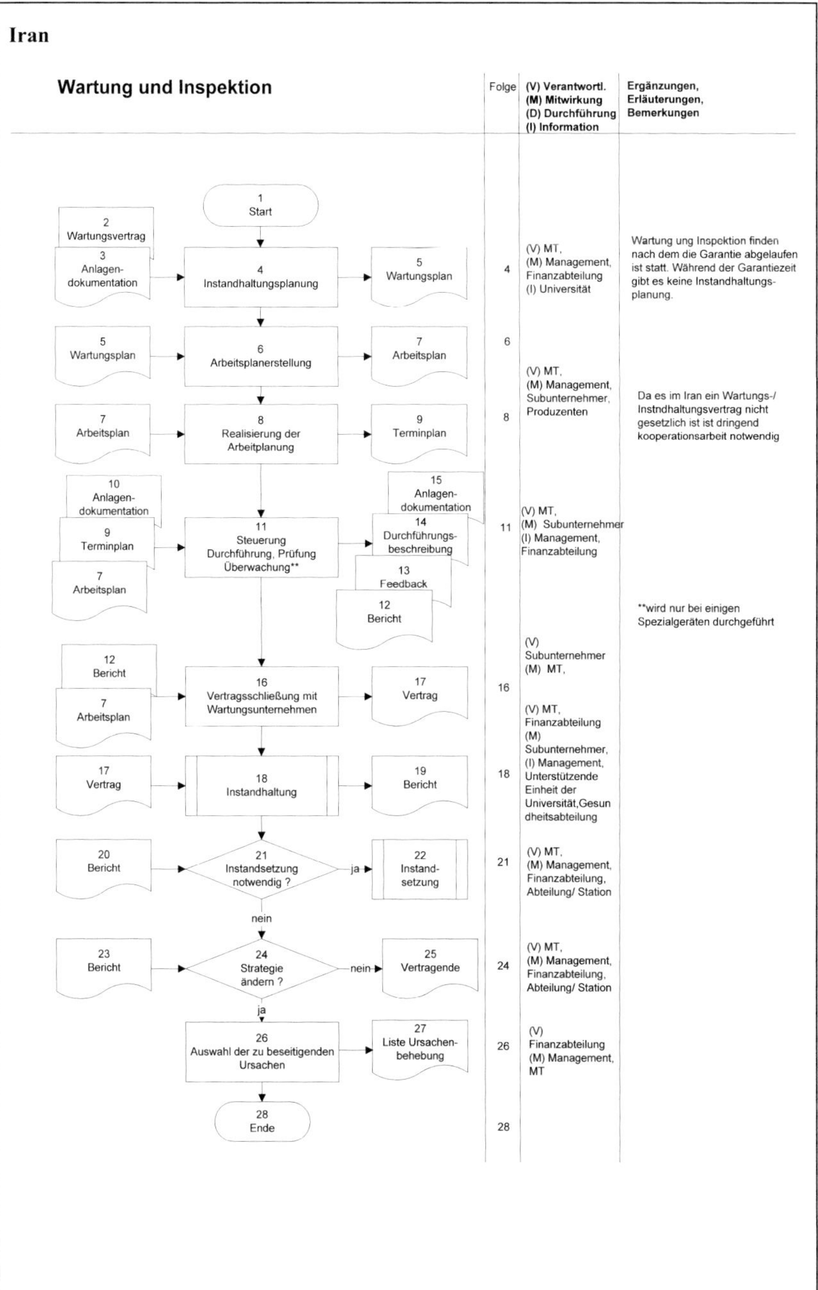

Deutschland

Durchführung Instandsetzung 1

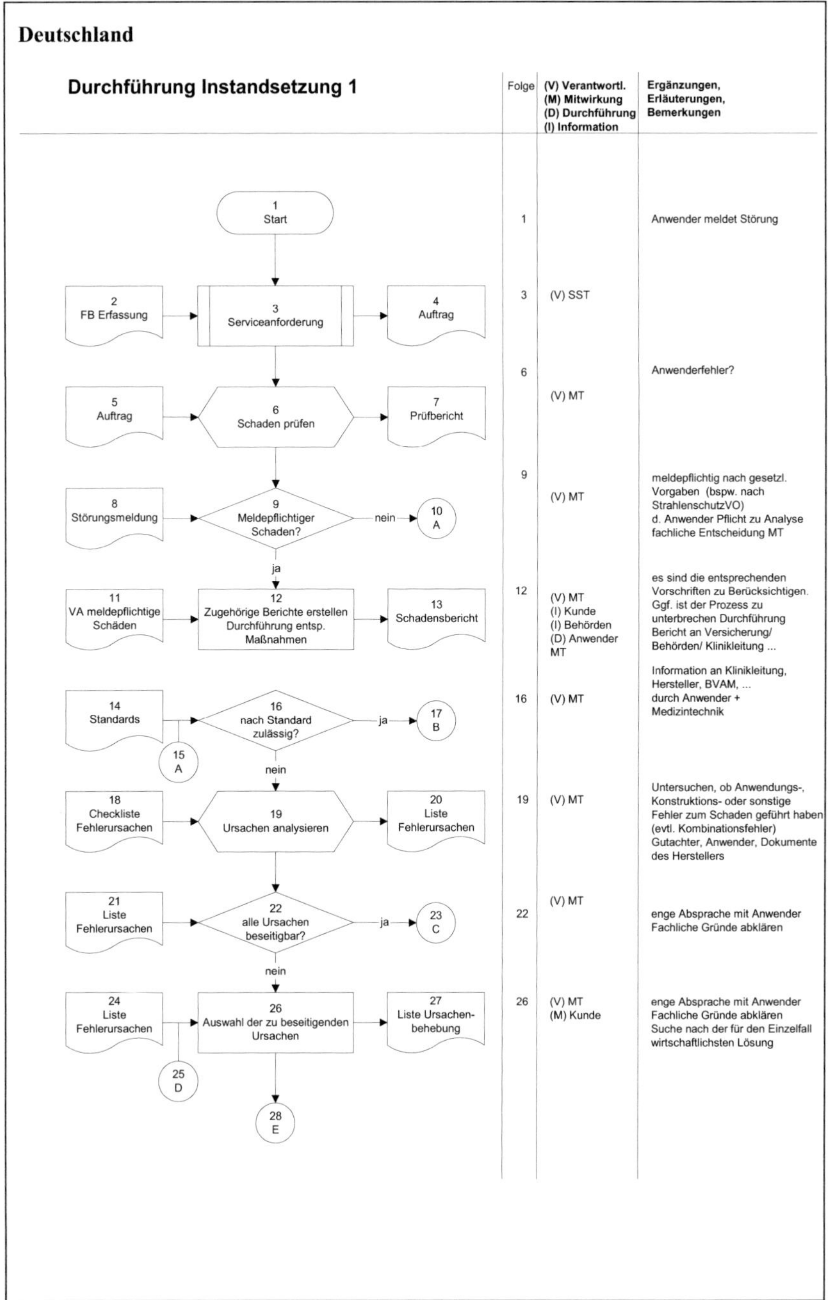

Deutschland

Durchführung Instandsetzung 2

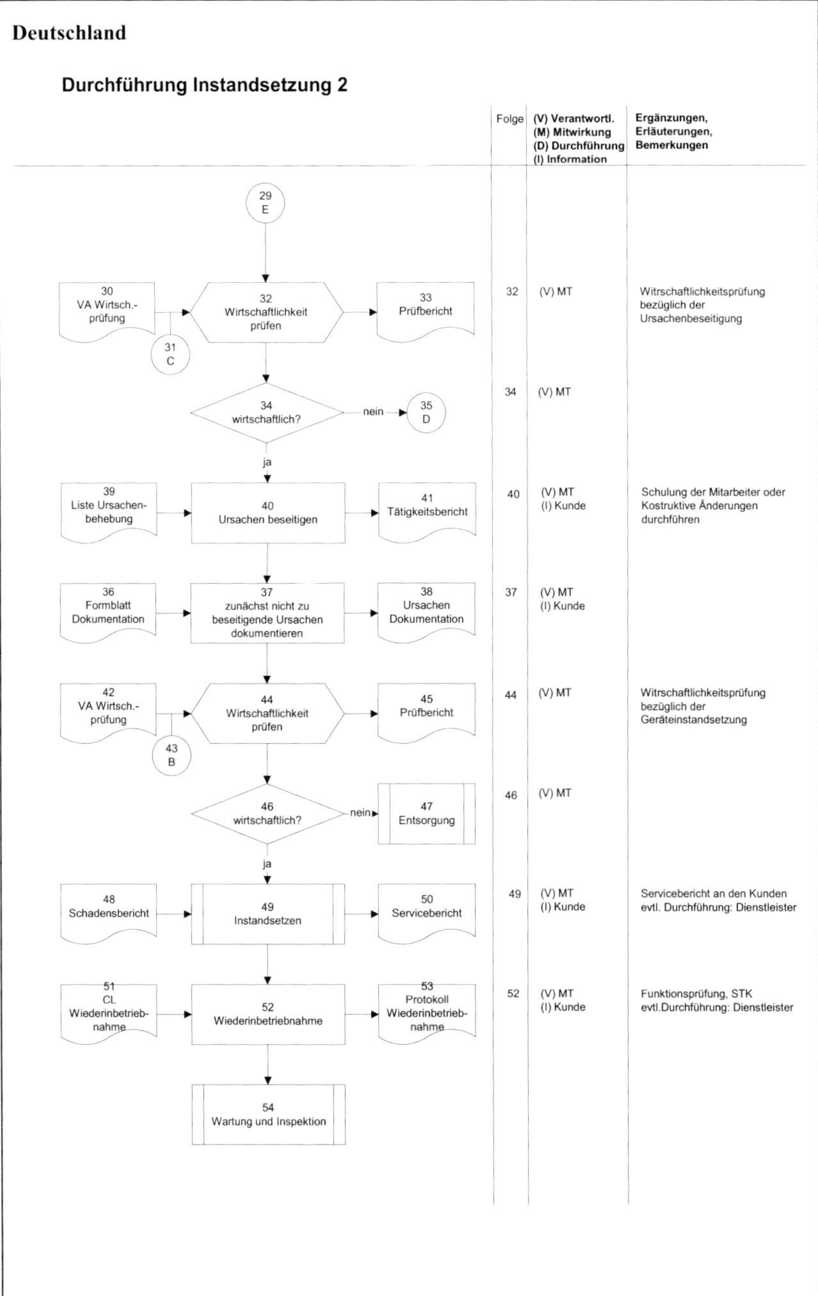

Iran

Durchführung Instandsetzung

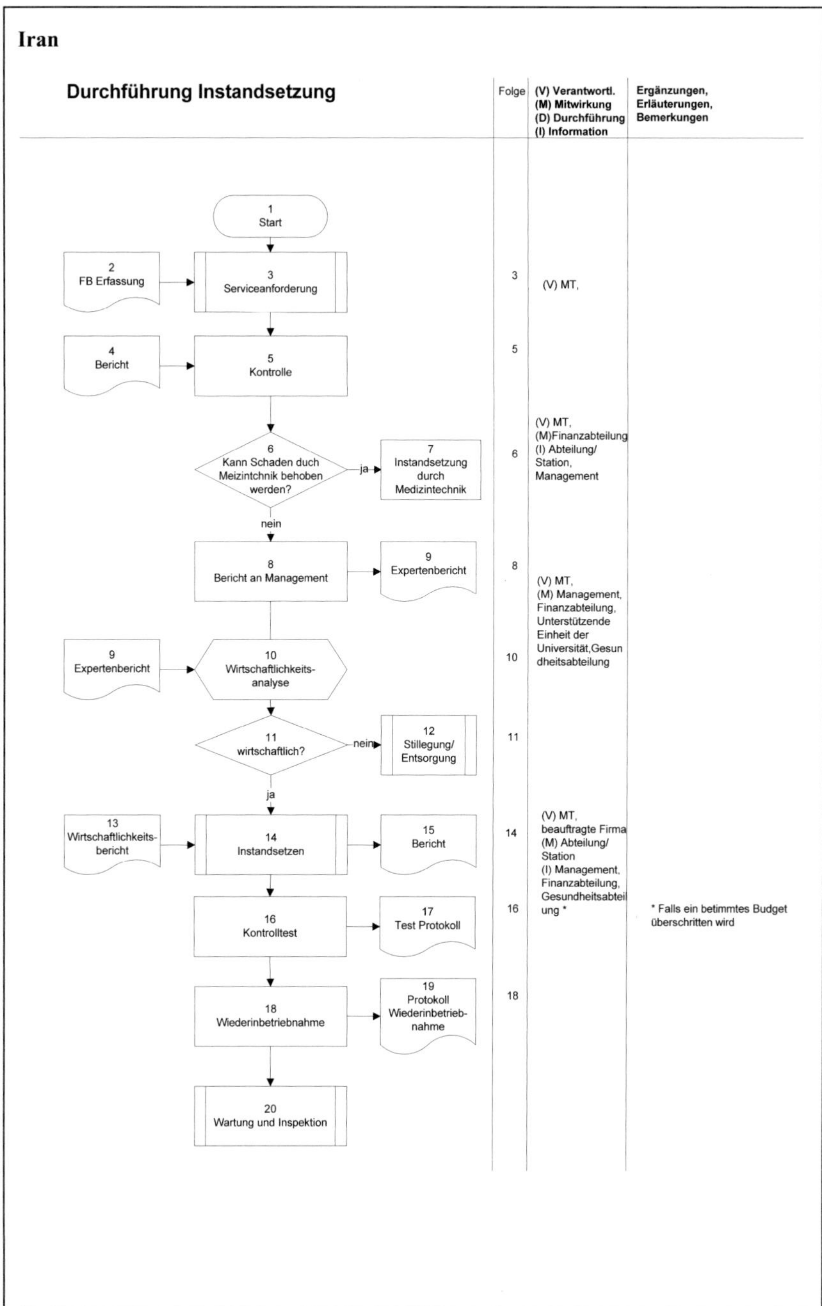

	Folge	(V) Verantwortl. (M) Mitwirkung (D) Durchführung (I) Information	Ergänzungen, Erläuterungen, Bemerkungen
	3	(V) MT,	
	5		
	6	(V) MT, (M)Finanzabteilung (I) Abteilung/ Station, Management	
	8	(V) MT, (M) Management, Finanzabteilung, Unterstützende Einheit der Universität,Gesundheitsabteilung	
	10		
	11		
	14	(V) MT, beauftragte Firma (M) Abteilung/ Station (I) Management, Finanzabteilung, Gesundheitsabteilung *	
	16		* Falls ein betimmtes Budget überschritten wird
	18		

Deutschland

Stilllegung / Entsorgung

Folge	(V) Verantwortl. (M) Mitwirkung (D) Durchführung (I) Information	Ergänzungen, Erläuterungen, Bemerkungen

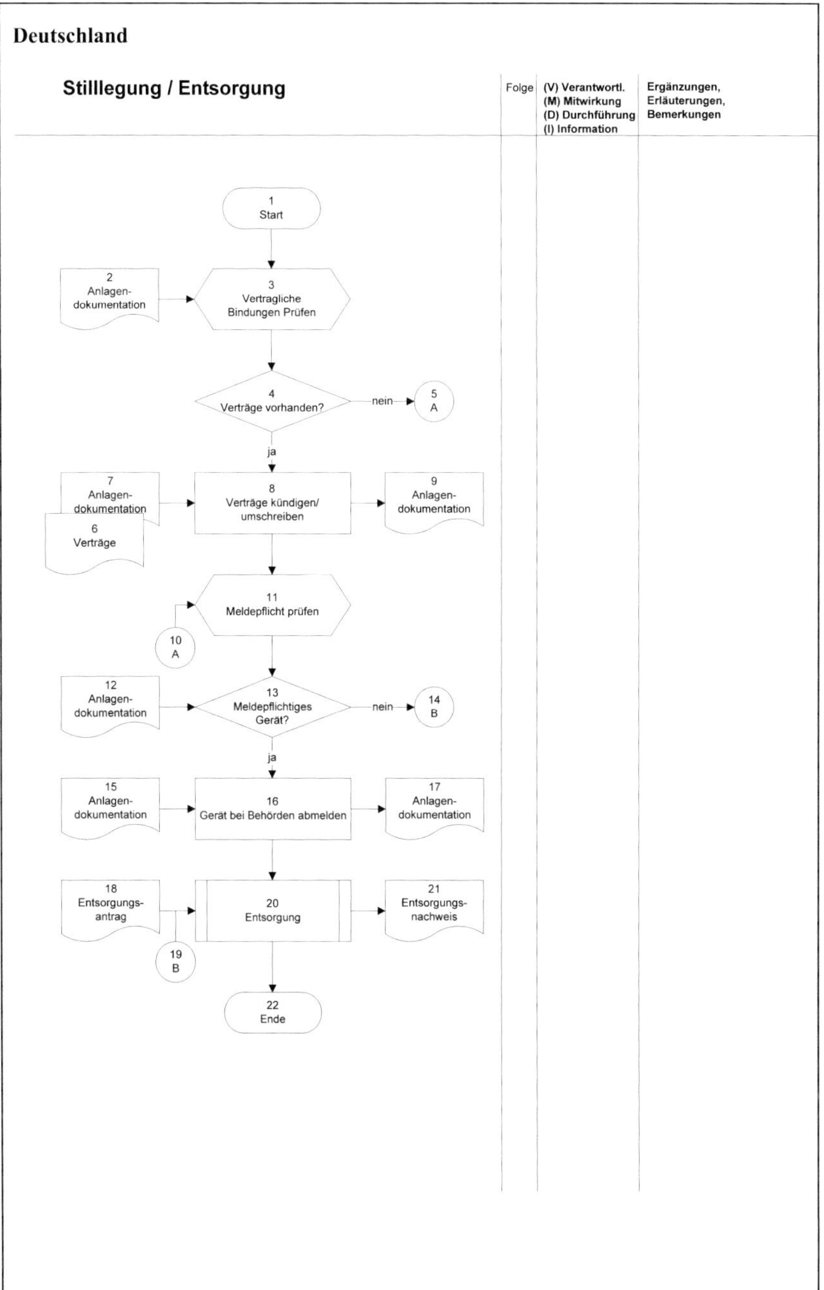

Iran

Stilllegung / Entsorgung	Folge	(V) Verantwortl. (M) Mitwirkung (D) Durchführung (I) Information	Ergänzungen, Erläuterungen, Bemerkungen

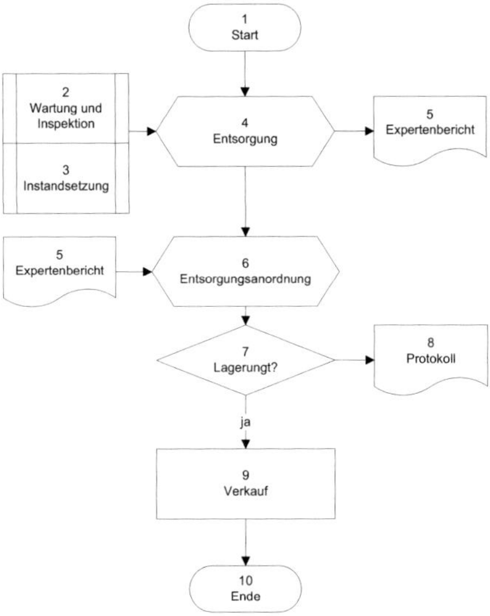

Flussdiagramm:

- 1 Start
- 2 Wartung und Inspektion
- 3 Instandsetzung
- 4 Entsorgung
- 5 Expertenbericht
- 6 Entsorgungsanordnung
- 7 Lagerungt?
- 8 Protokoll
- ja
- 9 Verkauf
- 10 Ende

Folge 4: (V)Finanzabteilung (M)MT, (I)Management, Krankenhausleitung, Haupt-Einkaufsleitung, Universität — Vor der Entsorgung muss das Lager alle staatlichen Stellen informieren.

Folge 6: (V)Finanzabteilung (M)MT, (I)Management, Krankenhausleitung, Haupt-Einkaufsleitung, Universität

Folge 7: — Entsorgungskriterien: -Nicht reparierbar -Veralterung der Technologie -Unwirtschaftlichkeit

Folge 9: (V)Finanzabteilung (M)MT, (I)Management, Krankenhausleitung, Haupt-Einkaufsleitung, Universität

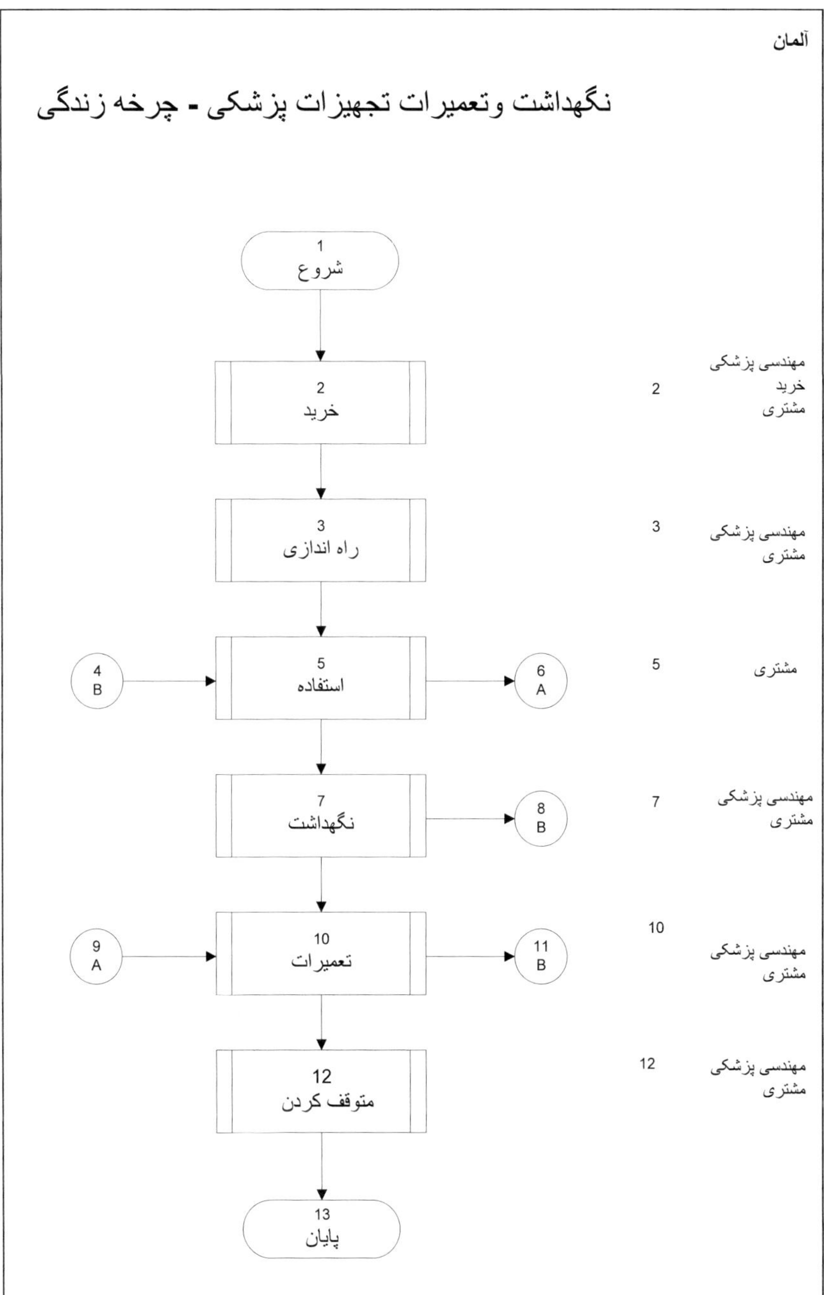

آلمان
نگهداشت وتعمیرات تجهیزات پزشکی ـ چرخه زندگی
1
شروع
2
خرید
3
راه اندازی
4
B
5
استفاده
6
A
7
نگهداشت
8
B
9
A
10
تعمیرات
11
B
12
متوقف کردن
13
پایان
2
مهندسی پزشکی
خرید
مشتَری
3
مهندسی پزشکی
مشتَری
5
مشتَری
7
مهندسی پزشکی
مشتَری
10
مهندسی پزشکی
مشتَری
12
مهندسی پزشکی
مشتَری

ایران

نگهداشت وتعمیرات تجهیزات پزشکی - چرخه زندگی

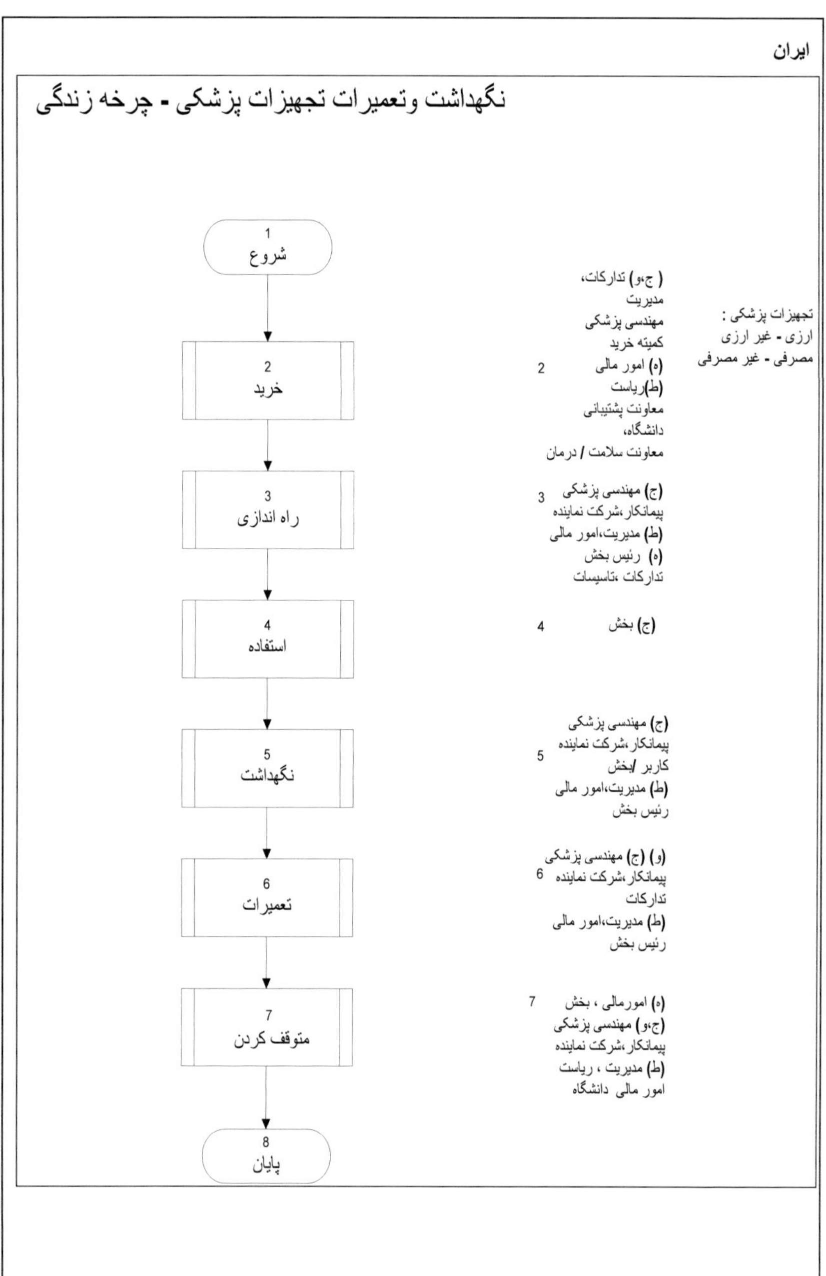

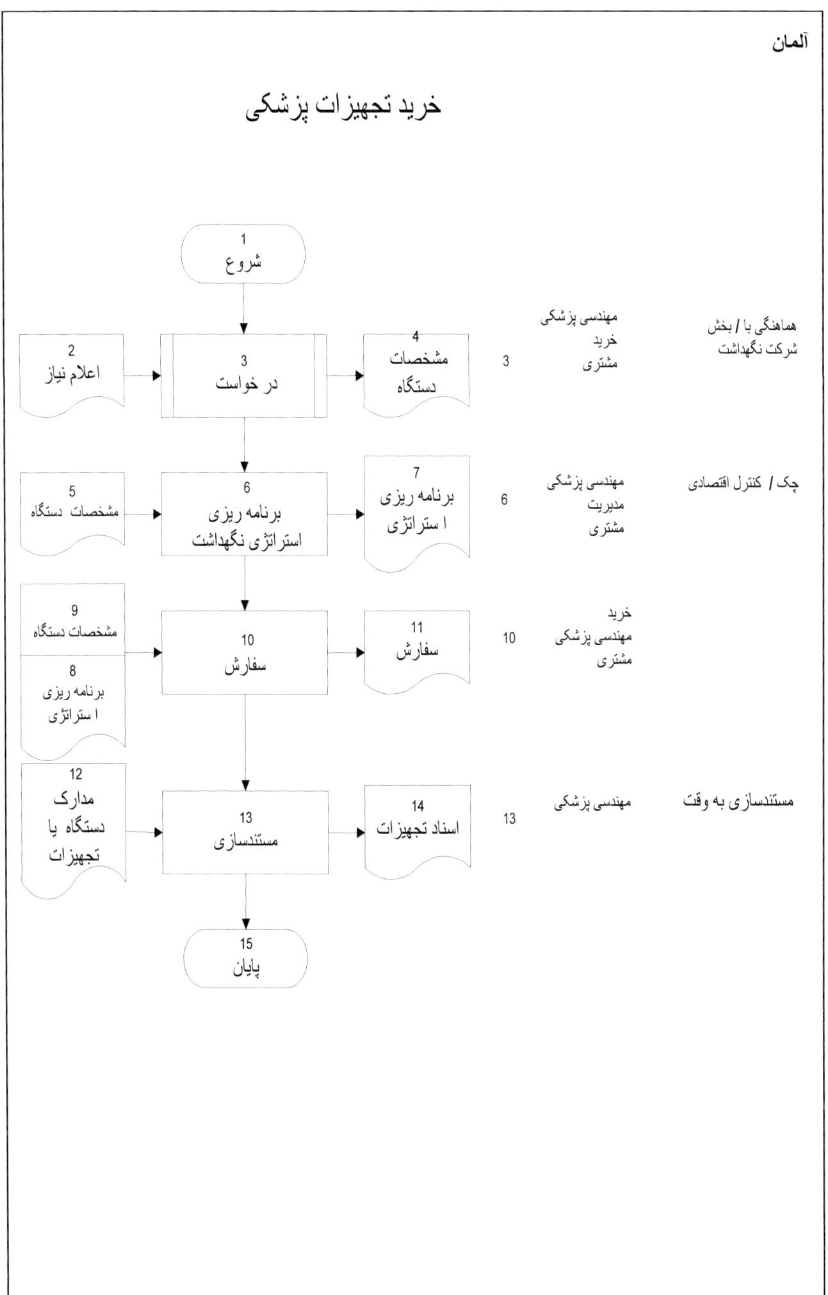
آلمان

خرید تجهیزات پزشکی

1
شروع

2
اعلام نیاز

3
در خواست

4
مشخصات دستگاه

3

مهندسی پزشکی
خرید
مشتری

هماهنگی با / بخش
شرکت نگهداشت

5
مشخصات دستگاه

6
برنامه ریزی استراتژی نگهداشت

7
برنامه ریزی ا ستراتژی

6

مهندسی پزشکی
مدیریت
مشتری

چک / کنترل اقتصادی

9
مشخصات دستگاه

8
برنامه ریزی ا ستراتژی

10
سفارش

11
سفارش

10

خرید
مهندسی پزشکی
مشتری

12
مدارک دستگاه یا تجهیزات

13
مستندسازی

14
اسناد تجهیزات

13

مهندسی پزشکی

مستندسازی به وقت

15
پایان

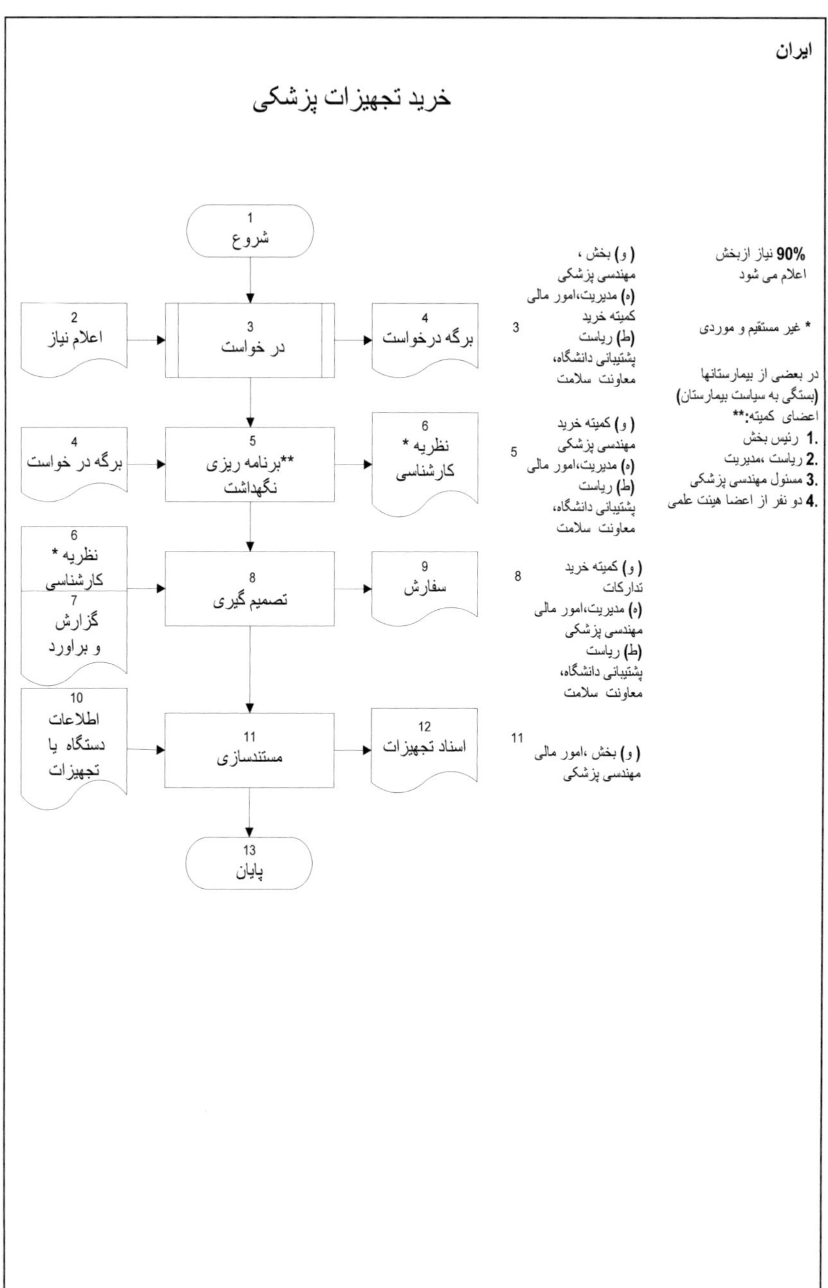
ایران

خرید تجهیزات پزشکی

1
شروع

2
اعلام نیاز

3
در خواست

4
برگه درخواست

4
برگه در خواست

5
**برنامه ریزی
نگهداشت

6
نظریه *
کارشناسی

6
نظریه *
کارشناسی
7
گزارش
و براورد

8
تصمیم گیری

9
سفارش

10
اطلاعات
دستگاه یا
تجهیزات

11
مستندسازی

12
اسناد تجهیزات

13
پایان

(و) بخش ،
مهندسی پزشکی
(ه) مدیریت،امور مالی
کمیته خرید
(ط) ریاست
پشتیبانی دانشگاه،
معاونت سلامت

3

(و) کمیته خرید
مهندسی پزشکی
(ه) مدیریت،امور مالی
(ط) ریاست
پشتیبانی دانشگاه،
معاونت سلامت

5

(و) کمیته خرید
تدارکات
(ه) مدیریت،امور مالی
مهندسی پزشکی
(ط) ریاست
پشتیبانی دانشگاه،
معاونت سلامت

8

(و) بخش ،امور مالی
مهندسی پزشکی

11

90% نیاز ازبخش
اعلام می شود

* غیر مستقیم و موردی

در بعضی از بیمارستانها
(بستگی به سیاست بیمارستان)
اعضای کمیته:**
1. رئیس بخش
2. ریاست ،مدیریت
3. مسئول مهندسی پزشکی
4. دو نفر از اعضا هیئت علمی

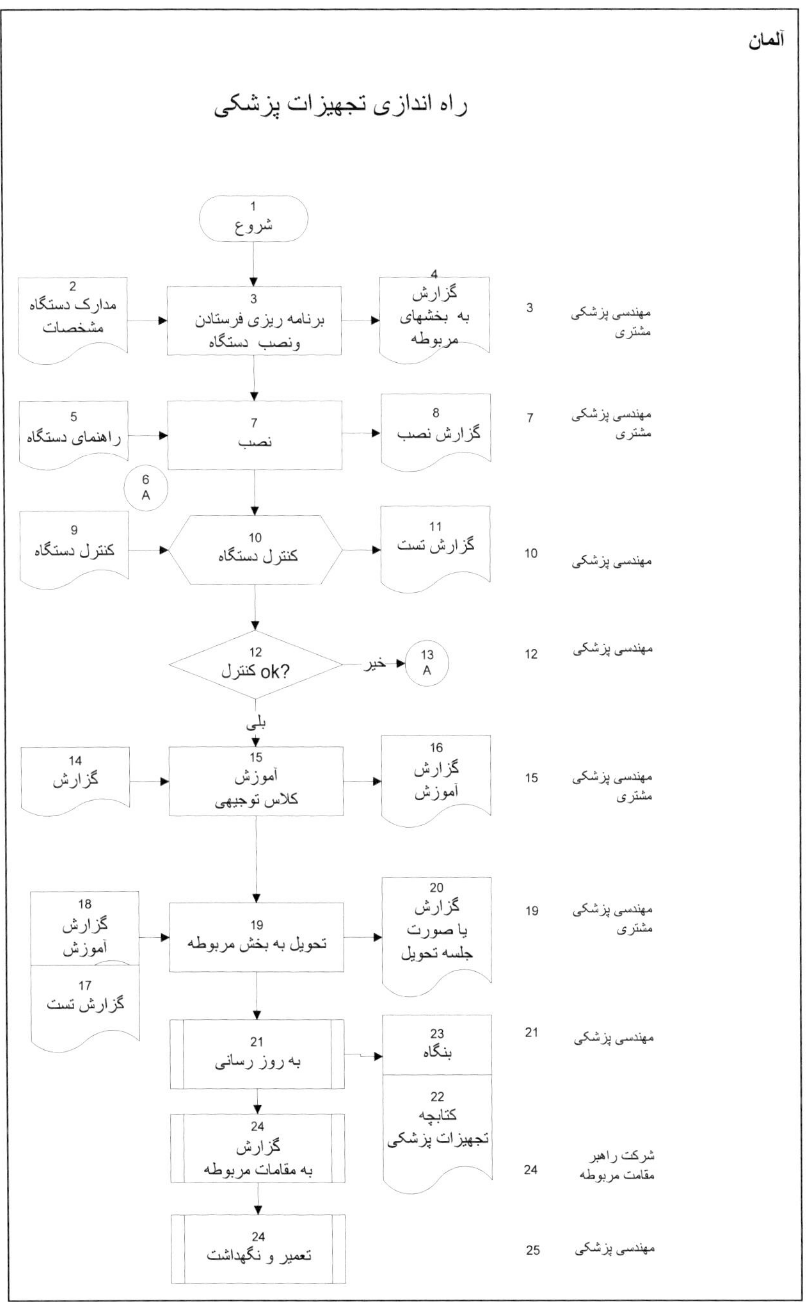
آلمان
راه اندازی تجهیزات پزشکی
1
شروع
2
مدارک دستگاه
مشخصات
3
برنامه ریزی فرستادن
ونصب دستگاه
4
گزارش
به بخشهای
مربوطه
3
مهندسی پزشکی
مشتری
5
راهنمای دستگاه
7
نصب
8
گزارش نصب
7
مهندسی پزشکی
مشتری
6
A
9
کنترل دستگاه
10
کنترل دستگاه
11
گزارش تست
10
مهندسی پزشکی
12
کنترل ok?
خیر
13
A
12
مهندسی پزشکی
بلی
14
گزارش
15
آموزش
کلاس توجیهی
16
گزارش
آموزش
15
مهندسی پزشکی
مشتری
18
گزارش
آموزش
17
گزارش تست
19
تحویل به بخش مربوطه
20
گزارش
یا صورت
جلسه تحویل
19
مهندسی پزشکی
مشتری
21
به روز رسانی
23
بنگاه
22
کتابچه
تجهیزات پزشکی
21
مهندسی پزشکی
24
گزارش
به مقامات مربوطه
شرکت راهبر
مقامت مربوطه
24
24
تعمیر و نگهداشت
25
مهندسی پزشکی

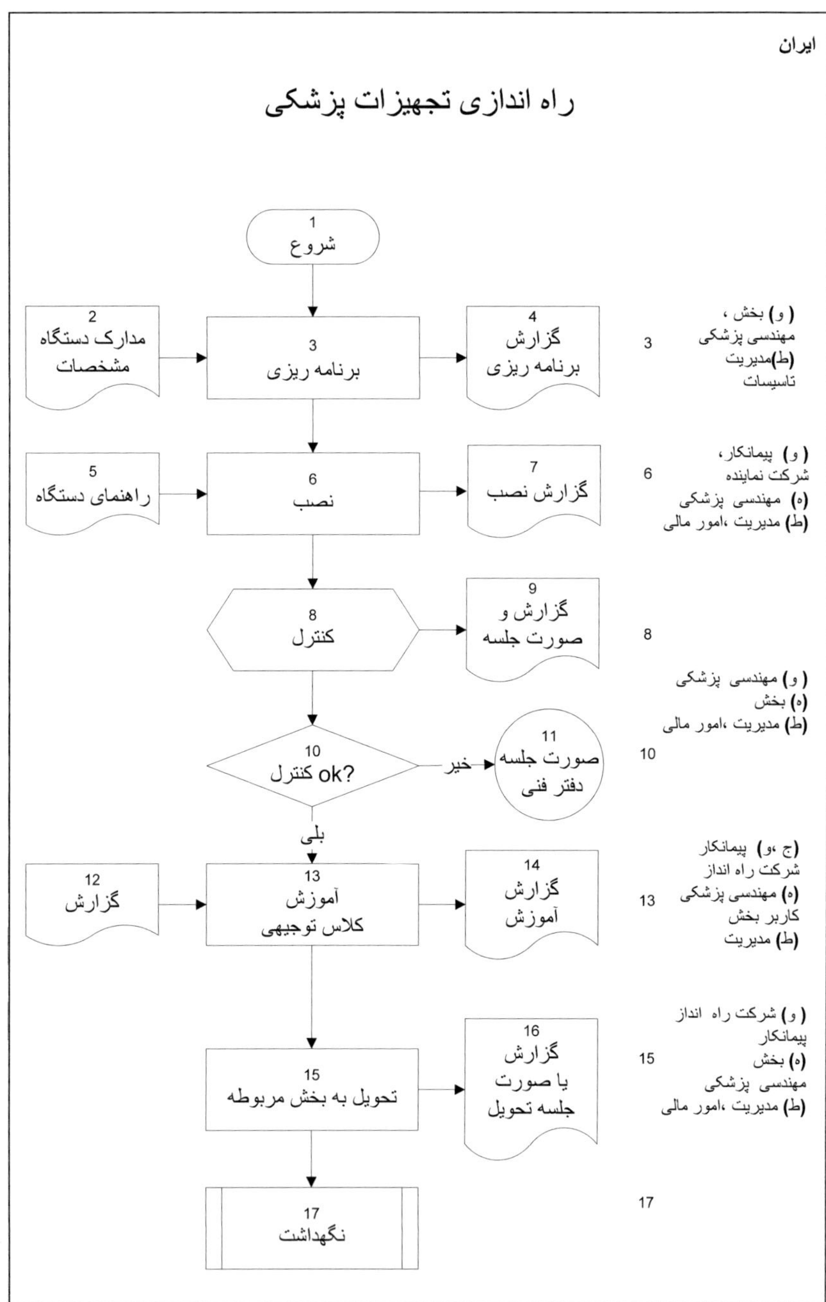
ایران

راه اندازی تجهیزات پزشکی

1
شروع

2
مدارک دستگاه
مشخصات

3
برنامه ریزی

4
گزارش
برنامه ریزی

3

(و) بخش ،
مهندسی پزشکی
(ط)مدیریت
تاسیسات

5
راهنمای دستگاه

6
نصب

7
گزارش نصب

6

(و) پیمانکار،
شرکت نماینده
(ه) مهندسی پزشکی
(ط) مدیریت ،امور مالی

8
کنترل

9
گزارش و
صورت جلسه

8

(و) مهندسی پزشکی
(ه) بخش
(ط) مدیریت ،امور مالی

10
کنترل ok؟

خیر

11
صورت جلسه
دفتر فنی

10

بلی

12
گزارش

13
آموزش
کلاس توجیهی

14
گزارش
آموزش

13

(ج ،و) پیمانکار
شرکت راه انداز
(ه) مهندسی پزشکی
کاربر بخش
(ط) مدیریت

15
تحویل به بخش مربوطه

16
گزارش
یا صورت
جلسه تحویل

15

(و) شرکت راه انداز
پیمانکار
(ه) بخش
مهندسی پزشکی
(ط) مدیریت ،امور مالی

17
نگهداشت

17

آلمان
نگهداشت تجهیزات پزشکی
1 شروع
4 مقررات کنترل
3 مستندات دستگاه
استراتژی نگهداشت
5 B
10 برنامه نگهداشت
11 A
6 برنامه ریزی استراتژیک
12 عملیاتی کردن برنامه ریزی کار
9 گزارش دستگاه
8 برنامه نگهداشت
7 استراتژی نگهداشت
13 برنامه کار
6
مهندسی پزشکی
مشتری
مقررات
کنترل بر اساس قوانین
12 (و) مهندسی پزشکی
در کل
مقایسه
خدمات خارجی یا داخلی
14 برنامه کار
15 عملیاتی کردن برنامه زمان
16 زمان بندی
15 (و) مهندسی پزشکی
(ه) مشتری
به عنوان منال
وظیفه ،زمان
19 برنامه زمان
18 دستور عمل دستگاه
17 برنامه کار
25 چک لیست اجرایی
28 بازخورد
20 هماهنگی و نظارت
24 مستندسازی
23 چک لیست اجرایی
22 بازخور
21 گزارش
20 (و) مهندسی پزشکی
(ه) مشتری
(ج) مقامات کنترل
خدمات خارجی یا داخلی
26 واگزارکردن هزینه
26 (ه) امور مالی
(و) مشتری
ریسد یا محاسبات داخلی
31 بازخورد
29 آنالیز مقایسه ای درخواست
27 رسید
30 گزارش آنالیز
29 (ه) مهندسی پزشکی
34 گزارش
32 آنالیز نگهداشت
33 گزارش
32 (و) مهندسی پزشکی
کنترل زمان و روند کار
35 نیاز تعمیر؟
بلی
36 تعمیرات
35 (و) مهندسی پزشکی
خیر
37 گزارش
38 تغییر استراتژی
خیر
39 A
38 (و) مهندسی پزشکی
هماهنگی دوره نگهداشت
بلی
40 B

ایران
نگهداشت تجهیزات پزشکی
1 شروع
2 قرارداد
3 مستندات دستگاه
4 برنامه ریزی
5 برنامه نگهداشت
4
(و) مهندسی پزشکی
(٥) مدیریت،امور مالی
(ط) دانشگاه
نگهداشت بعد از اتمام زمان گارانتی انجام می گیرد
5 برنامه نگهداشت
6 عملیاتی کردن برنامه ریزی کار
7 برنامه کار
6
در زمان گارانتی برای نگهداشت برنامه ای وجود ندارد .
7 برنامه کار
8 عملیاتی کردن برنامه زمان
9 زمان بندی
8
(و) مهندسی پزشکی
(٥) تولید کننده ، پیمانکار مدیریت
10 برنامه زمان
11 دستور عمل دستگاه
7 برنامه کار
12 ** نظارت و هماهنگی
13 مستندسازی
14 چک لیست اجرایی
15 *بازخور
16 گزارش
(و) مهندسی پزشکی
(٥) پیمانکار
(ط) مدیریت ، امور مالی
چون درایران قانونی تولید کننده را موظف به نظارت نمی کند به همکاری نیازاست
Feedback*
**
بعضی ازدستگاهای سرمایه ای به صورت موردی انجام می شود .
16 گزارش
7 برنامه کار
17 قرارداد نگهداشت با شرکت نماینده
18 قرارداد
17
(٥) مهندسی پزشکی
(و) پیمانکار
18 قرارداد
19 نگهداشت
20 گزارش نگهداشت
19
(٥) مهندسی پزشکی
امور مالی
(و) پیمانکار
(ط) بخش پشتیبانی
مدیریت ، معاونت سلامت
20 گزارش نگهداشت
19 نیاز تعمیر؟
بلی
20 تعمیرات
(و) مهندسی پزشکی
(٥) بخش
مدیریت ، امور مالی
خیر
21 گزارش
22 بازنگری در تصمیم نگهداشت
خیر
23 اتمام قرارداد
22
(و) مهندسی پزشکی
(٥) بخش
مدیریت ، امور مالی
بلی
24 پرداخت
25 اسناد مالی
26 فاکتور شرکت
27 تایید مهندسی پزشکی
28 تایید مدیریت
24
(و) حسابداری
(٥) مهندسی پزشکی
مدیریت
29 پایان

آلمان 1

تعمیرات تجهیزات پزشکی
1
شروع
2
بازخورد نگهداشت
3
درخواست تعمییر
4
سفارش
5
سفارش
6
چک تخریب
7
گزارش
8
خبر اختلال
9
اختلال باید گزارش شود؟
خیر
10
A
11
دستور گزارش اختلال
12
نوشتن گزارش مربوطه عملکرد مربوطه
13
گزارش تخریب
14
استاندارد
15
A
16
آیا با استاندارد هماهنگ است؟
بلی
17
B
18
چک لیست دلیل اختلال
19
آنالیز دلایل اختلال
20
لیست دلیل اختلال
21
لیست دلیل اختلال
22
تمام عوامل برطرف شده اند؟
بلی
23
C
24
لیست دلیل اختلال
25
D
26
گزینه برطرف دلیل اختلال
27
لیست برطرف کردن دلیل اختلال
28
E
1
استفاده کننده تخلیل را خبر می کند
3
(و) مهندسی پزشکی
اشتباه استفاده کننده ؟
6
(و) مهندسی پزشکی
آیا باید بر اساس قوانین به مقامات مربوطه گزارش داد؟
9
(و) مهندسی پزشکی
12
(ه) امور مالی
(و) مشتری
(و) مقامات مربوطه
(ج) استفاده کننده
مهندسی پزشکی
16
(و) مهندسی پزشکی
19
(و) مهندسی پزشکی
هماهنگی تنگاتنگ استفاده کننده
22
(و) مهندسی پزشکی
26
(و) مهندسی پزشکی
(ج) مشتری

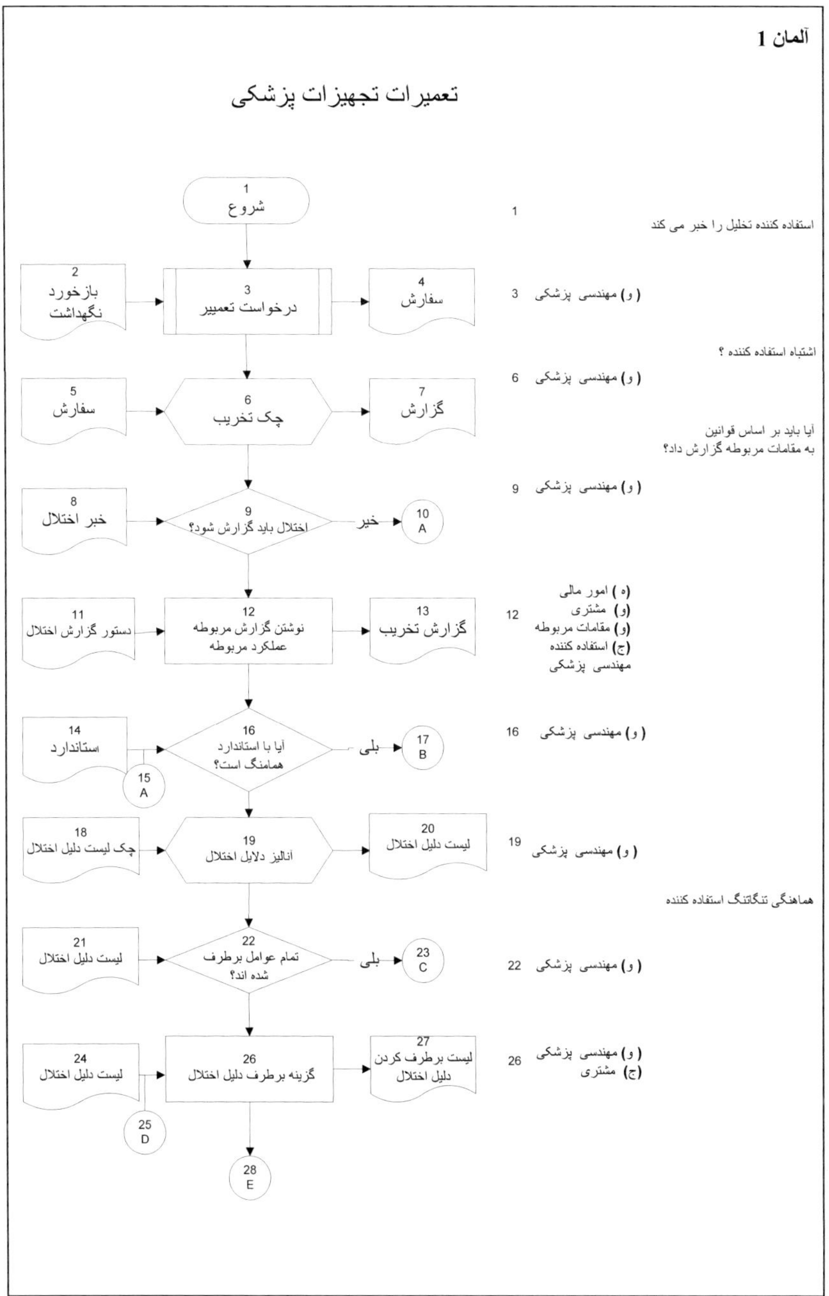

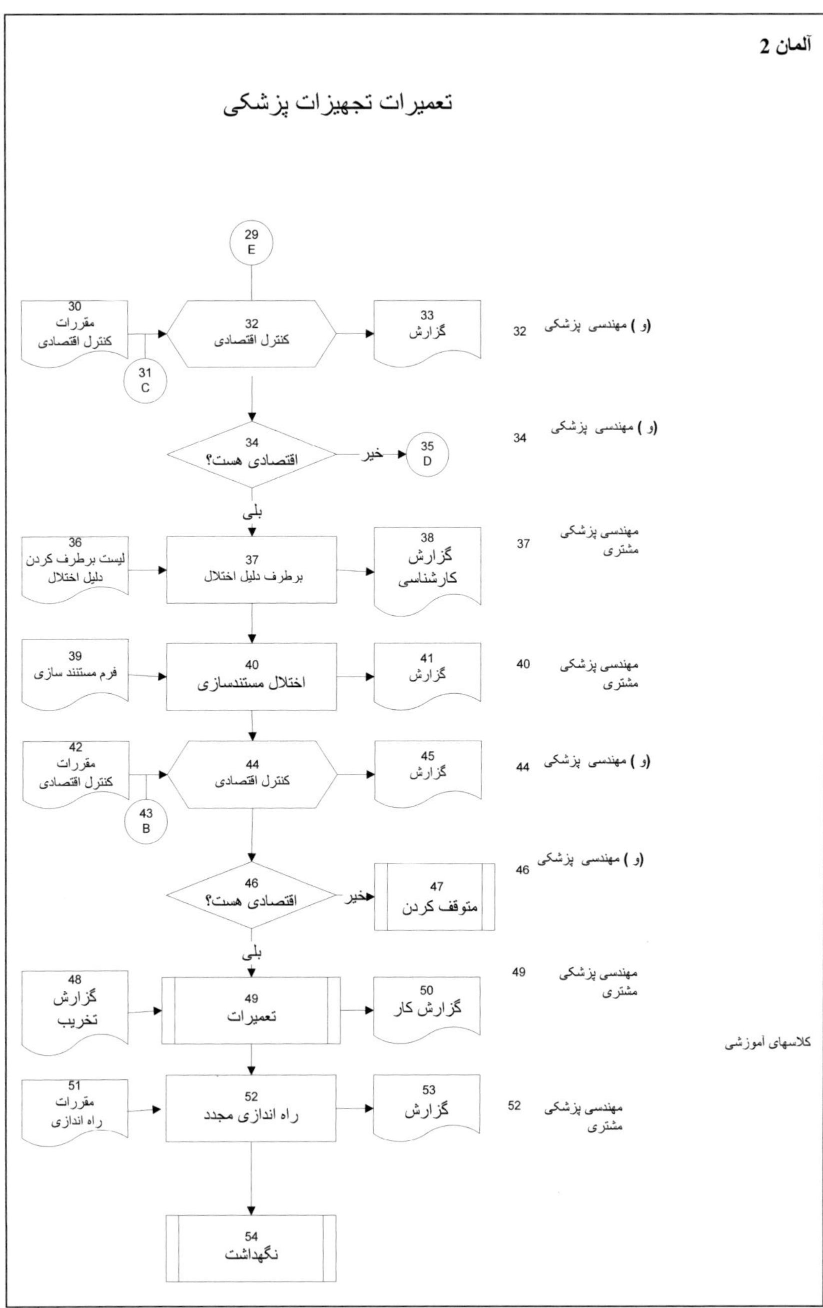
آلمان 2
تعمیرات تجهیزات پزشکی
29 E
30 مقررات کنترل اقتصادی
32 کنترل اقتصادی
33 گزارش
31 C
(و) مهندسی پزشکی 32
34 اقتصادی هست؟
خیر
35 D
بلی
(و) مهندسی پزشکی 34
36 لیست برطرف کردن دلیل اختلال
37 برطرف دلیل اختلال
38 گزارش کارشناسی
مهندسی پزشکی مشتری 37
39 فرم مستند سازی
40 اختلال مستندسازی
41 گزارش
مهندسی پزشکی مشتری 40
42 مقررات کنترل اقتصادی
44 کنترل اقتصادی
45 گزارش
43 B
(و) مهندسی پزشکی 44
46 اقتصادی هست؟
خیر
47 متوقف کردن
بلی
(و) مهندسی پزشکی 46
48 گزارش تخریب
49 تعمیرات
50 گزارش کار
مهندسی پزشکی مشتری 49
کلاسهای آموزشی
51 مقررات راه اندازی
52 راه اندازی مجدد
53 گزارش
مهندسی پزشکی مشتری 52
54 نگهداشت

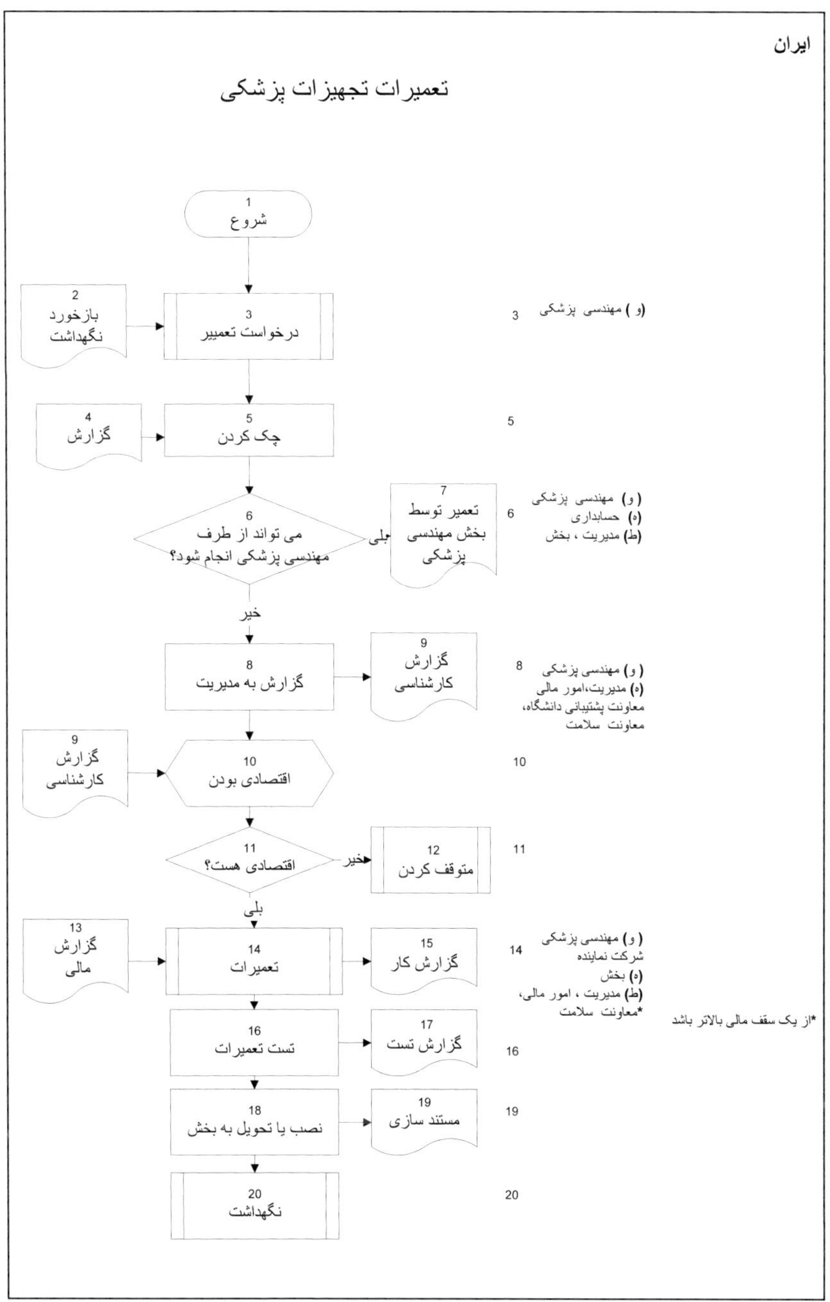
ایران

تعمیرات تجهیزات پزشکی

1
شروع

2
بازخورد نگهداشت

3
درخواست تعمیر

(و) مهندسی پزشکی   3

4
گزارش

5
چک کردن

5

6
می تواند از طرف مهندسی پزشکی انجام شود؟

بلی

7
تعمیر توسط بخش مهندسی پزشکی

(و) مهندسی پزشکی   6
(ه) حسابداری
(ط) مدیریت ، بخش

خیر

8
گزارش به مدیریت

9
گزارش کارشناسی

(و) مهندسی پزشکی   8
(ه) مدیریت،امور مالی
معاونت پشتیبانی دانشگاه،
معاونت سلامت

9
گزارش کارشناسی

10
اقتصادی بودن

10

11
اقتصادی هست؟

خیر

12
متوقف کردن

11

بلی

13
گزارش مالی

14
تعمیرات

15
گزارش کار

(و) مهندسی پزشکی   14
شرکت نماینده
(ه) بخش
(ط) مدیریت ، امور مالی،
معاونت سلامت*

*از یک سقف مالی بالاتر باشد

16
تست تعمیرات

17
گزارش تست

16

18
نصب یا تحویل به بخش

19
مستند سازی

19

20
نگهداشت

20

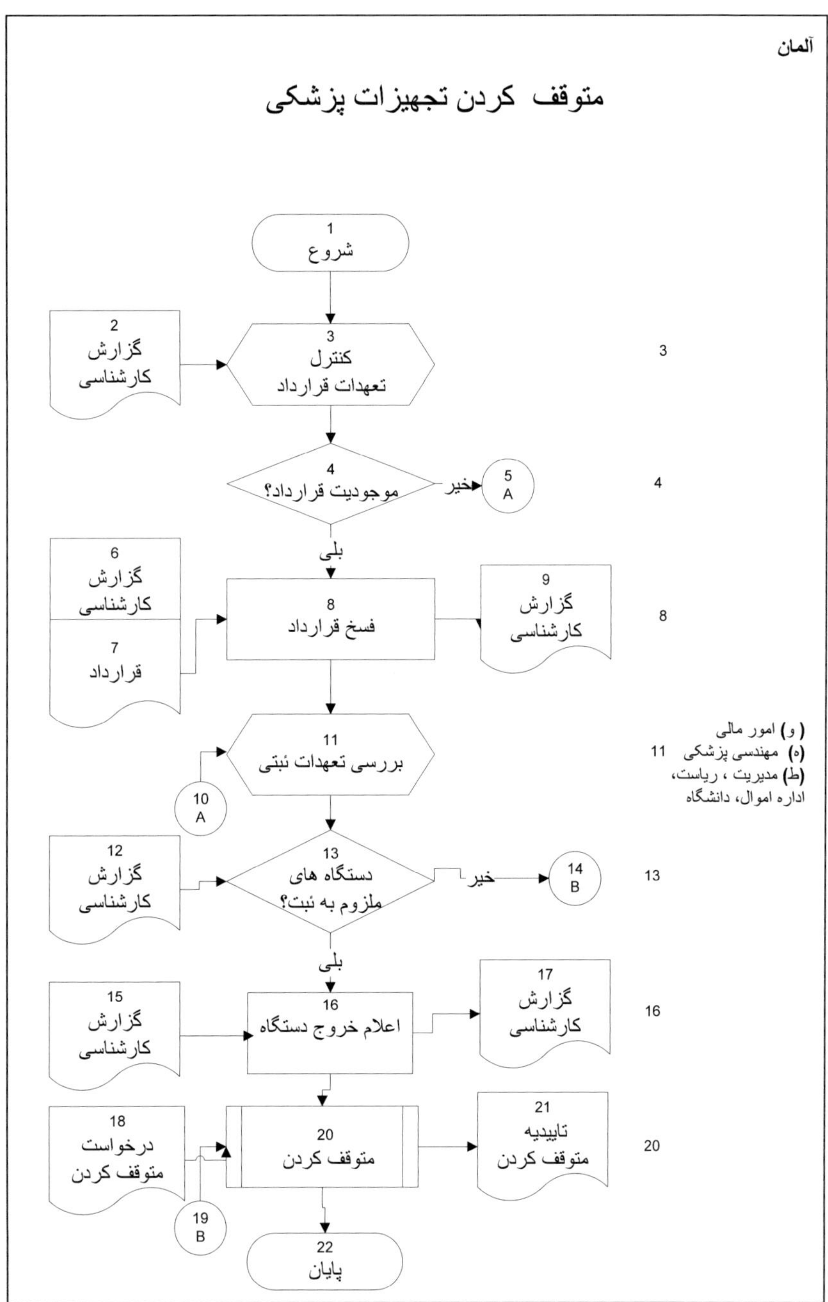
آلمان
متوقف کردن تجهیزات پزشکی
1
شروع
2
گزارش کارشناسی
3
کنترل تعهدات قرارداد
3
4
موجودیت قرارداد؟
خیر
5
A
4
بلی
6
گزارش کارشناسی
7
قرارداد
8
فسخ قرارداد
9
گزارش کارشناسی
8
11
بررسی تعهدات ئبتی
10
A
(و) امور مالی
(ه) مهندسی پزشکی   11
(ط) مدیریت ، ریاست،
اداره اموال، دانشگاه
12
گزارش کارشناسی
13
دستگاه های ملزوم به ئبت؟
خیر
14
B
13
بلی
15
گزارش کارشناسی
16
اعلام خروج دستگاه
17
گزارش کارشناسی
16
18
درخواست متوقف کردن
20
متوقف کردن
21
تاییدیه متوقف کردن
20
19
B
22
پایان

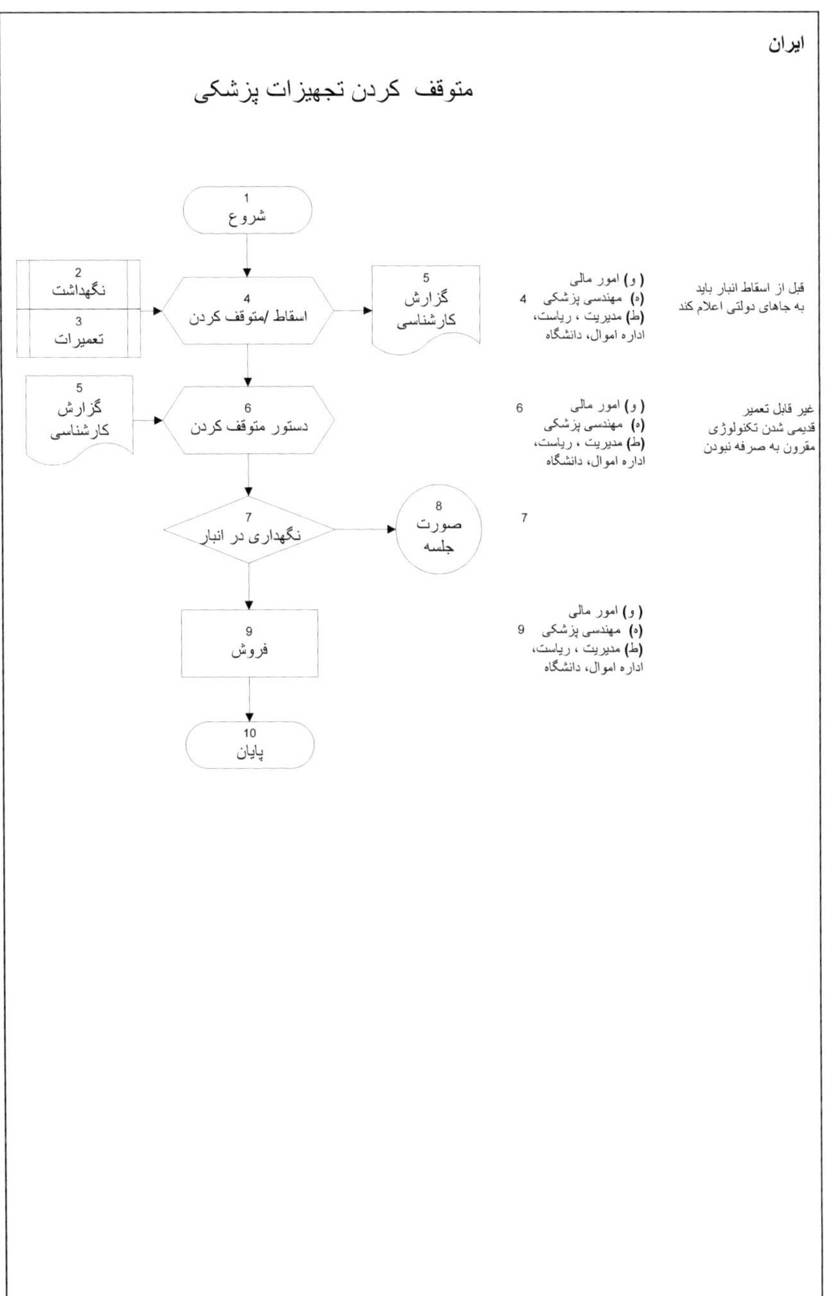
ایران

متوقف کردن تجهیزات پزشکی

1
شروع

2
نگهداشت

3
تعمیرات

4
اسقاط /متوقف کردن

5
گزارش کارشناسی

قبل از اسقاط انبار باید
به جاهای دولتی اعلام کند

(و) امور مالی
(ه) مهندسی پزشکی   4
(ط) مدیریت ، ریاست،
اداره اموال، دانشگاه

5
گزارش کارشناسی

6
دستور متوقف کردن

غیر قابل تعمیر
قدیمی شدن تکنولوژی
مقرون به صرفه نبودن

(و) امور مالی
(ه) مهندسی پزشکی   6
(ط) مدیریت ، ریاست،
اداره اموال، دانشگاه

7
نگهداری در انبار

8
صورت جلسه

7

9
فروش

(و) امور مالی
(ه) مهندسی پزشکی   9
(ط) مدیریت ، ریاست،
اداره اموال، دانشگاه

10
پایان

APPENDIX – A9

Interfaces of the process- Medical equipment

Deutschland 1

Prozess	Prozessschritt	Kostenfaktor[1]	Wertung	Qualitätsfaktor[2]	Wertung
Medizintechnik allgem.		Instandhaltungskosten bezogen auf Wiederbeschaffungswert Anschaffungskosten bezogen auf Anzahl der Anwendungen	A	Anzahl der meldepflichtigen Störungen an medizintechnischen Geräten	B
	Beschaffung	Anschaffungskosten bezogen auf Pflegetage Preis pro Gerät (gerätespezifisch)	B	Anzahl der beanstandeten Beschaffungen	B
	Inbetriebnahme	Schulungsaufwand pro Gerät (gerätespezifisch)	C	Anzahl der fehlgeschlagenen Testläufe	C
	Betrieb	Energieverbrauch pro Anwendung Materialkosten pro Anwendung	B	Auslastung der Geräte	A
	Wartung und Inspektion	Wartungskosten bezogen auf Wiederbeschaffungswert	A	Anzahl der meldepflichtigen Ereignisse	B
	Instandsetzung	Instandsetzungskosten bezogen auf Wiederbeschaffungswert	A	Dauer der Instandsetzung	A
	Stilllegung und Entsorgung	Entsorgungskosten bezogen auf Anschaffungswert	B	Anzahl der nicht dokumen-tierten Entsorgungen Dauer des Prozesses	C
Beschaffung Medizintechnik		Abstimmungsaufwand (Zeit) pro Beschaffung	B		B
	Beschaffungs-entscheidung	Abstimmungsaufwand (Zeit) pro Beschaffung	B	Anzahl der Kundenbeschwerden	A
	Festlegung Instandhaltungs-strategien	Abstimmungsaufwand (Zeit) pro Beschaffung	B	Anzahl der störungsbedingten Instandsetzungen	A
	Bestellung	Arbeitszeit pro Beschaffung	B		B
	Dokumentation der gerätebezogenen Vorgänge	Arbeitszeit pro Beschaffung	B	Anzahl der unvollständigen Datensätze nach Beschaffung	A
Inbetriebnahme Medizintechnik		Arbeitszeit pro Beschaffung	C	Gesamtdauer der Inbetriebnahme	B
	Planung und Festlegung der Anlieferung und Installation	Planungskosten bezogen auf Investitionskosten	A	Anzahl der Terminüberschreitungen Kundenzufriedenheit Anzahl der Nachbesserungen	A
	Installation	Arbeitszeit pro Beschaffung Materialkosten pro Auftrag	A	Dauer der Montage, Terminabweichungen	B
	Funktionsprüfung	Arbeitszeit pro Beschaffung Materialkosten pro Auftrag	B	Anzahl der Wiederholungen	B
	Durchführung von Einweisungen, Schulungen	Arbeitszeit pro Beschaffung, Schulungskosten pro Beschaffung	B	Anzahl anwenderverschuldeter Bedienfehler oder Störungen	A
	Übergabe	Arbeitszeit pro Beschaffung	B	Anzahl der nicht Dokumentierten Übergaben	B
	Meldung an zuständige Behörden	Genehmigungsgebühren pro Beschaffung	C		B

Deutschland 2

Prozess	Prozessschritt	Kostenfaktor[1]	Wertung	Qualitätsfaktor[2]	Wertung
Wartung und Inspektion Medizintechnik		Arbeitszeit pro Auftrag	A	Anzahl der nicht rechtzeitig erkannten Instandsetzungen	B
	Instandhaltungsplanung	Arbeitszeit pro Auftrag Hard-/Softwarekosten je Auftrag	B	Kapazitätsauslastung der Mitarbeiter	A
	Arbeitsplanerstellung	Arbeitszeit pro Auftrag	B	Abweichungen vom Arbeitsplan	B
	Terminplanung	Arbeitszeit pro Auftrag Telefonkosten pro Auftrag	B	Anzahl der Terminüberschreitungen Anzahl der Nachbesserungen	B
	Steuerung, Durchführung, Prüfung, Überwachung	Arbeitszeit pro Auftrag Materialkosten pro Auftrag Telefonkosten pro Auftrag	A	Anzahl der Nachbesserungen	B
	Kostenübermittlung		C		C
	Auftragsabweichungsanalyse	Arbeitszeit pro Auftrag	C	Anzahl der Abweichungen je Auftrag	B
	Inspektion analysieren	Arbeitszeit pro Auftrag	C	Anzahl der nicht rechtzeitig erkannten Instandsetzungen	B
Instandsetzung Medizintechnik			A		A
	Schaden prüfen	Arbeitszeit pro Auftrag	B	Anzahl der eingeleiteten Fehlmaßnahmen	A
	Zugehörige Berichte erstellen Durchführung entsp. Maßnahmen	Arbeitszeit pro Auftrag Telefonkosten pro Auftrag	B	Anzahl Beanstandungen durch zuständige Behörden	B
	Ursachen analysieren	Arbeitszeit pro Auftrag Hard-/Softwarekosten je Auftrag	C	Anzahl der Fehldiagnosen	B
	Auswahl der zu beseitigenden Ursachen	Arbeitszeit pro Auftrag Telefonkosten pro Auftrag	B	Anzahl der fehlgeschlagenen Behebungen	B
	zunächst nicht zu beseitigende Ursachen dokumentieren	Arbeitszeit pro Auftrag	C	Anzahl nicht dokumentierter Schwachstellen	B
	Ursachen beseitigen	Arbeitszeit pro Auftrag	B	Anzahl der Wiederholungen gleicher Ursachen	A
	Instandsetzen	Arbeitszeit pro Auftrag Materialkosten pro Auftrag Hard-/Softwarekosten je Auftrag	A	Anzahl der Nachbesserungen Kundenzufriedenheit	A
	Wiederinbetriebnahme	Arbeitszeit pro Auftrag	B	Anzahl der nicht dokumentierten STK'en	B

[1]Kostenfaktor:	Zähler: Welche Einheit bestimmt die Größe der Kosten Nenner: Welche Bezugsgröße kann zum Vergleich herangezogen werden
[2]Qualitätsfaktor:	Welches quantifizierbare Ereignis gibt die Abweichung von - oder die Erfüllung der geforderten Qualität wieder

Iran

Prozess	Prozessschritt	Kostenfaktor	Wertung	Qualitätsfaktor	Wertung
Beschaffung	Anfrage / Produktwahl	Abstimmungs- und Auswahlzeit	A	Meinung des Chefs und der Medizintechnik	A
	Planung der Instandsetzungsstrategie			Anzahl der Störungen	A
	Bestellung	Bestellungszeit	B	Anzahl der fehlerhaften Bestellungen	B
	Dokumentation	Arbeitszeit pro Gerät	B	Anzahl der unvollständigen Datensätze nach Beschaffung	B
Inbetriebnahme	Planung und Kontrolle des Geräts	Planungskosten bezogen auf Investitions-kosten	B	Qualität der Inbetriebnahme, Zeitaufwand, falls das Gerät nicht in Betrieb genommen werden kann, Kundenzufriedenheit	A
	Inbetriebnahme, Installation	Arbeitszeit, Materialkosten	B	Zeit für die Inbetriebnahme, Zeitaufwand, falls das Gerät nicht in Betrieb genommen werden kann	B
	Testbetrieb, Überprüfung	Arbeitszeit, Materialkosten	B	Anzahl der Wiederholungen	B
	Schulungen	Schulungszeit	A	Anzahl der anwenderbedingten Bedienfehler oder Störungen	B
	Übergabe an Abteilung	Arbeitszeit	B	Anzahl der Dokumentationen pro Gerät	B
Wartung und Inspektion	Planung der Wartung und Inspektion	Arbeitszeit	A	Auslastung zur Umsetzung der Wartung	B
	Arbeitsplaner-Stellung	Arbeitszeit	B	Einhalten des Arbeitsplans	A
	Terminplanung	Arbeitszeit, Verwaltungs-kosten	B	Anzahl der Terminüberschreitungen, Anzahl der Nachbesserungen	A
	Überwachung, Steuerung, Koordinierung	Arbeitszeit, Materialkosten, Verwaltungs-kosten	B	Anzahl der Nachbesserungen	B
	Kostenübermittlung	Verwaltungszeit	B	Genauigkeit der Informationen	B
Instandsetzung	Schaden prüfen	Arbeitszeit	A	Anzahl der Fehldiagnosen	B
	Dokumentation	Arbeitszeit, Verwaltungskosten	B	Anzahl der eingeleiteten Fehlmaßnahmen	B
	Instandsetzung	Arbeitszeit, Verwaltungskosten	B	Anzahl der Nachbesserungen, Kundenzufriedenheit	A
	Überwachung, Steuerung	Arbeitszeit	B	Genauigkeit der Dokumentation	B

آلمان

فرایند	قدم فرایند	فاکتور های هزینه	ارزش	فاکتور های کیفیت	ارزش
مهندسی پزشکی عمومی		هزینه های تعمیر و نگهداری بر حسب ارزش تهیه مجددبر حسب تعداد کاربرد	A	تعداد معایب دستگاههای پزشکی لازم به اعلام	B
	خرید	هزینه اکتساب بر حسب روز های مراقبت قیمت بازا دستگاه	B	تعداد خرید های ایراد دار	B
	راه اندازی	مخارج آموزش بازا دستگاه	C	تعداد تستهای نا مو فق	C
	بهره برداری	مصرف انرژی و مواد بازا کاربرد	B	بهره برداری از ظرفیت دستگاه	A
	نگهداری و بازرسی	هزینه های نگهداری بر حسب ارزش تهیه مجدد	A	تعداد اتفاقات لازم به اعلام	B
	تعمیر	هزینه های تعمیر بر حسب ارزش تهیه مجدد	A	مدت تعمیر	A
	خاتمه بهره برداری و اسفاط	هزینه های اسفاط بر حسب قیمت خرید	B	تعداد اسفاط ثبت نشده، مدت فرایند	C
خرید تجهیزات پزشکی		مدت هماهنگی بازا خرید	B		B
	تصمیم خرید	مدت هماهنگی بازا خرید	B	تعداد شکایات	A
	تعیین استراتژی نگهداری	مدت هماهنگی بازا خرید	B	تعداد تعمیرات بعلت اختلال	A
	سفارش	مدت کار بازا خرید	B		B
	مستند سازی مراحل مربوط به دستگاه	مدت کار بازا خرید	B	تعداد نگاشته های ناکامل بعد از خرید	A
راه اندازی تجهیزات پزشکی		مدت کار بازا تجهیزات	C	مدت کامل راه اندازی	B
	بر نامه ریزی و تعین ارسال و نصب	هزینه های برنامه ریزی بازا هزینه های سرمایه گذاری	A	تعداد عدم رعایت موعد، رضایت بخش، تعداد تصحیحات	A
	نصب	زمان کار بازا تجهیزات، هزینه مواد بازا سفارش	A	تعداد مونتاژ، انحراف از موعد	B
	آزمون کارکرد	زمان کار بازا تجهیزات، هزینه مواد بازا سفارش	B	تعداد تکرار	B
	ایجاد دستور کار، آموزش	زمان کار بازا تجهیزات، هزینه آموزش بازا تجهیزات،	B	تعداد اشتباهات ناشی از اپراتور	A
	تحویل	مدت کار بازا تجهیزات	B	تعداد تحویلات ثبت نشده	B

فرایند	قلم فرایند	فاکتور های هزینه	ارزش	فاکتور های کیفیت	ارزش
نگهداری و بازرسی تجهیزات پزشکی		زمان کار بازا سفارش	A	تعداد تعمیراتی که سر موقع مشخص نشده اند	B
	برنامه ریزی نگهداری	زمان کار بازا سفارش، هزینه نرم و سخت افزار کار بازا سفارش	B	ظرفیت کاربری کارکنان	A
	ایجاد برنامه ریزی کار	زمان کار بازا سفارش	B	انحراف از برنامه ریزی کار	B
	برنامه ریزی زمان	زمان کار بازا سفارش، هزینه طفی بازا سفارش	B	تعداد عدم رعایت موعد، تعداد تصحیحات	B
	کنترل، اجرا، تست، نظارت	زمان کار بازا سفارش، هزینه مواد بازا سفارش، هزینه طفی بازا سفارش	A	تعداد تصحیحات	B
	برآورد هزینه		C		C
	بررسی انحرافت از قرارداد	زمان کار بازا سفارش	C	تعداد انحرافت بازا سفارش	B
	بررسی نتایج بازرسی	زمان کار بازا سفارش	C	تعداد تعمیراتی که سر موقع مشخص نشده اند	B
تعمرات تجهیزات پزشکی			A		A
	بررسی خسارات	زمان کار بازا سفارش	B	تعداد اقامات انجام شد علد	A
	تهیه گزارشات مربوطه، اجرای اقامات لازم	زمان کار بازا سفارش، هزینه طفی بازا سفارش	B	تعداد ایرادات از طریق ادارات مربوطه	B
	بررسی علها	زمان کار بازا سفارش، هزینه نرم و سخت افزار کار بازا سفارش	C	تعداد تشخیصات اشتبه	B
	انتخاب علل لازم به رفع	زمان کار بازا سفارش، هزینه طفی بازا سفارش	B	تعداد اشکال زدایی های اشتبه	B
	ابتدا خرج علل غیرلازم به رفع	زمان کار بازا سفارش	C	تعداد نقطه ضعفهای ثبت نشد	B
	رفع علل	زمان کار بازا سفارش	B	تعداد تکرار علل مشابه	A
	تعمیر	زمان کار بازا سفارش، هزینه مواد بازا سفارش، هزینه نرم و سخت افزار کار بازا سفارش	A	تعداد تصحیحات، رضایت بخش	A
	راه اندازی مجدد	زمان کار بازا سفارش	B	تعداد قطعات ثبت نشد	B
1 فاکتور های هزینه	صورتکه چه واحدی برای میزان هزینه ها تعیین کننده است؟ مخرج چه مقیاسی میواند برای مقایسه در نظر گرفته شود؟				
2 فاکتور های کیفیت	چه رویداد قابل اندازه گیری میواندنشنگر انحراف یا برآوردگی کیفیت باشد؟				

246

ایران

فرایند	قدم فرایند	فاکتورهای هزینه	فاکتورهای کیفیت	فاکتورهای زمان
جهیزات پزشکی				
خرید				
	اعلام نیاز/انتخاب کالا	زمان اطلاع و زمان انتخاب (روند اداری)	نظر رئیس و مهندسی پزشکی	
	برنامه ریزی تعمیرات		تعداد خرابی ها	
	سفارش	زمان سفارش	تعداد سفارش های مشکل دار	
	مستند سازی	زمان کار برای هر دستگاه	تکمیل نمودن اطلاعات بعد از خرید	
راه اندازی				
	برنامه ریزی و چک کردن دستگاه	هزینه های برنامه ریزی در مقایسه با هزینه های سرمایه گزاری	- کیفیت راه اندازی - اتلاف وقت اگر در زمان مقرر دستگاه راه اندازی نشود - رضایت بخش	
	نصب	زمان نصب هزینه مواد	زمان نصب،اتلاف وقت به شرط راه اندازی نشدن دستگاه	
	چک کردن	زمان نصب هزینه مواد	تکرار کار	
	آموزش	زمان آموزش	مواردی که اپراتور به دلیل ضعف آموزش یاد نگرفته است	
	تحویل به بخش	زمان	مستندات تحویل دستگاه	

فرایند	قدم فرایند	فاکتورهای هزینه	فاکتورهای کیفیت	فاکتورهای زمان
نگهداشت				
	برنامه ریزی نگهداشت	زمان	میزان مشغولیت عوامل اجرایی نگهداشت	
	برنامه ریزی کار	زمان	انجام برنامه کاری در زمان پیش بینی شده	
	برنامه ریزی زمان	زمان کاری هزینه های اداری(تلفن)	. انجام نشدن فرایند در زمان تعیین شده . لزوم تکرار انجام فرایند	
	نظارت و هماهنگی	زمان کاری هزینه کالا/مواد هزینه های اداری	لزوم تکرار انجام فرایند	
	حسابرسی	زمان سیکل اداری	دقت اطلاعات	
تعمیرات				
	چک کردن خرابی یا مسئله دستگاه	زمان کاری	خطای ناشی از تشخیص غلط	
	تولید گزارش/مستندسازی	زمان کاری هزینه های اداری	خطای درج فرایند تعمیرات	
	تعمیرات	زمان کاری هزینه های اداری	لزوم تکرار انجام فرایندرضایت بخش	
	نظارت و هماهنگی	زمان کاری	دقت در درج مستندات	

APPENDIX – A10

Identity values –Medical equipment

Deutschland

Prozess	Prozessschritt	Sender	Information/Inhalt	Empfänger
Medizintechnik allgem.	Beschaffung			
	Inbetriebnahme			
	Wartung und Inspektion			
	Instandsetzung			
	Stilllegung und Entsorgung			
Beschaffung Medizintechnik	Beschaffungs-entscheidung	Kunde (Besteller)	Gerätespezifikationen	Einkauf
	Beschaffungs-entscheidung	Medizintechnik	Zustimmung Gerät	Einkauf
	Beschaffungs-entscheidung	Einkauf	Kosten, Gerätespezifikationen, Aufstellungsort, begleitende Maßnahmen	Medizintechnik
	Festlegung Instandhaltungs-strategien	Medizintechnik	Strategiefestlegungen, Wirtschaftlichkeitsinformationen	Geschäftsführung
	Bestellung	Einkauf	Kosten, Gerätespezifikationen, Aufstellungsort, begleitende Maßnahmen	Lieferant
	Dokumentation der gerätebezogenen Vorgänge	Einkauf	Gerätedaten	Medizintechnik
Inbetriebnahme Medizintechnik	Planung und Festlegung der Anlieferung und Installation	Kunde (Besteller)	Ort, Zeitpunkt, Art der Montage	Medizintechnik
	Planung und Festlegung der Anlieferung und Installation	Medizintechnik	Ort, Zeitpunkt, Art der Montage	beteiligte Organisations-einheiten
	Planung und Festlegung der Anlieferung und Installation	Medizintechnik	Ort, Zeitpunkt, Art der Montage	Lieferant
	Durchführung der Installation	Medizintechnik	Ort, Zeitpunkt, Art der Montage	Kunde (Nutzer)
	Funktionsprüfung	Lieferant	Testbericht, Fehlerdokumentation	Medizintechnik
	Funktionsprüfung	Medizintechnik	Terminbestätigung oder Terminänderung	Kunde (Nutzer)
	Durchführung von Einweisungen, Schulungen	Lieferant	Bedienungshinweise	Kunde (Nutzer)
	Durchführung von Einweisungen, Schulungen	Lieferant	Servicehinweise, Bedienungshinweise	Medizintechnik
	Übergabe	Medizintechnik	Gerät, Bedienungsanleitung	Kunde (Verantwortlicher)
	Meldung an zuständige Behörden	Betreiber	Anmeldung, Anzeige Geräteinformationen	zuständige Behörden

Iran

Prozess	Prozessschritt	Sender	Information/Inhalt	Empfänger
Beschaffung	Bedarfserklärung	Abteilung	Geräteinformation	Management
	Bedarfserklärung	Abteilung	Geräteinformation	Verwaltungabtl.
	Beschaffungsentscheidung	Verwaltungsabtl.	Gerätespezifikation	Medizintechnik*
	Beschaffungsentscheidung	Medizintechnik	Informationsbeschaffung und technische Beratung	Management
	Beschaffung	Management	Einkaufsbefehl/ dessen Bestätigung	Verwaltungsabtl.
	Beschaffung	Verwaltungsabtl.	Bestellungsinformation	Lieferant/ Niederlassung
	Warenannahme	Lieferant/ Verwaltungsabtl.	Geräteinformation	Medizintechnik
Inbetriebnahme	Planung und Entscheidung	Medizintechnik	Zeitpunkt, Ort, Art der Inbetriebnahme	Management**
	Planung und Entscheidung	Management	Zeitpunkt, Ort, Art der Montage	Alle involvierten Abteilungen
	Testbetrieb	Lieferant	Testbericht	Medizintechnik
	Testbetrieb	Medizintechnik	Bestätigung des Nutzers, Abnahme	Abteilung
	Schulungen	Lieferant/ Niederlassung	Bedienungsanleitung	Medizintechnik
	Schulungen	Lieferant/ Niederlassung , Medizintechnik	Bedienungsanleitung	Abteilung
	Übergabe an Abteilung	Medizintechnik	Gerät, Bedienungsanleitung	Abteilung

Deutschland

Prozess	Prozessschritt	Sender	Information/Inhalt	Empfänger
Wartung und Inspektion Medizintechnik	Instandhaltungs-planung	Medizintechnik	Wartungsplan	Kunde (Verantwortlicher)
	Arbeitsplanerstellung	Medizintechnik	Arbeitsplan	Medizintechnik
	Terminplanung	Medizintechnik	Termine	Kunde (Nutzer)
	Steuerung, Durchführung, Überwachung	Medizintechnik	Termine, Qualitäten	Kunde (Besteller)
	Kostenübermittlung	Medizintechnik	Kosten (Rechnung)	Kunde (Besteller)
	Auftragsabweichungs-analyse	Medizintechnik	alle Auftragsdaten	Medizintechnik
	Inspektion analysieren	Medizintechnik	Inspektionsdaten	Medizintechnik
Instandsetzung Medizintechnik	Serviceanforderung	Kunde (Nutzer)	Schadensbeschreibung	Servicestelle
	Serviceanforderung	Servicestelle	Schadensbeschreibung	Medizintechnik
	Schaden prüfen	Medizintechnik	Prüfbericht	Kunde (Verantwortlicher)
	Zugehörige Berichte erstellen Durchführung entsp. Maßnahmen	Medizintechnik	Meldung über Vorkommnis	Betreiber
	Zugehörige Berichte erstellen Durchführung entsp. Maßnahmen	Betreiber	Meldung über Vorkommnis	Behörden etc.
	Ursachen analysieren	ggf. Gutachter	Schadensbild	Medizintechnik
	Auswahl der zu beseitigenden Ursachen	Medizintechnik	Ursachen mit Behebungsvorschlägen	Kunde (Verantwortlicher)
	Auswahl der zu beseitigenden Ursachen	Kunde (Verantwortlicher)	zu beseitigende Ursachen	Medizintechnik
	Ursachen beseitigen	Medizintechnik	Schulung, Änderungsinformation	Kunde (Nutzer)
	zunächst nicht zu beseitigende Ursachen dokumentieren	Medizintechnik	Auflistung nicht zu beseitigender Ursachen	Medizintechnik
	Instandsetzen	Medizintechnik	Servicebericht	Kunde (Besteller)
	Wiederinbetriebnahme	Medizintechnik	Protokoll, Testbericht	Betreiber
	Wiederinbetriebnahme	Betreiber	Anmeldung, Anzeige Geräteinformationen	zuständige Behörden

Iran

Prozess	Prozessschritt	Sender	Information/Inhalt	Empfänger
Wartung und Inspektion Medizintechnik	Instandhaltungs-planung	Medizintechnik	Wartungsplan	Kunde (Verantwortlicher)
	Arbeitsplanerstellung	Medizintechnik	Arbeitsplan	Medizintechnik
	Terminplanung	Medizintechnik	Termine	Kunde (Nutzer)
	Steuerung, Durchführung, Überwachung	Medizintechnik	Termine, Qualitäten	Kunde (Besteller)
	Kostenübermittlung	Medizintechnik	Kosten (Rechnung)	Kunde (Besteller)
	Auftragsabweichungs-analyse	Medizintechnik	alle Auftragsdaten	Medizintechnik
	Inspektion analysieren	Medizintechnik	Inspektionsdaten	Medizintechnik
Instandsetzung Medizintechnik	Serviceanforderung	Kunde (Nutzer)	Schadensbeschreibung	Servicestelle
	Serviceanforderung	Servicestelle	Schadensbeschreibung	Medizintechnik
	Schaden prüfen	Medizintechnik	Prüfbericht	Kunde (Verantwortlicher)
	Zugehörige Berichte erstellen Durchführung entsp. Maßnahmen	Medizintechnik	Meldung über Vorkommnis	Betreiber
	Zugehörige Berichte erstellen Durchführung entsp. Maßnahmen	Betreiber	Meldung über Vorkommnis	Behörden etc.
	Ursachen analysieren	ggf. Gutachter	Schadensbild	Medizintechnik
	Auswahl der zu beseitigenden Ursachen	Medizintechnik	Ursachen mit Behebungsvorschlägen	Kunde (Verantwortlicher)
	Auswahl der zu beseitigenden Ursachen	Kunde (Verantwortlicher)	zu beseitigende Ursachen	Medizintechnik
	Ursachen beseitigen	Medizintechnik	Schulung, Änderungsinformation	Kunde (Nutzer)
	zunächst nicht zu beseitigende Ursachen dokumentieren	Medizintechnik	Auflistung nicht zu beseitigender Ursachen	Medizintechnik
	Instandsetzen	Medizintechnik	Servicebericht	Kunde (Besteller)
	Wiederinbetriebnahme	Medizintechnik	Protokoll, Testbericht	Betreiber
	Wiederinbetriebnahme	Betreiber	Anmeldung, Anzeige Geräteinformationen	zuständige Behörden

آلمان

گیرنده	اطلاعات	فرستنده	گام فرایند	فرایند
			توصیف	تجهیزات پزشکی
			راه‌اندازی	
			نگهداری و بازرسی	
			تعمیر	
			خاتمه بهره‌برداری، رفع	
خرید	مشخصات دستگاه	متقاضی دهند	تصمیم خرید	خرید تجهیزات پزشکی
خرید	تصویب دستگاه	مهندسی پزشکی	تصمیم خرید	
مهندسی پزشکی	هزینه، مشخصات دستگاه، محل نصب، الزامات جنبی	خرید	تصمیم خرید	
مدیریت	تعیین استراتژی، بازدهی اقتصادی	مهندسی پزشکی	تعیین استراتژی نگهداری	
فروشنده	هزینه، مشخصات دستگاه، محل نصب، الزامات جنبی	خرید	متقاضی	
مهندسی پزشکی	اطلاعات دستگاه	خرید	مستند سازی الزامات مربوط به دستگاه	
مهندسی پزشکی	زمان، مکان، نحوه نصب	متقاضی دهند	برنامه‌ریزی و تعیین تحویل و نصب دستگاه	راه‌اندازی تجهیزات پزشکی
واحد مربوطه	زمان، مکان، نحوه نصب	مهندسی پزشکی	برنامه‌ریزی و تعیین تحویل و نصب دستگاه	
فروشنده	زمان، مکان، نحوه نصب	مهندسی پزشکی	برنامه‌ریزی و تعیین تحویل و نصب دستگاه	
استفاده کننده	زمان، مکان، نحوه نصب	مهندسی پزشکی	نصب	
مهندسی پزشکی	گزارش تست، ثبت اشکالات	فروشنده	تست کارکرد	
استفاده کننده	تایید با تعمیر موعد مقرر	مهندسی پزشکی	تست کارکرد	
استفاده کننده	راهنمای عملکرد	فروشنده	تهیه راهنمای کاربردی، آموزش	
مهندسی پزشکی	راهنمای عملکرد و سرویس	فروشنده	تهیه راهنمای کاربردی، آموزش	
استفاده کننده	راهنمای عملکرد دستگاه	مهندسی پزشکی	تحویل	
ادارات مربوطه	اعلام و ثبت اطلاعات دستگاه	گرداننده	ثبت در ادارات مربوطه	

ایران

تجهیزات پزشکی

فرایند	گام فرایند	فرستنده	اطلاعات	گیرنده
خرید				
	اعلام نیاز	بخش	اطلاعات دستگاه	مدیریت
	اعلام نیاز	بخش	اطلاعات دستگاه	کارپردازی
	تصمیم گیری	کارپردازی	معرفی کالا و پیش فاکتور	مهندسی پزشکی*
	تصمیم گیری	مهندسی پزشکی	اطلاعات و مشاورت فنی	مدیریت
	خرید	مدیریت	دستور خرید/تایید	کارپردازی
	خرید	کارپردازی	سفارشی	شرکت تولیدکننده یا نماینده
	تحویل کالا	شرکت تولید کننده ویا کارپردازی	اطلاعات دستگاه	مهندسی پزشکی
راه اندازی برنامه ریزی				
	بررسی و تصمیم گیری	مهندسی پزشکی	زمان،مکان،نحوه نصب	مدیریت**
	بررسی و تصمیم گیری	مدیریت	زمان،مکان،نحوه انجام	کل واحدهای درگیر
	تست	شرکت	گزارش تست راه اندازی	مهندسی پزشکی
	تست	مهندسی پزشکی	تایید بهره برداری	بخش
	آموزش	شرکت	نحوه کاربری	مهندسی پزشکی
	آموزش	شرکت،مهندسی پزشکی	نحوه کاربری	بخش
	تحویل به بخش	مهندسی پزشکی	دستگاه،کتابچه راهنمای کاربری	بخش

آلمان

گیرنده	اطلاعات	فرستنده	گام فرایند	فرایند
بخش	برنامه نگهداری	مهندسی پزشکی	برنامه ریزی نگهداری	نگهداری و بازرسی تجهیزات پزشکی
مهندسی پزشکی	برنامه کار	مهندسی پزشکی	برنامه ریزی کار	
بخش	زمان	مهندسی پزشکی	برنامه ریزی زمان	
بخش	زمان و سطح کیفیت	مهندسی پزشکی	کنترل، اجرا، تست، نظارت	
بخش	هزینه، صورت حساب	مهندسی پزشکی	برآورد هزینه	
مهندسی پزشکی	اطلاعات کامل دستور کار	مهندسی پزشکی	بررسی انحرافات از قرارداد	
مهندسی پزشکی	اطلاعات بازرسی	مهندسی پزشکی	بررسی نتایج بازرسی	
قسمت سرویس	توصیف خسارت	بخش	درخواست تعمیر بخش	تعمیر تجهیزات پزشکی
مهندسی پزشکی	توصیف خسارت	قسمت سرویس	درخواست تعمیر	
بخش مسئول	گزارش بررسی	مهندسی پزشکی	بررسی خسارت	
اپراتور	اعلام حوادث	مهندسی پزشکی	تهیه گزارشات مربوطه، اجرای اقدامات لازم	
ادارات و غیره	اعلام حوادث	اپراتور	تهیه گزارشات مربوطه، اجرای اقدامات لازم	
مهندسی پزشکی	لیست خسارات	کارشناس	بررسی علتها	
بخش مسئول	گزارش و پیشنهاد راه حل	مهندسی پزشکی	انتخاب عال لازم به رفع	
مهندسی پزشکی	عال لازم به رفع	بخش مسئول	انتخاب عال لازم به رفع	
بخش مسئول	آموزش، اطلاعات راجع به تغییرات	مهندسی پزشکی	رفع عال	
مهندسی پزشکی	لیست عال غیرلازم به رفع	مهندسی پزشکی	ابتدادرج عال غیرلازم به رفع	
بخش مسئول	گزارش سرویس	مهندسی پزشکی	تعمیر	
اپراتور	گزارش تست	مهندسی پزشکی	راه اندازی مجدد	
ادارات مربوطه	ثبت و معرفی اطلاعات دستگاه	اپراتور	راه اندازی مجدد	

آلمان

ایران

نگهداشت و بازرسی			
برنامه ریزی	مهندسی پزشکی	برنامه	مدیریت مهندسی پزشکی
عملیاتی کردن برنامه ریزی	مهندسی پزشکی	پیشنهاد قرارداد	معاونت پشتیبانی، دایره حقوقی
عملیاتی کردن برنامه ریزی	معاونت پشتیبانی	تایید	مهندسی پزشکی
عملیاتی کردن برنامه ریزی	مهندسی پزشکی	تایید	مدیریت
برنامه زمان	مهندسی پزشکی	برنامه زمان بندی شده	بخش، مدیریت
نظارت و هماهنگی	مهندسی پزشکی	زمان و سطح کیفیت	بخش
پرداخت هزینه	مهندسی پزشکی	صورت حساب	مدیریت، امور مالی

تعمیرات			
درخواست	درخواست تعمیر بخش	اعلام خرابی	مهندسی پزشکی
چک کردن	مهندسی پزشکی	گزارش و پیشنهاد راه حل	مدیریت،بخش
تعمیر	مهندسی پزشکی	تصمیم گیری تعمیر	مهندس پزشکی،شرکت
مستندسازی	مهندس پزشکی	علت ارجاع به شرکت	مهندس پزشکی،مدیریت
تعمیرات	مهندس پزشکی	گزارش، صورت حساب	مدیریت،مهندس پزشکی ،امور مالی
تحویل به بخش	مهندس پزشکی	دستگاه، کل اطلاعات	بخش

اسقاط			
	بهروردار	اعلام اسقاط	مدیریت، تدارکات، حسابداری
	مدیریت	اعلام اسقاط	مهندس پزشکی
	مهندس پزشکی	گزارش کارشناسی	مدیریت
	مدیریت	دستور اسقاط	امور مالی

* در بعضی از بیمارستانها جای فرستنده و گیرنده می تواند عوض شود.

** کالای سرمایه ای

در کل مستند سازی شفائی خیلی زعیف می باشد. در بیمارستانهای مختلف سیستمهای مختلفی حاکم هستند.

APPENDIX - A11

Questionnaire Technical Facilities – Maintenance

OPIK

Optimierung und Analyse von Prozessen in Krankenhäusern

Wartung/Inspektion technische Anlagen

Anlagenhersteller:	
Anlagenbezeichnung:	
Standort:	
Wartungsintervall	
Baujahr:	
Auftragskurzbeschreibung:	
Tätigkeitsbeschreibung	

Anlagentyp:
- Versorgungstechnik (Heizung, Klima, Lüftung)
- Elektrotechnik
- Bautechnik
- sonstige

Auftragsnummer	Ereignisnummer	Meldungsnummer	Datum	eigen ☐ / fremd ☐
				[bei Fremdvergabe bitte Lieferschein beilegen]

Zeiten (in Minuten)

Nr.	Bezeichnung	Wegezeit	Informationszeit	Arbeitszeit
1.	Instandhaltungsplanung			
2.	Arbeitsplanerstellung			
3.	Terminplanung			
4.	Durchführung der Wartung			
5.	Dokumentationsaufwand			
6.	Auftragsabweichungs-analyse			
7.	Inspektionsanalyse			

Material

Nr.	Anzahl	Bezeichnung	Preis [€]
1.			
2.			
3.			
4.			
5.			
6.			
7.			

Qualität (Durch den Erbringer der Leistungen auszufüllen)

1. Wurde Auftrag termingerecht durchgeführt? ☐ ja ☐ nein
2. Wie viele Anläufe wurden benötigt, um den Auftrag erfolgreich abzuschließen? ___ Anläufe
3. Ist die Erzeugung eines Instandsetzungsauftrages notwendig? ☐ ja ☐ nein

---

بهينه سازى و تجزيه و تحليل فرآيندهاى بيمارستانى   **OPIK**

مراقبت و نگهداشت تجهيزات تاسيسات فنى

	نوع تاسيسات	
توليد كننده دستگاه:		
سال توليد:		تاسيسات (گرما، سرما، هوا)
اسم \ نام دستگاه:	برق :	
مكان :	ساختمان :	
ريتم نگهداشت:	غيره :	

توضيح نعاليتها به طور مختصر/ مشكل دستگاه چه بوده است؟

توضيح عملكرد/ تعمير دستگاه چگونه انجام شده است؟

☐ خارج از بيمارستان   ☐ داخل بيمارستان   تاريخ:

اگر از خارج از بيمارستان نگهداشت انجام مى گيرد لطفا توضيح دهيد

زمان (دقيقه)

زمان كار	زمان اطلاع رسانى تا اقدام	زمان اطلاع رسانى	زمان اطلاع رسانى	فعاليت	شماره
				برنامه ريزى نگهداشت	1.
				تهييه برنامه ريزى كارى	2.
				برنامه ريزى زمانى	3.
				انجام نگهداشت	4.
				مستندسازى	5.
				تجزيه و تحليل فعاليت	6.
				نظارت و بازرسى	7.

لوازم مصرف

هزينه (تومان)	تعداد	نام كالا	شماره
			1.
			2.
			3.
			4.
			5.
			6.
			7.

كيفيت

1.	ايا سفارش در زمان در نظر گرفته انجام گرديده؟
2.	تا انجام نهايى كار چند بار مراجعه لازم بوده؟ ___ مراجعه — خبر ☐ بلى ☐
3.	ايا نياز به تعميرات است؟ — خبر ☐ بلى ☐

APPENDIX - A12

Questionnaire Technical Facilities - Repair

	Optimierung und Analyse von Prozessen in Krankenhäusern

Instandsetzung technischer Anlagen

Daten der Anlage

Anlagen- hersteller:		Anlagentyp:	Versorgungs- technik *(Heizung, Klima, Lüftung)*	
Anlagenbez.:			Elektro- technik	
Standortort:				
Leistungsbereich:			Bautechnik	
Baujahr:			sonstige	

Auftragskurz- beschreibung:	
Tätigkeits- beschreibung	

Auftragsnummer	Ereignisnummer	Meldungsnummer	Datum

Zeiten (in Minuten)

Nr.	Bezeichnung		Wegezeit	Information	Arbeitszeit
1.	Schadensprüfung				
2.	Berichte erstellen				
3.	Ursachen- u. Schwachstellen- analyse Ursachenauswahl, Ursachenbeseitigung	*(für Ursachen, die bspw. nach erfolgter Instandsetzung wieder als Störung auftreten würden)*			
4.	Instandsetzung				
5.	Wiederinbetriebnahme				

Material

Nr.	Anzahl	Bezeichnung	Preis in €
1.			
2.			
3.			
4.			
5.			
6.			

Qualität (Durch den Erbringer der Leistungen auszufüllen)

1. Ist ein Fehler gleicher Ursache bereits aufgetreten   ☐ ja ☐ nein
2. Wie viele Anläufe wurden benötigt, um den Auftrag erfolgreich abzuschließen?   _____ Anläufe
3. Wie beurteilen Sie die Qualität der Information über den Auftrag   1 2 3 4 5 6   ☐ ☐ ☐ ☐ ☐ ☐
4. Verbesserungsvorschläge:

بهینه سازی و تجزیه و تحلیل فرایندهای بیمارستانی

CPIK

تعمیرات و بازرسی تجهیزات تاسیسات فنی

		نوع تاسیسات	تولید کننده دستگاه :
	تاسیسات (گرما، سرما، هوا)		سال تولید :
	الکترونیک :		اسم \ نام دستگاه :
	سازه :		مکان :
	غیره :		بخش کاری :

توضیح فعالیتها به طور مختصر/ مشکل دستگاه چه بوده است؟

توضیح عملکرد/ تعمیر دستگاه چگونه انجام شده است؟

لوازم مصرف

زمان (دقیقه)

زمان کار	زمان اطلاع رسانی تا اقدام	زمان راه	مشخص سازی	شماره		هزینه (تومان)	توضیحات	تعداد	شماره
			تعیین اشکل	1					1.
			تهیه گزارش	2.					2.
			تجزیه و تحلیل علت	3.					3.
			تعمیر	4.					4.
			کار گیری دوباره	5.					5.

کیفیت

1.	ایا تابه حل اشتباهی با چنین حالتی پیش امده است ؟ خیر ☐ بلی ☐ 2 تا انجام نهایی کار چندبار مراجعه لازم بوده؟	مراجعه____
3.	کیفیت اطلاع رسانی این درخواست را ارزشیابی فرمایید؟ ☐1 ☐2 ☐3 ☐4 ☐5 ☐6 1 خیلی خوب 6 نارسا	
4.	پیشنهادهای اصلاحی	

O Itfverd ISH arbruhe TH	دانشگاه علوم پزشکی تهران	تاسیسات فنی / پرستنامه 2
Facility Management TUMS	تاریخ	1/1

APPENDIX - A13

Process Comparison- Technical Facilities

Deutschland

Lebenszyklus Gebäudetechnik

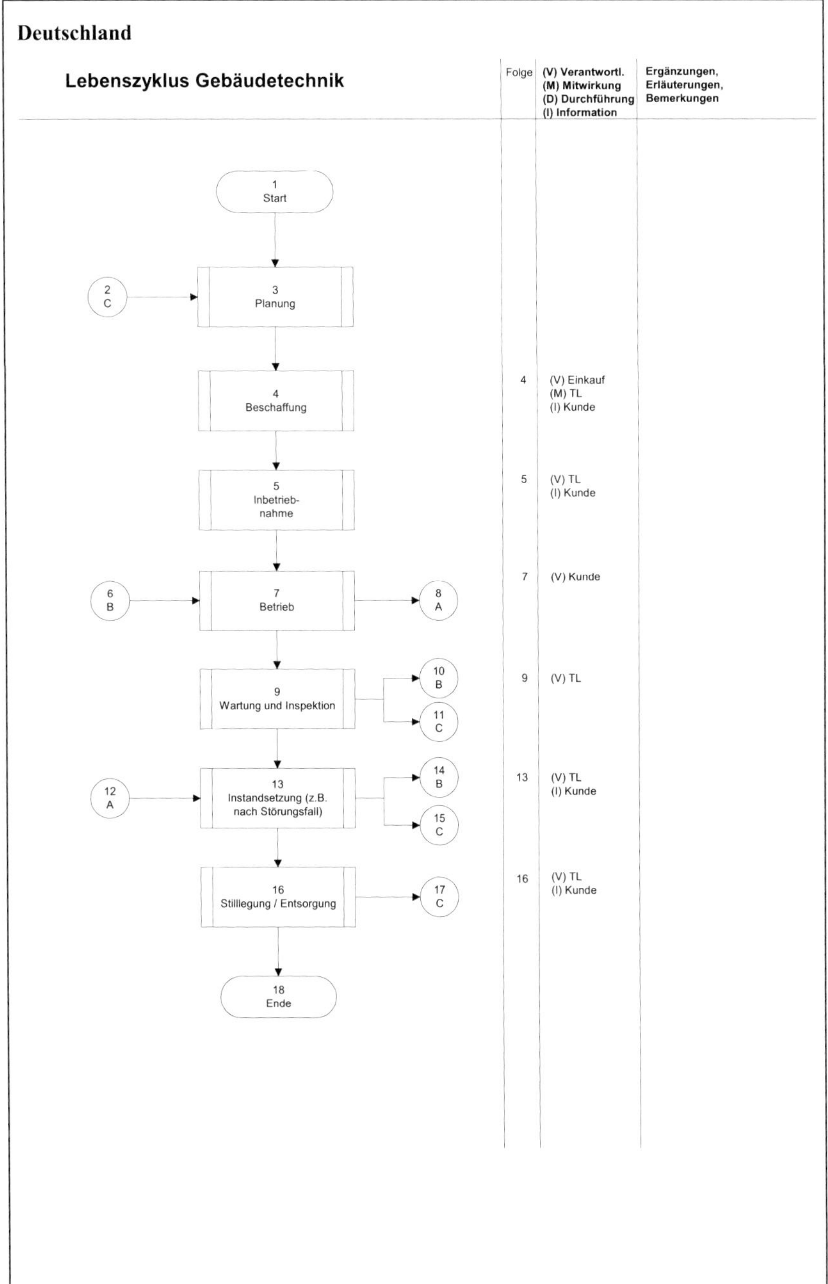

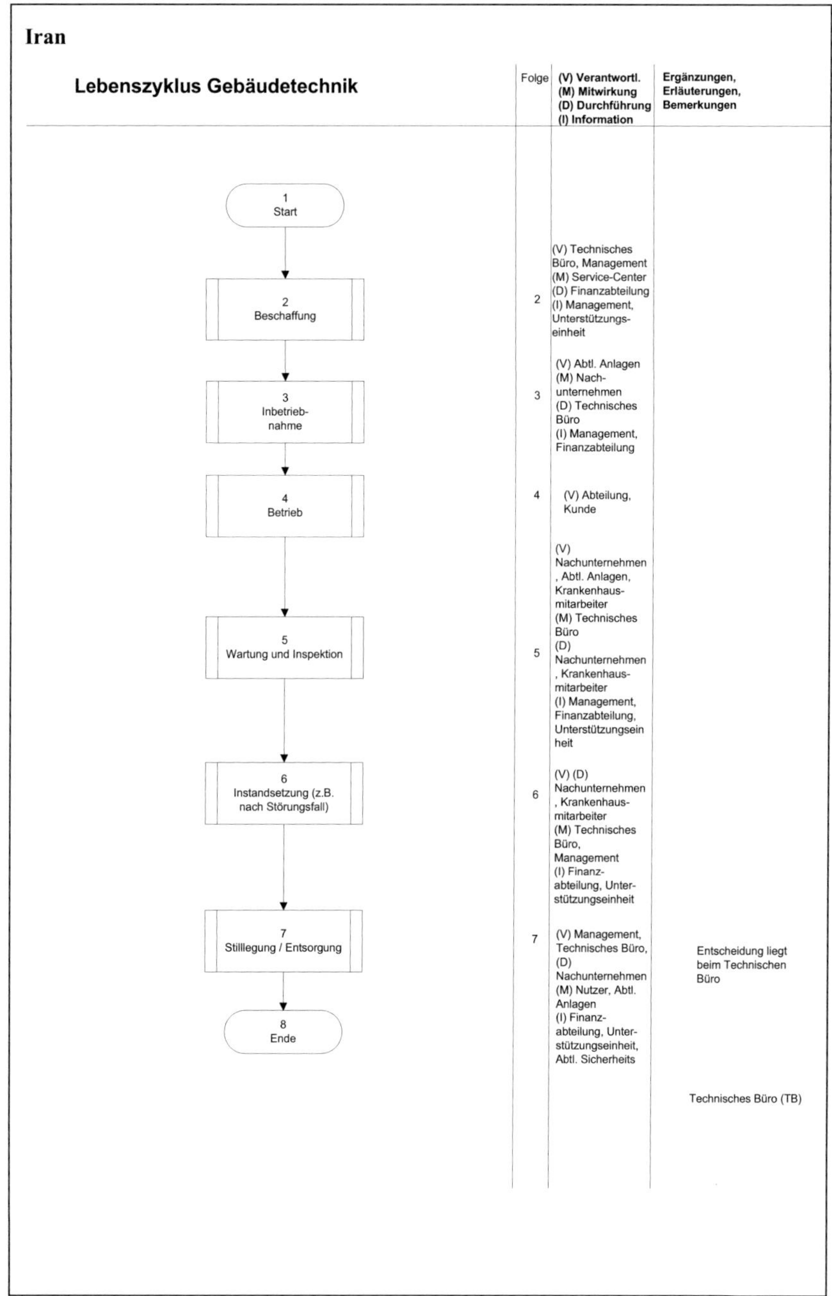
Iran

Lebenszyklus Gebäudetechnik

Folge
(V) Verantwortl.
(M) Mitwirkung
(D) Durchführung
(I) Information

Ergänzungen,
Erläuterungen,
Bemerkungen

1
Start

2
Beschaffung

2
(V) Technisches
Büro, Management
(M) Service-Center
(D) Finanzabteilung
(I) Management,
Unterstützungs-
einheit

3
Inbetrieb-
nahme

3
(V) Abtl. Anlagen
(M) Nach-
unternehmen
(D) Technisches
Büro
(I) Management,
Finanzabteilung

4
Betrieb

4
(V) Abteilung,
Kunde

5
Wartung und Inspektion

5
(V)
Nachunternehmen
, Abtl. Anlagen,
Krankenhaus-
mitarbeiter
(M) Technisches
Büro
(D)
Nachunternehmen
, Krankenhaus-
mitarbeiter
(I) Management,
Finanzabteilung,
Unterstützungsein
heit

6
Instandsetzung (z.B.
nach Störungsfall)

6
(V) (D)
Nachunternehmen
, Krankenhaus-
mitarbeiter
(M) Technisches
Büro,
Management
(I) Finanz-
abteilung, Unter-
stützungseinheit

7
Stilllegung / Entsorgung

7
(V) Management,
Technisches Büro,
(D)
Nachunternehmen
(M) Nutzer, Abtl.
Anlagen
(I) Finanz-
abteilung, Unter-
stützungseinheit,
Abtl. Sicherheits

Entscheidung liegt
beim Technischen
Büro

8
Ende

Technisches Büro (TB)

Deutschland

Planung technischer Anlagen
(Gebäudetechnik)

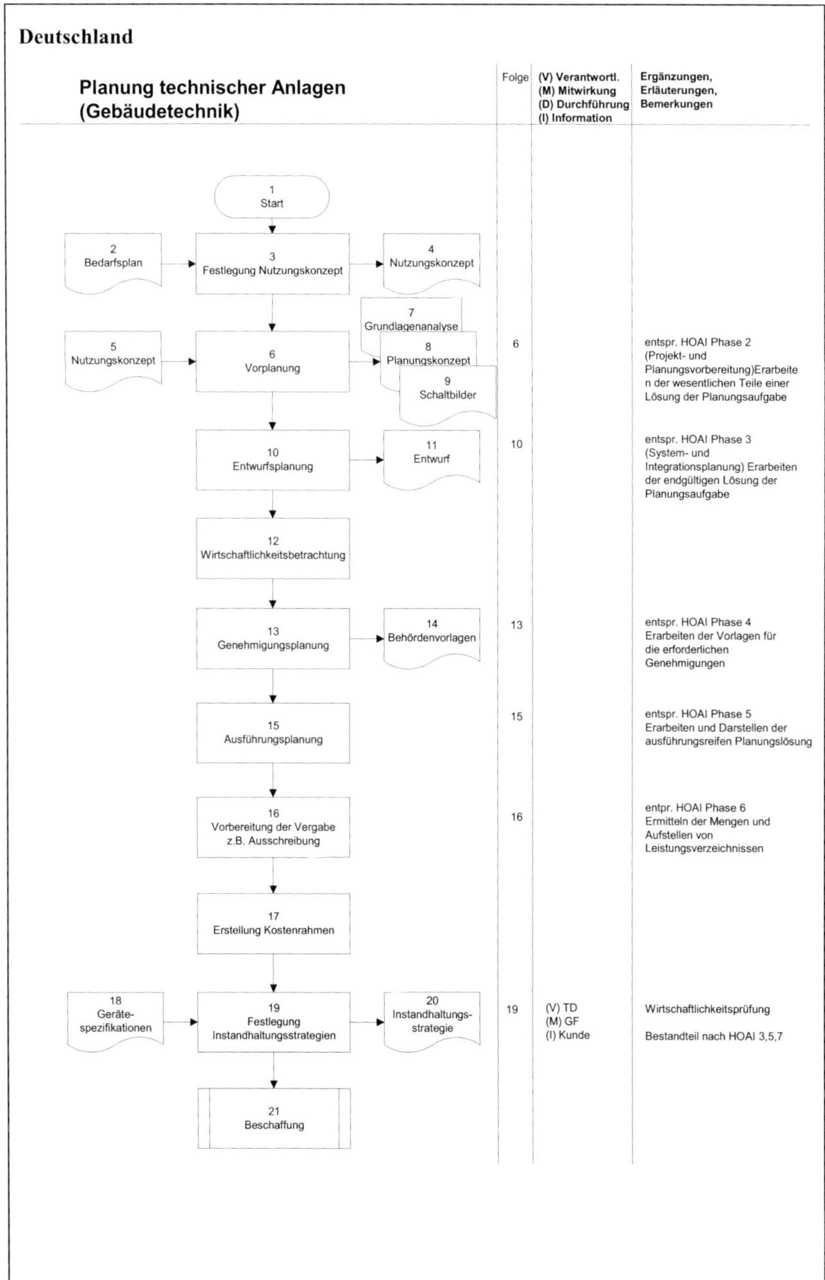

Deutschland

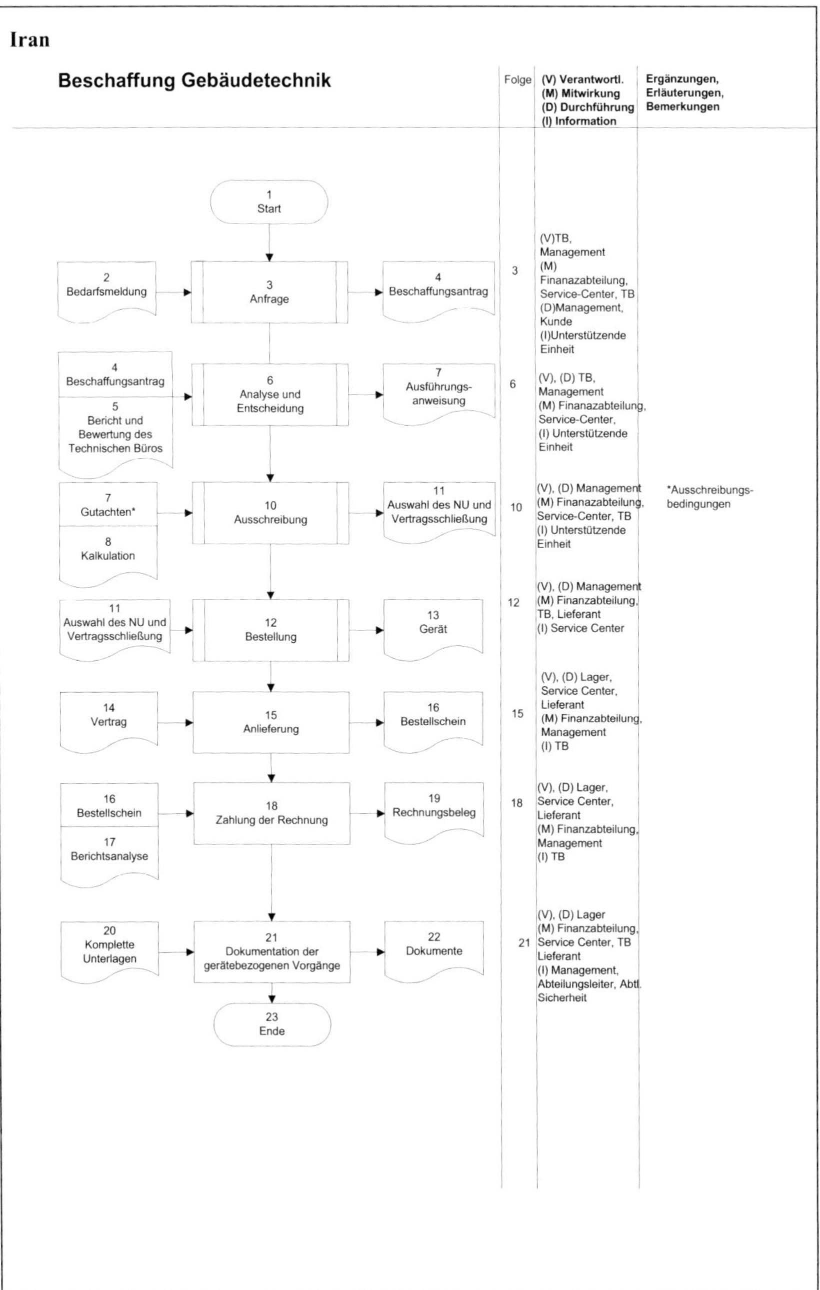

Iran
Beschaffung Gebäudetechnik
Folge
(V) Verantwortl.
(M) Mitwirkung
(D) Durchführung
(I) Information
Ergänzungen, Erläuterungen, Bemerkungen
1 Start
2 Bedarfsmeldung
3 Anfrage
4 Beschaffungsantrag
3
(V)TB, Management
(M) Finanazabteilung, Service-Center, TB
(D)Management, Kunde
(I)Unterstützende Einheit
4 Beschaffungsantrag
5 Bericht und Bewertung des Technischen Büros
6 Analyse und Entscheidung
7 Ausführungs-anweisung
6
(V), (D) TB, Management
(M) Finanazabteilung, Service-Center,
(I) Unterstützende Einheit
7 Gutachten*
8 Kalkulation
10 Ausschreibung
11 Auswahl des NU und Vertragsschließung
10
(V), (D) Management
(M) Finanazabteilung, Service-Center, TB
(I) Unterstützende Einheit
*Ausschreibungs-bedingungen
11 Auswahl des NU und Vertragsschließung
12 Bestellung
13 Gerät
12
(V), (D) Management
(M) Finanzabteilung, TB, Lieferant
(I) Service Center
14 Vertrag
15 Anlieferung
16 Bestellschein
15
(V), (D) Lager, Service Center, Lieferant
(M) Finanzabteilung, Management
(I) TB
16 Bestellschein
17 Berichtsanalyse
18 Zahlung der Rechnung
19 Rechnungsbeleg
18
(V), (D) Lager, Service Center, Lieferant
(M) Finanzabteilung, Management
(I) TB
20 Komplette Unterlagen
21 Dokumentation der gerätebezogenen Vorgänge
22 Dokumente
21
(V), (D) Lager
(M) Finanzabteilung, Service Center, TB Lieferant
(I) Management, Abteilungsleiter, Abtl. Sicherheit
23 Ende

Deutschland

Inbetriebnahme
Gebäudetechnik

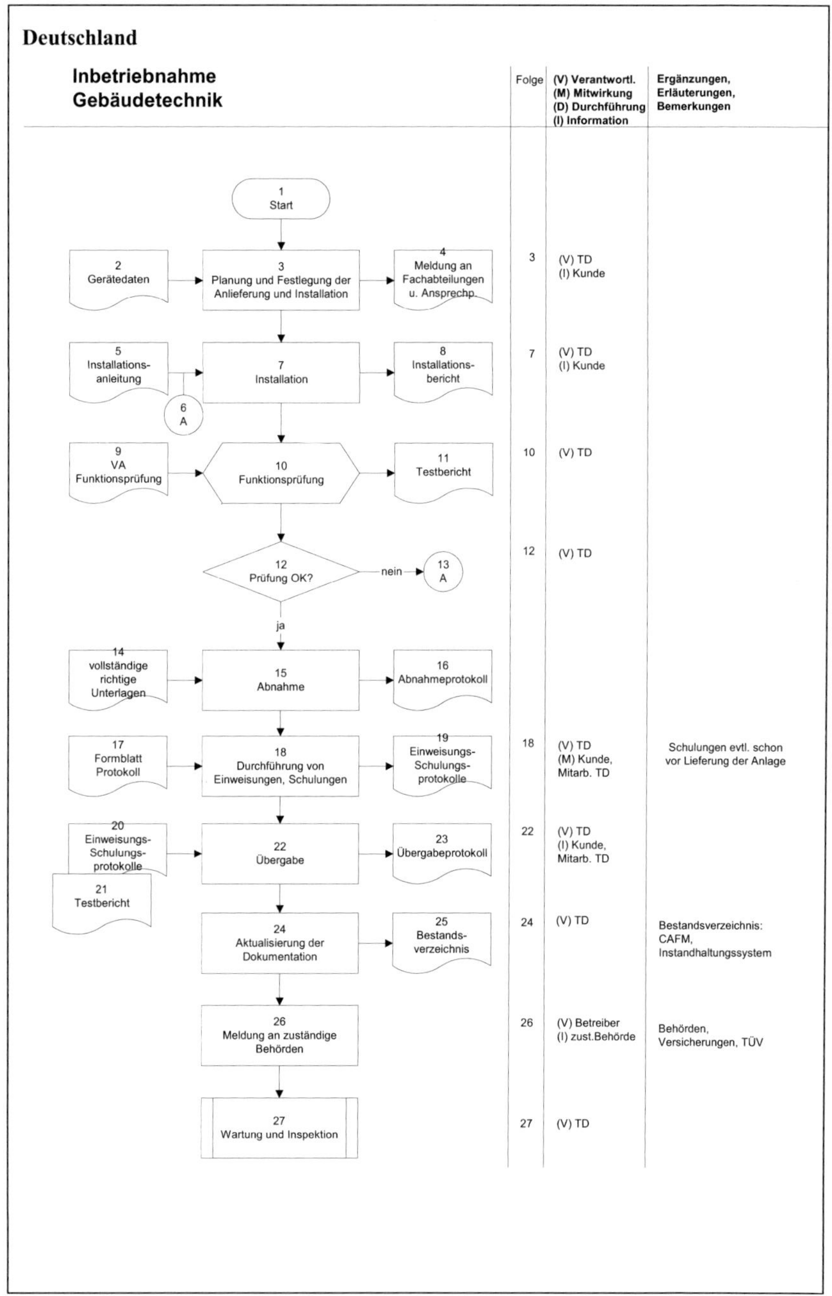

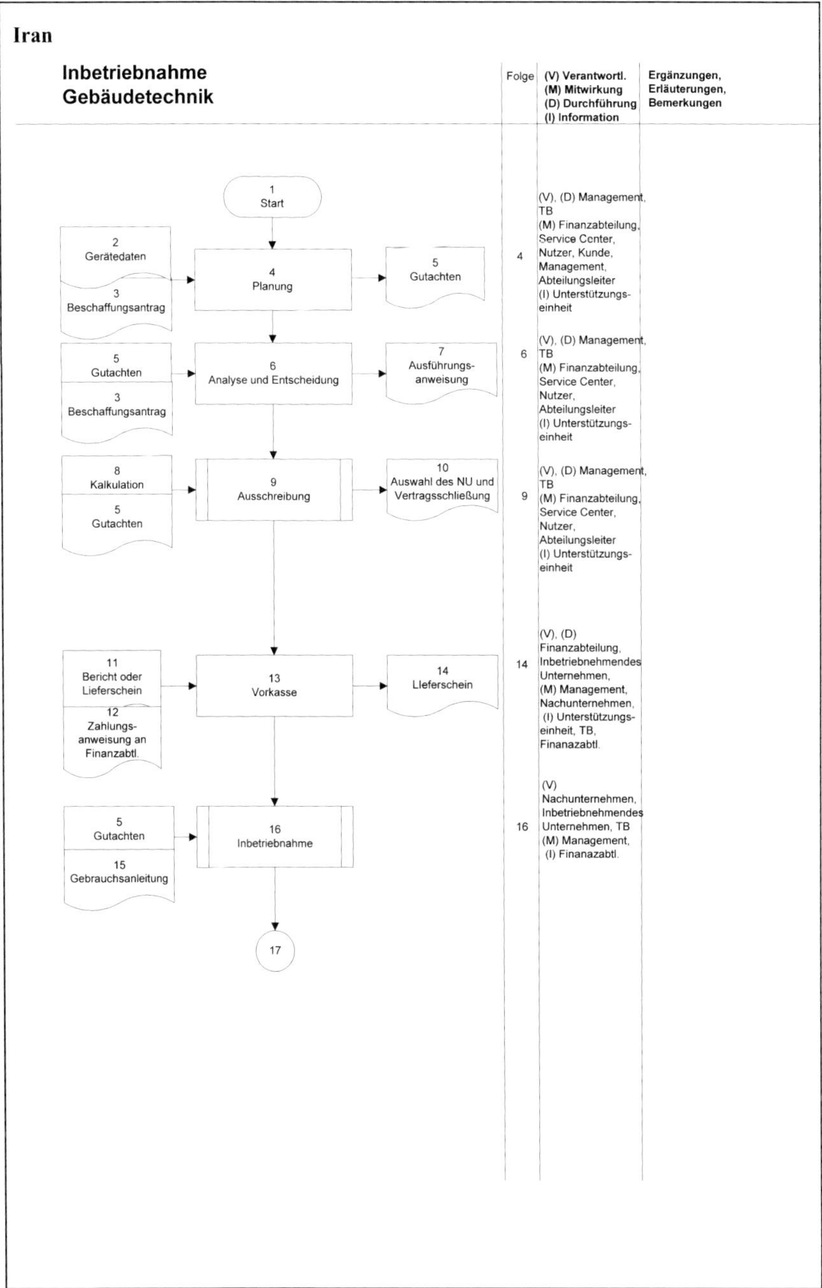
Iran

Inbetriebnahme
Gebäudetechnik

Folge
(V) Verantwortl.
(M) Mitwirkung
(D) Durchführung
(I) Information

Ergänzungen,
Erläuterungen,
Bemerkungen

1
Start

2
Gerätedaten

3
Beschaffungsantrag

4
Planung

5
Gutachten

4

(V), (D) Management,
TB
(M) Finanzabteilung,
Service Center,
Nutzer, Kunde,
Management,
Abteilungsleiter
(I) Unterstützungs-
einheit

5
Gutachten

3
Beschaffungsantrag

6
Analyse und Entscheidung

7
Ausführungs-
anweisung

6

(V), (D) Management,
TB
(M) Finanzabteilung,
Service Center,
Nutzer,
Abteilungsleiter
(I) Unterstützungs-
einheit

8
Kalkulation

5
Gutachten

9
Ausschreibung

10
Auswahl des NU und
Vertragsschließung

9

(V), (D) Management,
TB
(M) Finanzabteilung,
Service Center,
Nutzer,
Abteilungsleiter
(I) Unterstützungs-
einheit

11
Bericht oder
Lieferschein

12
Zahlungs-
anweisung an
Finanzabtl.

13
Vorkasse

14
Lieferschein

14

(V), (D)
Finanzabteilung,
Inbetriebnehmendes
Unternehmen,
(M) Management,
Nachunternehmen,
(I) Unterstützungs-
einheit, TB,
Finanazabtl.

5
Gutachten

15
Gebrauchsanleitung

16
Inbetriebnahme

16

(V)
Nachunternehmen,
Inbetriebnehmendes
Unternehmen, TB
(M) Management,
(I) Finanazabtl.

17

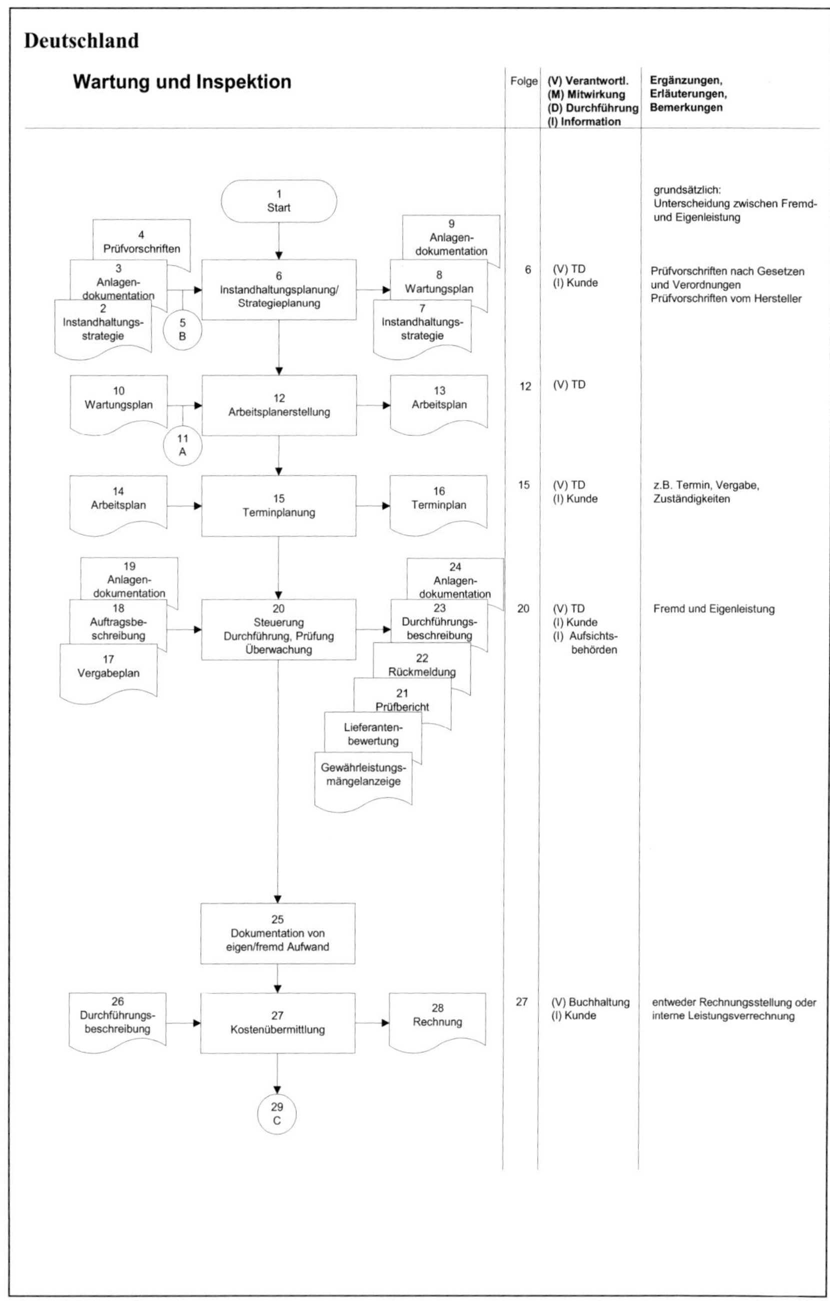

Deutschland
Wartung und Inspektion
Folge
(V) Verantwortl.
(M) Mitwirkung
(D) Durchführung
(I) Information
Ergänzungen, Erläuterungen, Bemerkungen
1 Start
4 Prüfvorschriften
3 Anlagendokumentation
2 Instandhaltungsstrategie
5 B
6 Instandhaltungsplanung/ Strategieplanung
9 Anlagendokumentation
8 Wartungsplan
7 Instandhaltungsstrategie
6
(V) TD
(I) Kunde
grundsätzlich: Unterscheidung zwischen Fremd- und Eigenleistung
Prüfvorschriften nach Gesetzen und Verordnungen
Prüfvorschriften vom Hersteller
10 Wartungsplan
11 A
12 Arbeitsplanerstellung
13 Arbeitsplan
12
(V) TD
14 Arbeitsplan
15 Terminplanung
16 Terminplan
15
(V) TD
(I) Kunde
z.B. Termin, Vergabe, Zuständigkeiten
19 Anlagendokumentation
18 Auftragsbeschreibung
17 Vergabeplan
20 Steuerung Durchführung, Prüfung Überwachung
24 Anlagendokumentation
23 Durchführungsbeschreibung
22 Rückmeldung
21 Prüfbericht
Lieferantenbewertung
Gewährleistungsmängelanzeige
20
(V) TD
(I) Kunde
(I) Aufsichtsbehörden
Fremd und Eigenleistung
25 Dokumentation von eigen/fremd Aufwand
26 Durchführungsbeschreibung
27 Kostenübermittlung
28 Rechnung
27
(V) Buchhaltung
(I) Kunde
entweder Rechnungsstellung oder interne Leistungsverrechnung
29 C

Iran

**Inbetriebnahme
Gebäudetechnik 2**

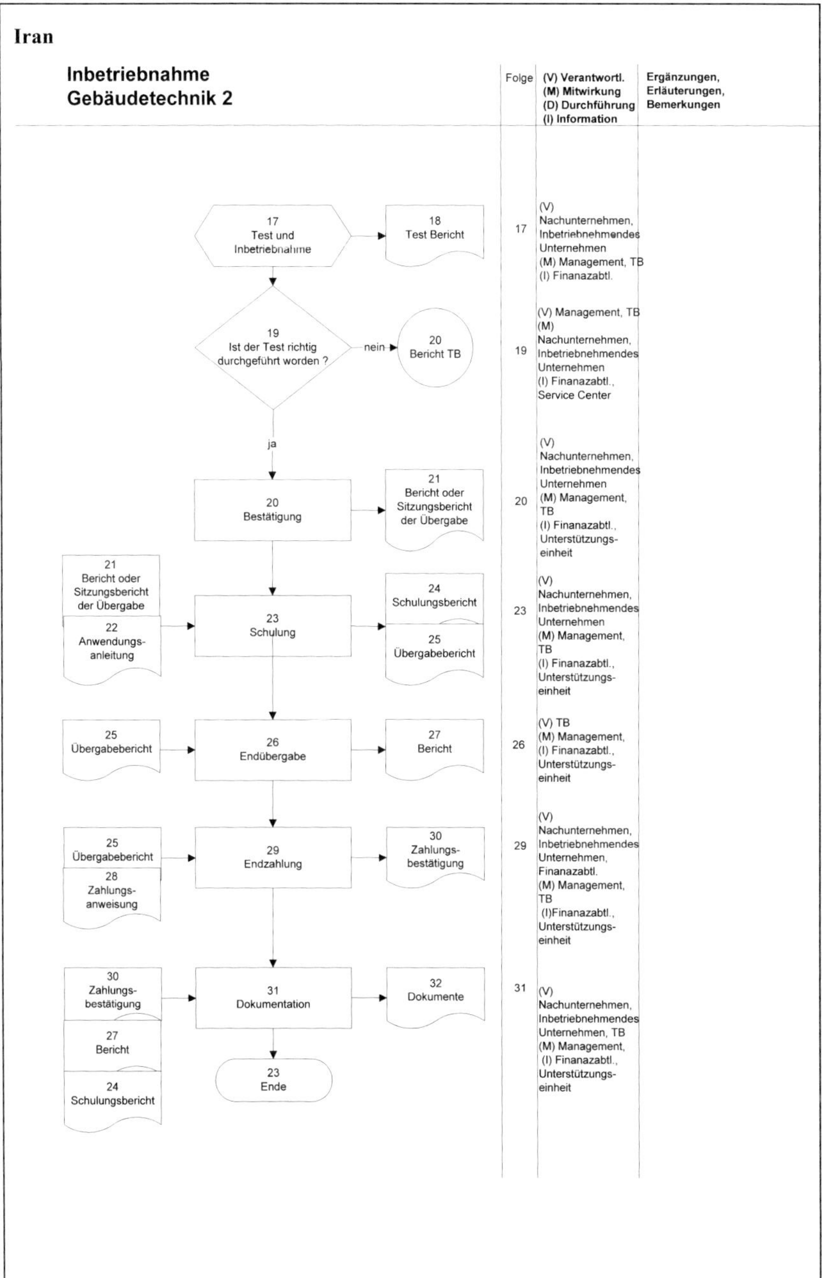

Deutschland

Durchführung Instandsetzung 1

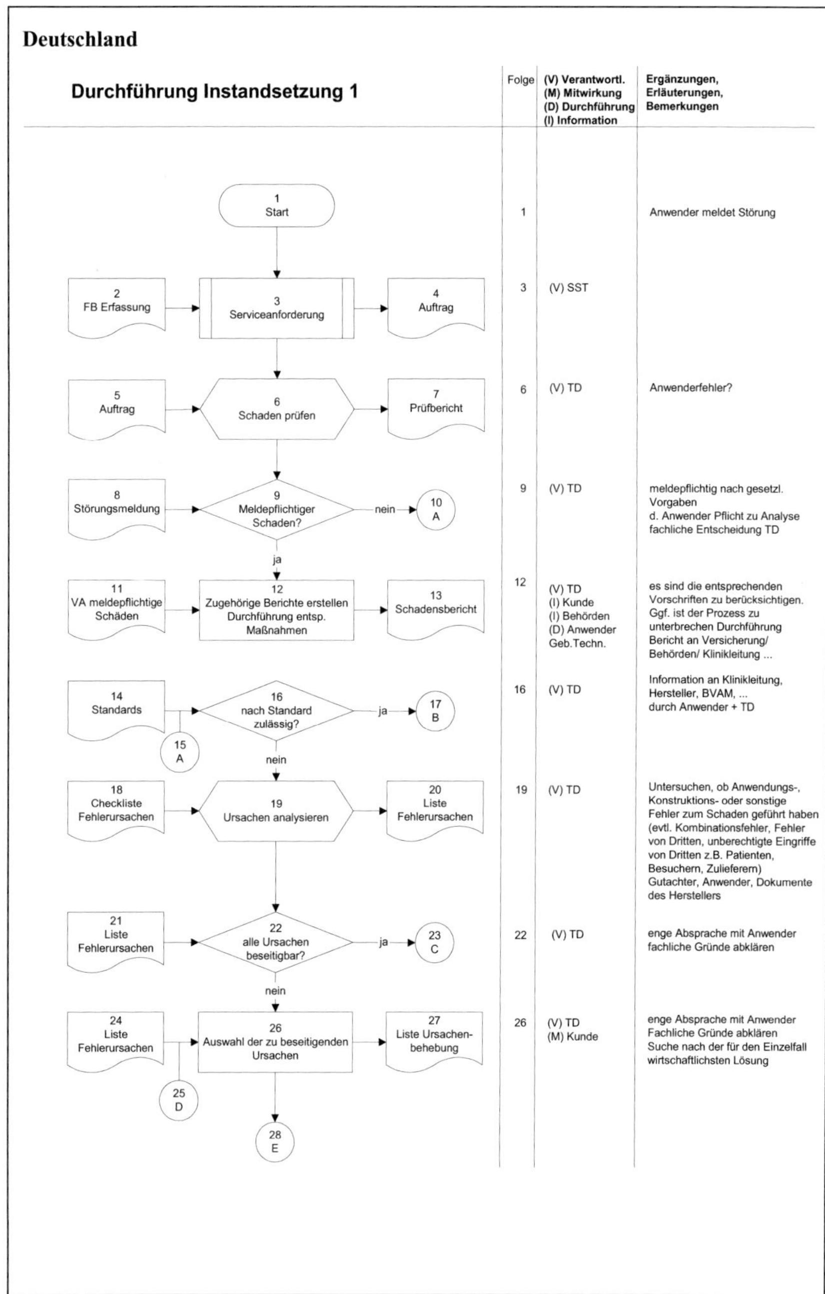

Iran

Wartung 2

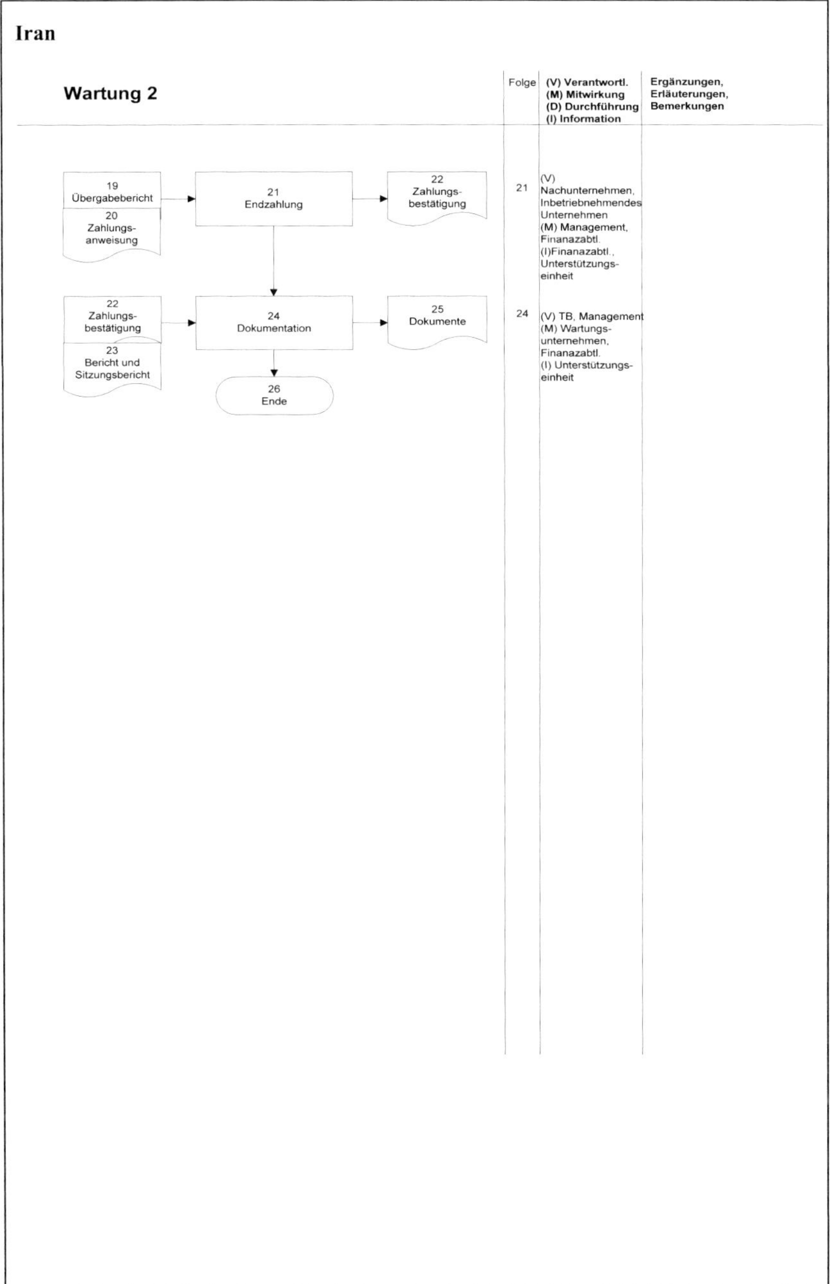

Folge	(V) Verantwortl. (M) Mitwirkung (D) Durchführung (I) Information	Ergänzungen, Erläuterungen, Bemerkungen
21	(V) Nachunternehmen, Inbetriebnehmendes Unternehmen (M) Management, Finanazabtl. (I)Finanazabtl., Unterstützungs- einheit	
24	(V) TB, Management (M) Wartungs- unternehmen, Finanazabtl. (I) Unterstützungs- einheit	

Deutschland

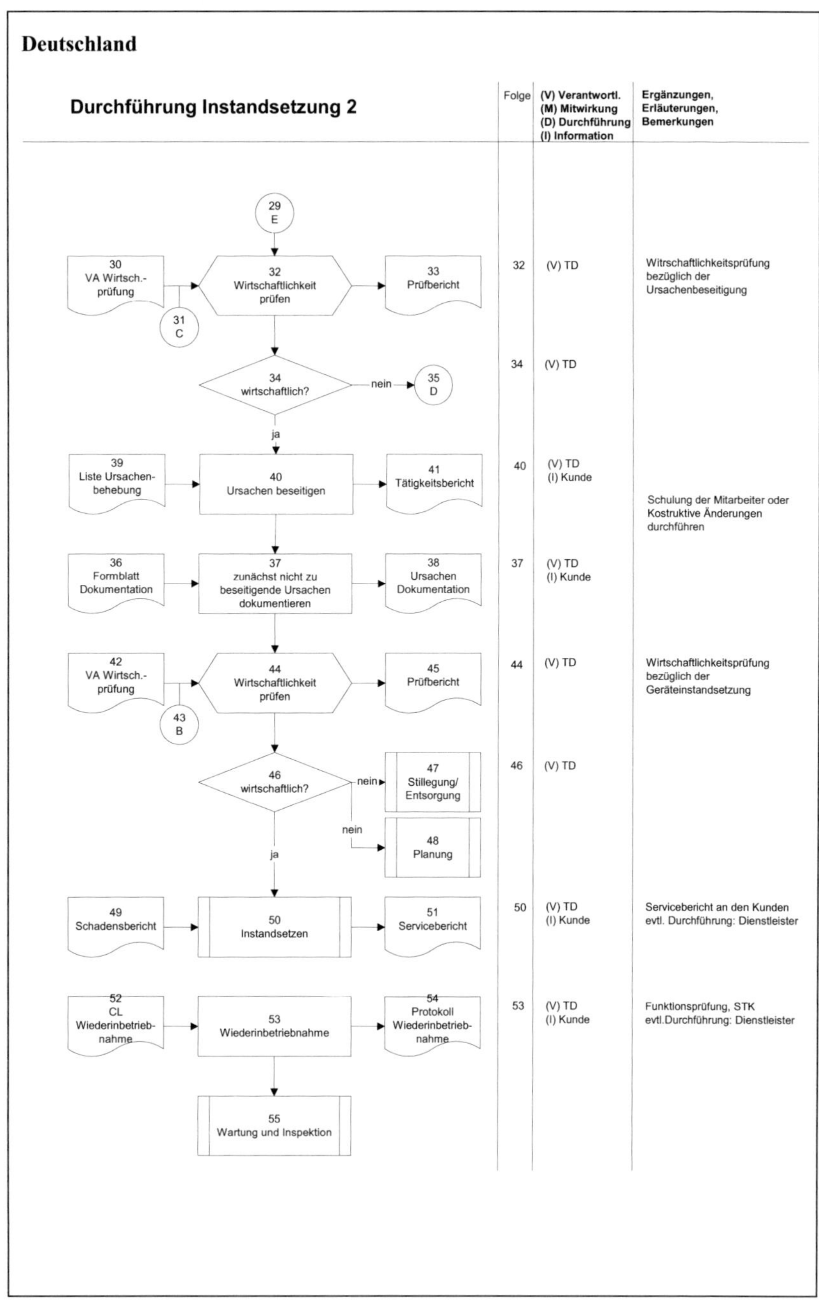

Iran

Instandsetzung

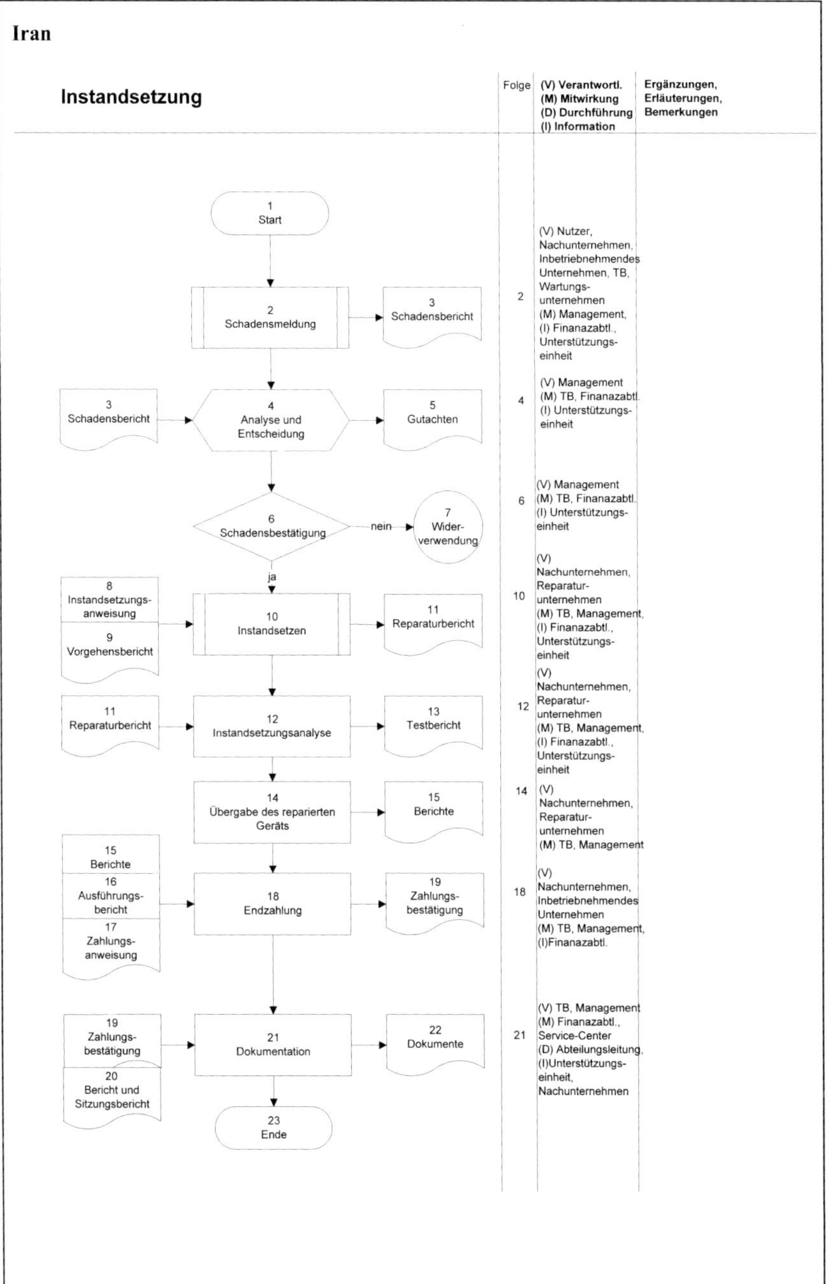

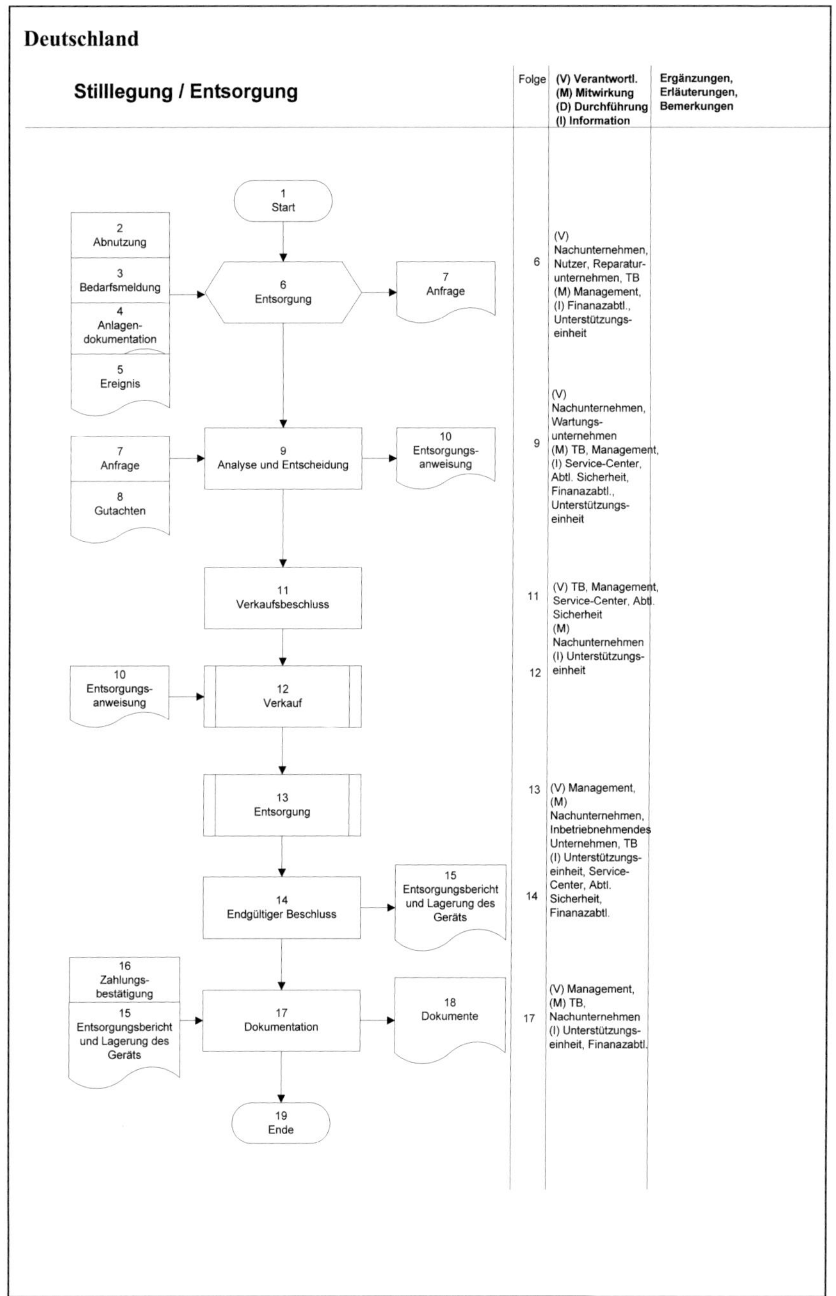

Deutschland
Stilllegung / Entsorgung
Folge
(V) Verantwortl.
(M) Mitwirkung
(D) Durchführung
(I) Information
Ergänzungen, Erläuterungen, Bemerkungen
1 Start
2 Abnutzung
3 Bedarfsmeldung
4 Anlagendokumentation
5 Ereignis
6 Entsorgung
7 Anfrage
6
(V) Nachunternehmen, Nutzer, Reparaturunternehmen, TB (M) Management, (I) Finanazabtl., Unterstützungseinheit
7 Anfrage
8 Gutachten
9 Analyse und Entscheidung
10 Entsorgungsanweisung
9
(V) Nachunternehmen, Wartungsunternehmen (M) TB, Management, (I) Service-Center, Abtl. Sicherheit, Finanazabtl., Unterstützungseinheit
11 Verkaufsbeschluss
11
(V) TB, Management, Service-Center, Abtl. Sicherheit (M) Nachunternehmen (I) Unterstützungseinheit
10 Entsorgungsanweisung
12 Verkauf
12
13 Entsorgung
13
(V) Management, (M) Nachunternehmen, Inbetriebnehmendes Unternehmen, TB (I) Unterstützungseinheit, Service-Center, Abtl. Sicherheit, Finanazabtl.
14 Endgültiger Beschluss
15 Entsorgungsbericht und Lagerung des Geräts
14
16 Zahlungsbestätigung
15 Entsorgungsbericht und Lagerung des Geräts
17 Dokumentation
18 Dokumente
17
(V) Management, (M) TB, Nachunternehmen (I) Unterstützungseinheit, Finanazabtl.
19 Ende

Iran

Stilllegung / Entsorgung

	Folge	(V) Verantwortl. (M) Mitwirkung (D) Durchführung (I) Information	Ergänzungen, Erläuterungen, Bemerkungen

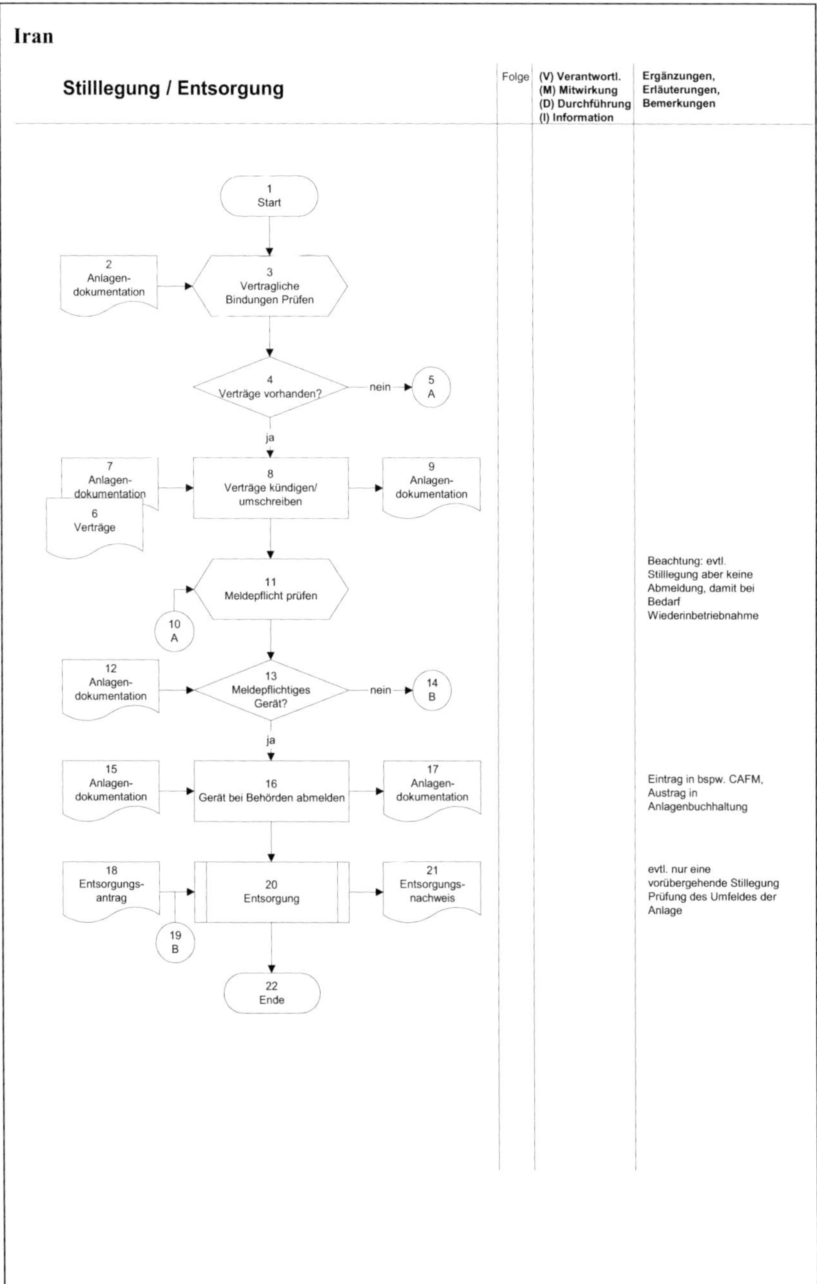

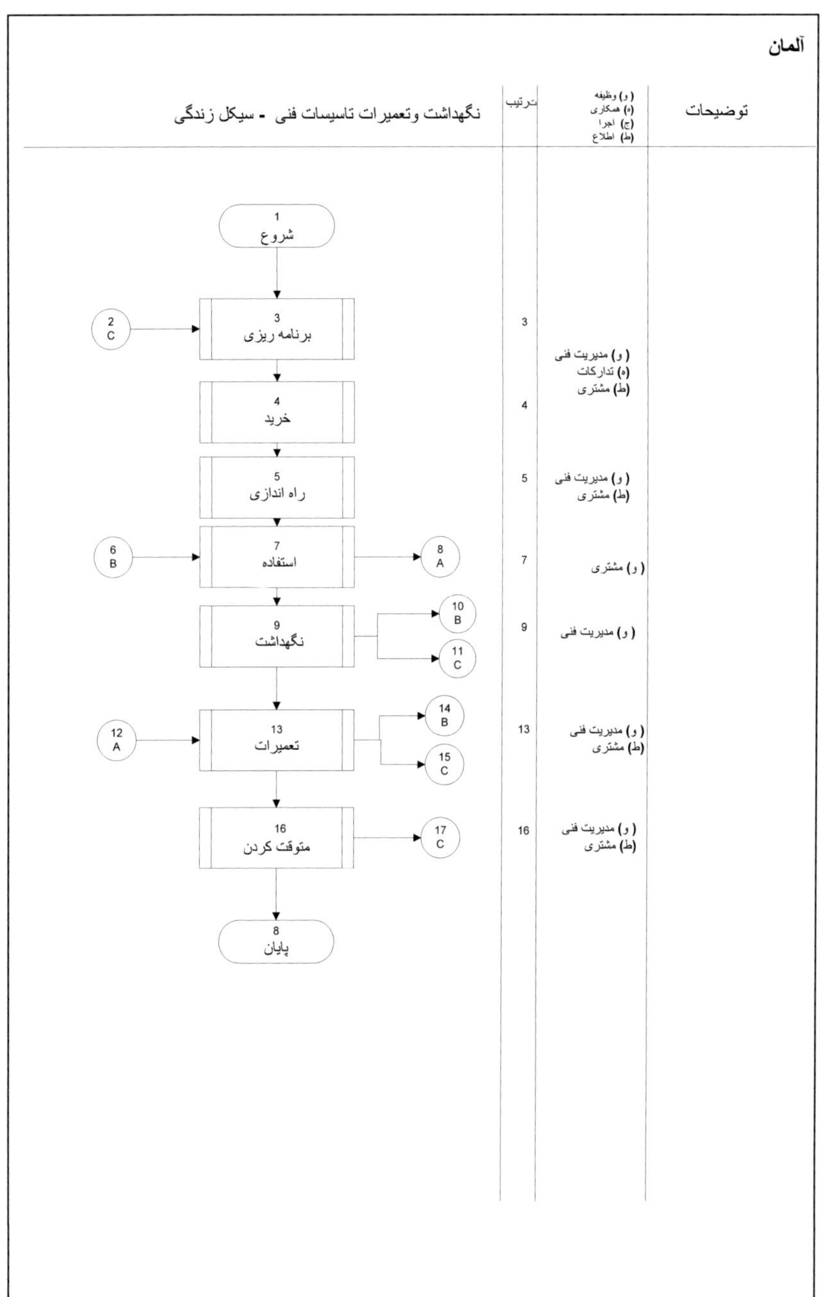

آلمان

توضیحات

(و) وظیفه
(ه) همکاری
(ج) اجرا
(ط) اطلاع

ترتیب

نگهداشت وتعمیرات تاسیسات فنی - سیکل زندگی

1
شروع

2
C

3
برنامه ریزی

3

(و) مدیریت فنی
(ه) تدارکات
(ط) مشتری

4
خرید

4

5
راه اندازی

5

(و) مدیریت فنی
(ط) مشتری

6
B

7
استفاده

8
A

7

(و) مشتری

9
نگهداشت

10
B

11
C

9

(و) مدیریت فنی

12
A

13
تعمیرات

14
B

15
C

13

(و) مدیریت فنی
(ط) مشتری

16
متوقت کردن

17
C

16

(و) مدیریت فنی
(ط) مشتری

8
پایان

ایران

نگهداشت وتعمیرات تاسیسات فنی

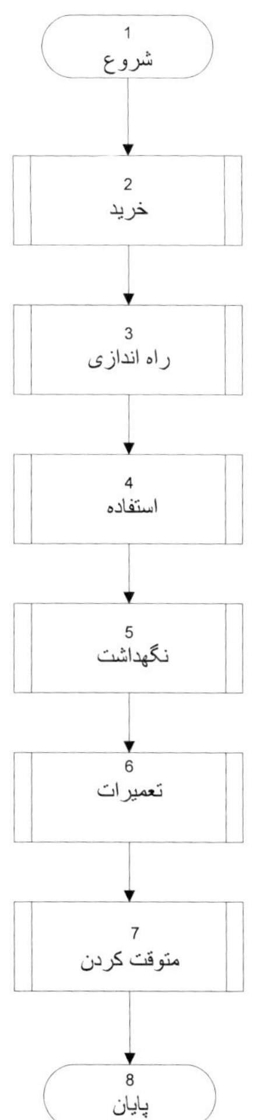

(و) دفتر فنی *
مدیریت
(ه)تدارکات
(ج)امور مالی   2
(ط) مدیریت و
معاونت پشتیبانی

(و) تاسیسات
(ه) پیمانکار   3
(ج) دفترفنی
(ط) مدیریت،امور مالی

4
ج)   بخش
مشتری

(و)   پیمانکارتاسیسات
، همکاران بیمارستان
(ه) دفترفنی   5
(ج) پیمانکار
، همکاران بیمارستان
(ط) مدیریت،امور مالی
معاونت پشتیبانی

(و) (ج) پیمانکار   6
، همکاران بیمارستان
(ه) دفترفنی ، مدیریت
(ط) امور مالی
معاونت پشتیبانی

7
(ج) پیمانکار
(و) مدیریت ،دفترفنی   تصمیمگیری دفترفنی
(ه)بهره بردار،تاسیسات
(ط) امور مالی
معاونت پشتیبانی
حراست

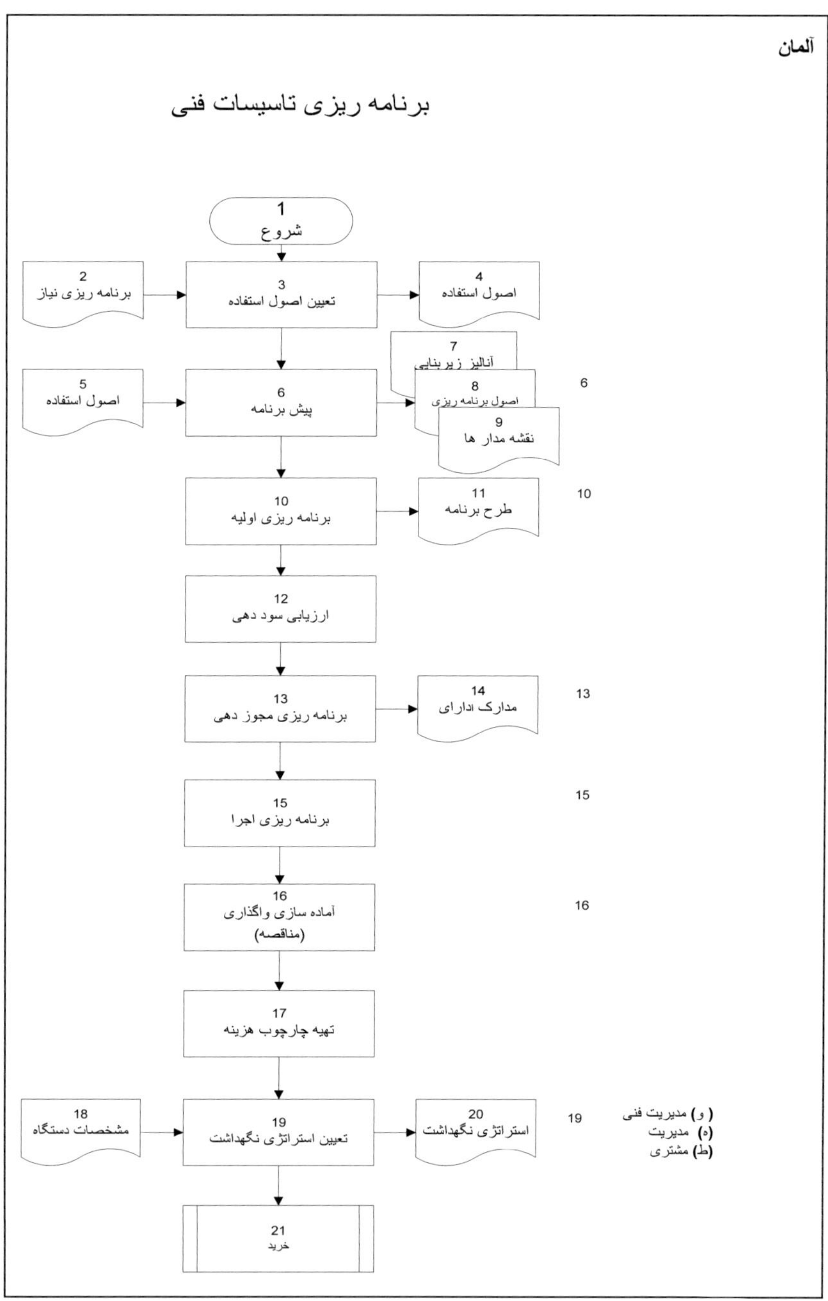
آلمان

برنامه ریزی تاسیسات فنی

1
شروع

2
برنامه ریزی نیاز

3
تعیین اصول استفاده

4
اصول استفاده

5
اصول استفاده

6
پیش برنامه

7
آنالیز زیربنایی

8
اصول برنامه ریزی

9
نقشه مدار ها

6

10
برنامه ریزی اولیه

11
طرح برنامه

10

12
ارزیابی سود دهی

13
برنامه ریزی مجوز دهی

14
مدارک اdarای

13

15
برنامه ریزی اجرا

15

16
آماده سازی واگذاری
(مناقصه)

16

17
تهیه چارچوب هزینه

18
مشخصات دستگاه

19
تعیین استراتژی نگهداشت

20
استراتژی نگهداشت

19

(و) مدیریت فنی
(ه) مدیریت
(ط) مشتری

21
خرید

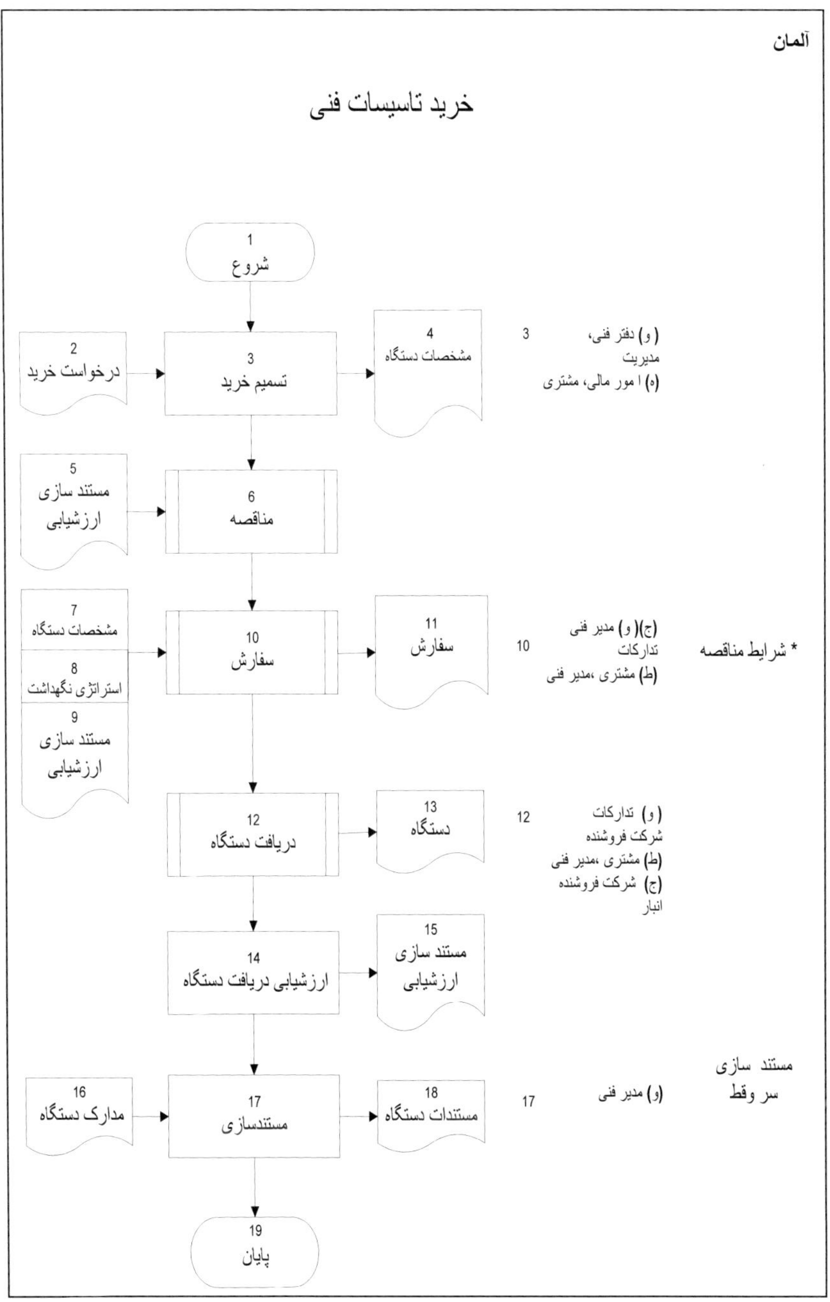
آلمان

خرید تاسیسات فنی

1
شروع

2
درخواست خرید

3
تسمیم خرید

4
مشخصات دستگاه

3
(و) دفتر فنی،
مدیریت
(ه) ا مور مالی، مشتری

5
مستند سازی
ارزشیابی

6
مناقصه

7
مشخصات دستگاه

8
استراتژی نگهداشت

9
مستند سازی
ارزشیابی

10
سفارش

11
سفارش

10
(ج)(و) مدیر فنی
تدارکات
(ط) مشتری ،مدیر فنی

* شرایط مناقصه

12
دریافت دستگاه

13
دستگاه

12
(و) تدارکات
شرکت فروشنده
(ط) مشتری ،مدیر فنی
(ج) شرکت فروشنده
انبار

14
ارزشیابی دریافت دستگاه

15
مستند سازی
ارزشیابی

16
مدارک دستگاه

17
مستندسازی

18
مستندات دستگاه

17
(و) مدیر فنی

مستند سازی
سر وقط

19
پایان

ایران

خرید تاسیسات فنی

1
شروع

2
اعلام نیاز

3
در خواست

4
برگه
در خواست

(و) دفتر فنی، مدیریت
(ه) ا مور مالی، تدارکات
(ج) واحد مدیریت، مشتری
(ط) معاونت پشتیبانی
3

4
برگه در خواست

5
گزارش و براورد دفتر فنی

6
بررسی و تصمیمگیری

7
دستور اجرا به بخش مالی

(ج) (و) دفتر فنی، مدیریت
(ه) ا مور مالی، تدارکات
(ط) معاونت پشتیبانی
6

8
گزارش کارشناسی *

9
برآورد هزینه

10
مناقصه

11
انتخاب پیمانکار و عقد قرارداد

(ج)(و) مدیریت
(ه) ا مور مالی، تدارکات
دفتر فنی
(ط) معاونت پشتیبانی
10

* شرایط مناقصه

11
انتخاب پیمانکار و عقد قرارداد

12
سفارش

13
دستگاه

(ج)(و) مدیریت
(ه) ا مور مالی،
دفتر فنی،
شرکت فروشنده
(ط) تدارکات
12

14
قرارداد

15
دریافت دستگاه

16
قبض انبار

(ج)(و) انبار ، تدارکات
شرکت فروشنده
(ه) ا مور مالی، مدیریت
(ط) دفتر فنی
15

16
قبض انبار

17 بررسی
گزارش

18
پرداخت هزینه

19
قبض رسید

(ج)(و) انبار ، تدارکات
شرکت فروشنده
(ه) ا مور مالی، مدیریت
(ط) دفتر فنی
18

20
کل سوابق

21
مستندسازی

22
مستندات

(ج)(و) انبار
(ه) ا مور مالی، تدارکات
دفتر فنی، شرکت فروشنده
(ط) مدیریت مدیر واحد ، حراست
21

23
پایان

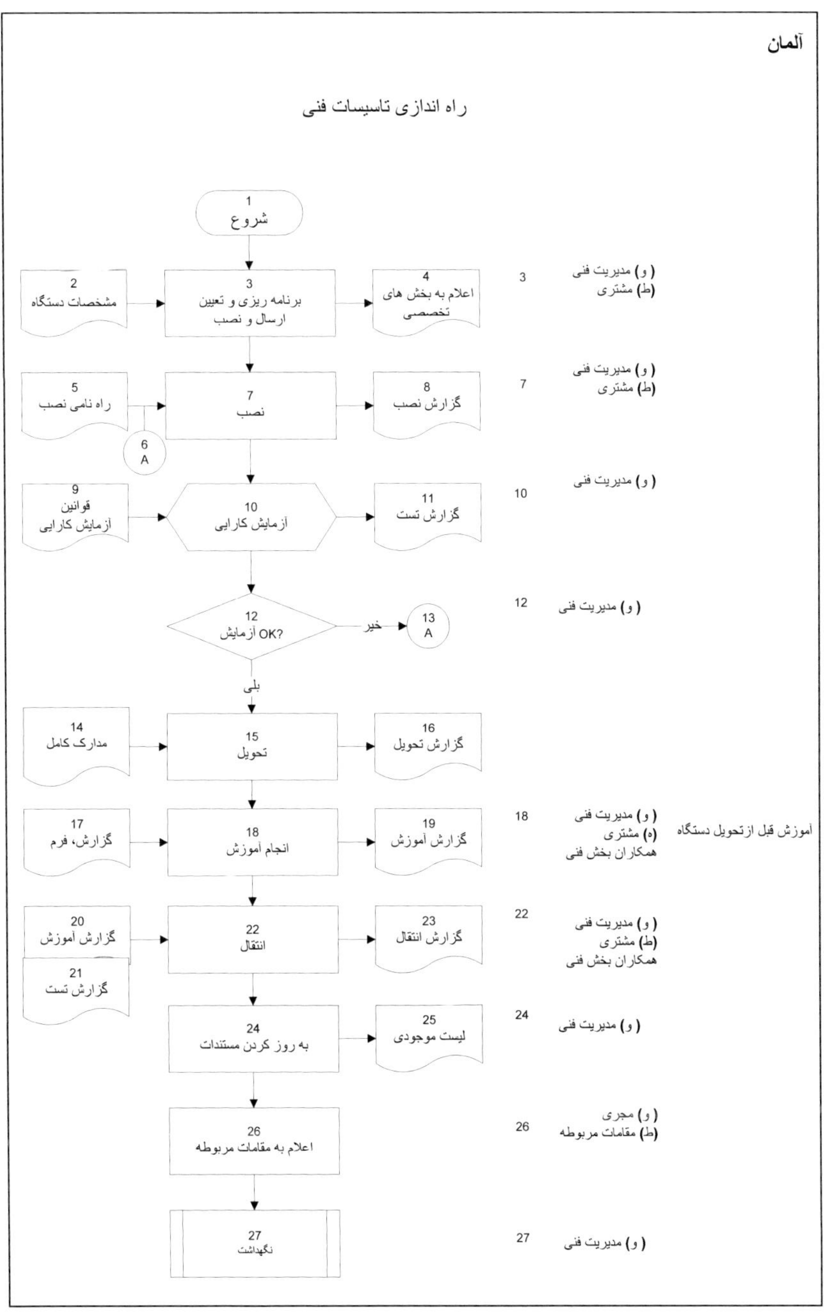
آلمان
راه اندازی تاسیسات فنی
1
شروع
2
مشخصات دستگاه
3
برنامه ریزی و تعیین ارسال و نصب
4
اعلام به بخش های تخصصی
3
(و) مدیریت فنی
(ط) مشتری
5
راه نامی نصب
6
A
7
نصب
8
گزارش نصب
7
(و) مدیریت فنی
(ط) مشتری
9
قوانین آزمایش کارایی
10
آزمایش کارایی
11
گزارش تست
10
(و) مدیریت فنی
12
آزمایش OK?
خیر
13
A
بلی
12
(و) مدیریت فنی
14
مدارک کامل
15
تحویل
16
گزارش تحویل
17
گزارش، فرم
18
انجام آموزش
19
گزارش آموزش
18
(و) مدیریت فنی
(ه) مشتری
همکاران بخش فنی
آموزش قبل از تحویل دستگاه
20
گزارش آموزش
21
گزارش تست
22
انتقال
23
گزارش انتقال
22
(و) مدیریت فنی
(ط) مشتری
همکاران بخش فنی
24
به روز کردن مستندات
25
لیست موجودی
24
(و) مدیریت فنی
26
اعلام به مقامات مربوطه
26
(و) مجری
(ط) مقامات مربوطه
27
نگهداشت
27
(و) مدیریت فنی

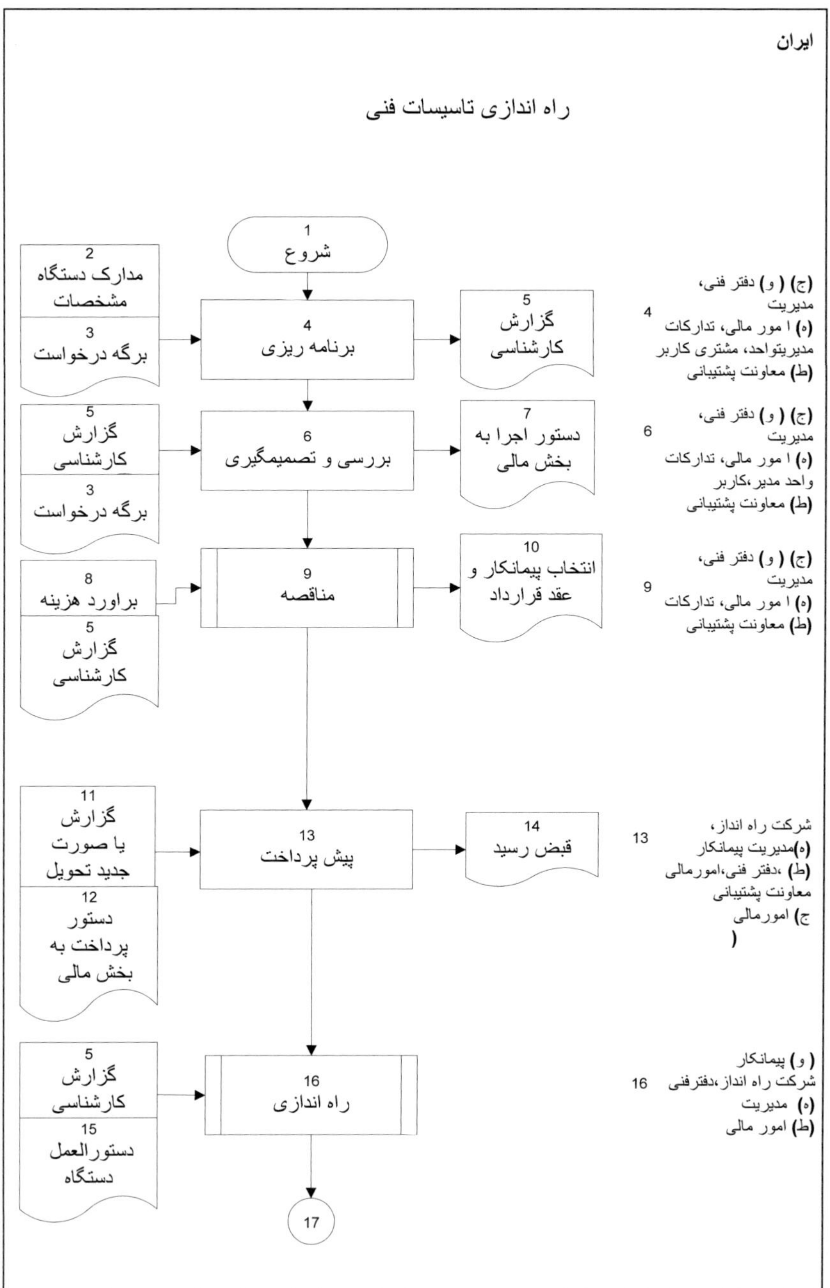
ایران
راه اندازی تاسیسات فنی
1 شروع
2 مدارک دستگاه مشخصات
3 برگه درخواست
4 برنامه ریزی
5 گزارش کارشناسی
(ج) (و) دفتر فنی، مدیریت
(ه) ا امور مالی، تدارکات مدیریتواحد، مشتری کاربر
(ط) معاونت پشتیبانی
4
5 گزارش کارشناسی
3 برگه درخواست
6 بررسی و تصمیمگیری
7 دستور اجرا به بخش مالی
(ج) (و) دفتر فنی، مدیریت
(ه) ا امور مالی، تدارکات واحد مدیر،کاربر
(ط) معاونت پشتیبانی
6
8 براورد هزینه
5 گزارش کارشناسی
9 مناقصه
10 انتخاب پیمانکار و عقد قرارداد
(ج) (و) دفتر فنی، مدیریت
(ه) ا امور مالی، تدارکات
(ط) معاونت پشتیبانی
9
11 گزارش یا صورت جدید تحویل
12 دستور پرداخت به بخش مالی
13 پیش پرداخت
14 قبض رسید
شرکت راه انداز،
(ه)مدیریت پیمانکار
(ط) ،دفتر فنی،امورمالی معاونت پشتیبانی
ج) امورمالی
(
13
5 گزارش کارشناسی
15 دستورالعمل دستگاه
16 راه اندازی
(و) پیمانکار
شرکت راه انداز ،دفترفنی
(ه) مدیریت
(ط) امور مالی
16
17

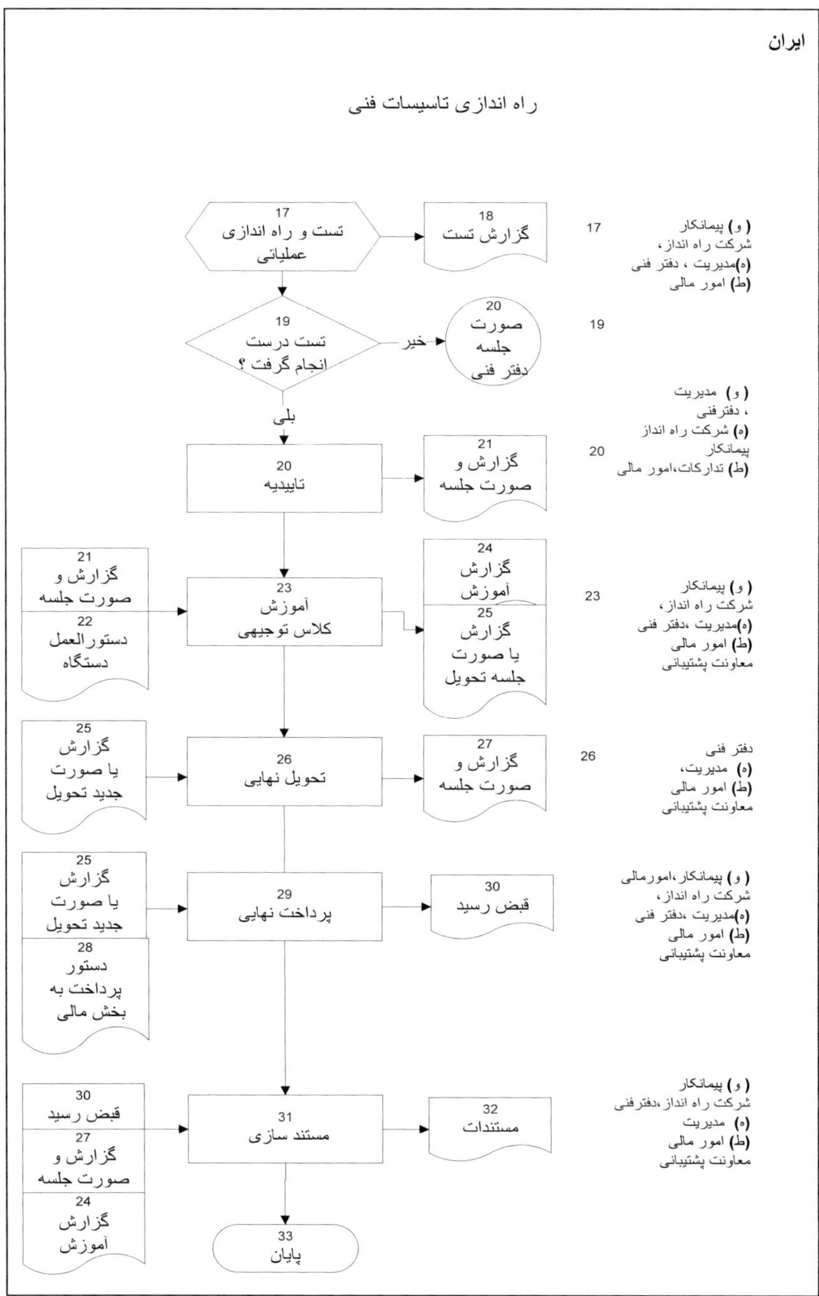
ایران
راه اندازی تاسیسات فنی
17
تست و راه اندازی عملیاتی
18
گزارش تست
17
(و) پیمانکار
شرکت راه انداز ،
(ه)مدیریت ، دفتر فنی
(ط) امور مالی
19
تست درست انجام گرفت ؟
خیر
20
صورت جلسه دفتر فنی
19
بلی
(و) مدیریت
، دفترفنی
(ه) شرکت راه انداز
پیمانکار
(ط) تدارکات،امور مالی
20
تاییدیه
21
گزارش و صورت جلسه
20
21
گزارش و صورت جلسه
22
دستورالعمل دستگاه
23
آموزش کلاس توجیهی
24
گزارش آموزش
25
گزارش یا صورت جلسه تحویل
23
(و) پیمانکار
شرکت راه انداز ،
(ه)مدیریت ،دفتر فنی
(ط) امور مالی
معاونت پشتیبانی
25
گزارش یا صورت جدید تحویل
26
تحویل نهایی
27
گزارش و صورت جلسه
26
دفتر فنی
(ه) مدیریت،
(ط) امور مالی
معاونت پشتیبانی
25
گزارش یا صورت جدید تحویل
28
دستور پرداخت به بخش مالی
29
پرداخت نهایی
30
قبض رسید
(و) پیمانکار ،امورمالی
شرکت راه انداز ،
(ه)مدیریت ،دفتر فنی
(ط) امور مالی
معاونت پشتیبانی
30
قبض رسید
27
گزارش و صورت جلسه
24
گزارش آموزش
31
مستند سازی
32
مستندات
(و) پیمانکار
شرکت راه انداز ،دفترفنی
(ه) مدیریت
(ط) امور مالی
معاونت پشتیبانی
33
پایان

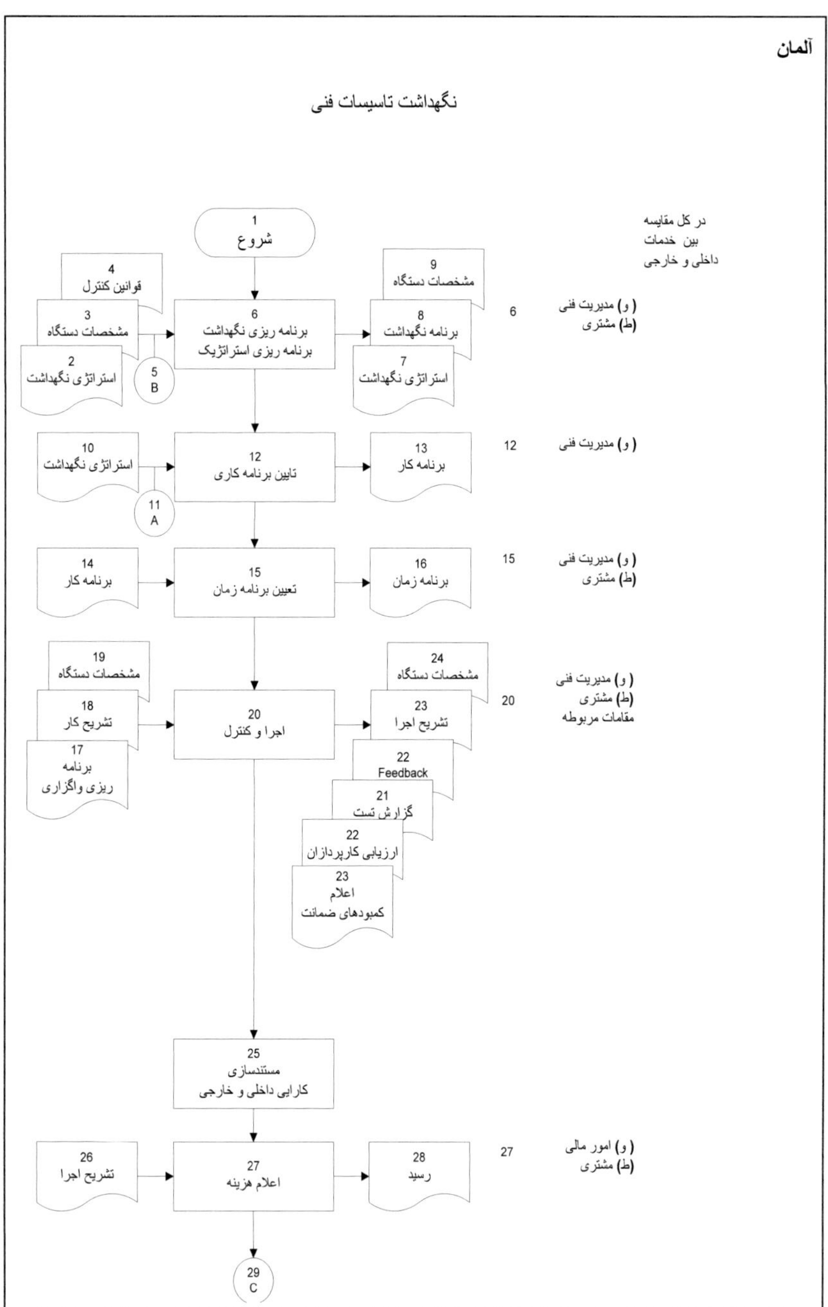

آلمان
نگهداشت تاسیسات فنی
1 شروع
در کل مقایسه بین خدمات داخلی و خارجی
4 قوانین کنترل
3 مشخصات دستگاه
2 استراتژی نگهداشت
5 B
6 برنامه ریزی نگهداشت برنامه ریزی استراتژیک
9 مشخصات دستگاه
8 برنامه نگهداشت
7 استراتژی نگهداشت
6
(و) مدیریت فنی
(ط) مشتری
10 استراتژی نگهداشت
11 A
12 تایین برنامه کاری
13 برنامه کار
12
(و) مدیریت فنی
14 برنامه کار
15 تعیین برنامه زمان
16 برنامه زمان
15
(و) مدیریت فنی
(ط) مشتری
19 مشخصات دستگاه
18 تشریح کار
17 برنامه ریزی واگزاری
20 اجرا و کنترل
24 مشخصات دستگاه
23 تشریح اجرا
22 Feedback
21 گزارش تست
22 ارزیابی کارپردازان
23 اعلام کمبودهای ضمانت
20
(و) مدیریت فنی
(ط) مشتری
مقامات مربوطه
25 مستندسازی کارایی داخلی و خارجی
26 تشریح اجرا
27 اعلام هزینه
28 رسید
27
(و) امور مالی
(ط) مشتری
29 C

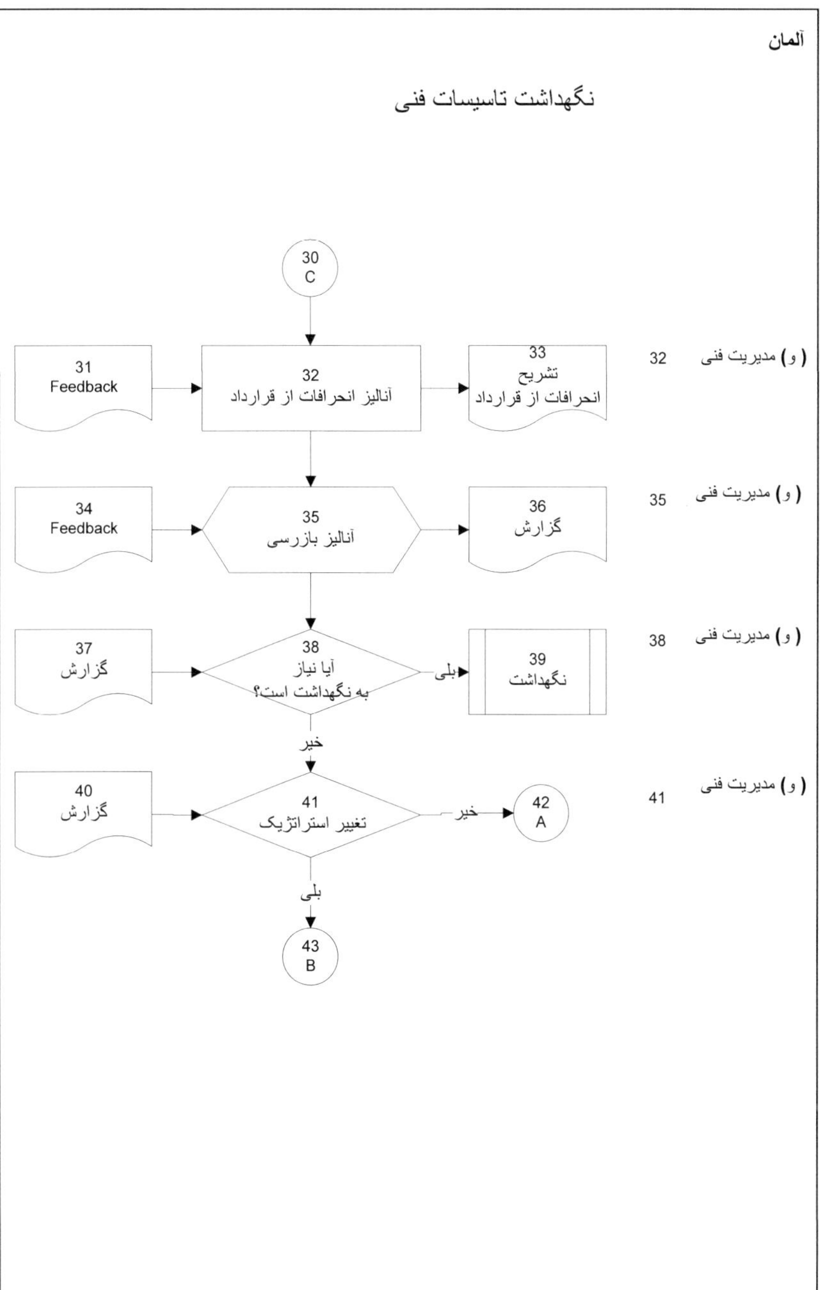
آلمان

نگهداشت تاسیسات فنی

30
C

31
Feedback

32
آنالیز انحرافات از قرارداد

33
تشریح
انحرافات از قرارداد

(و) مدیریت فنی   32

34
Feedback

35
آنالیز بازرسی

36
گزارش

(و) مدیریت فنی   35

37
گزارش

38
آیا نیاز
به نگهداشت است؟

بلی

39
نگهداشت

(و) مدیریت فنی   38

خیر

40
گزارش

41
تغییر استراتژیک

خیر

42
A

(و) مدیریت فنی   41

بلی

43
B

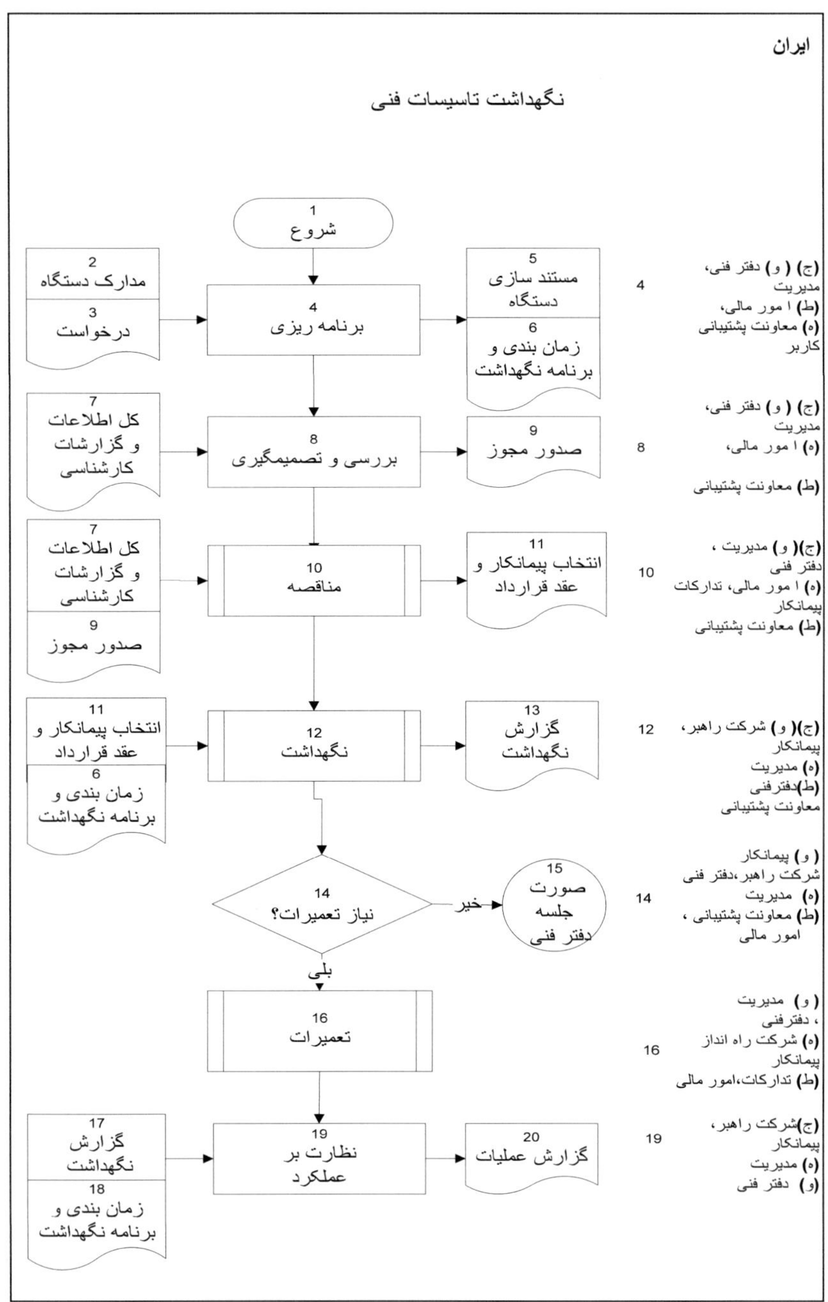
ایران
نگهداشت تاسیسات فنی
۱ شروع
۲ مدارک دستگاه
۳ درخواست
۴ برنامه ریزی
۵ مستند سازی دستگاه
۶ زمان بندی و برنامه نگهداشت
۷ کل اطلاعات و گزارشات کارشناسی
۸ بررسی و تصمیمگیری
۹ صدور مجوز
۷ کل اطلاعات و گزارشات کارشناسی
۹ صدور مجوز
۱۰ مناقصه
۱۱ انتخاب پیمانکار و عقد قرارداد
۱۱ انتخاب پیمانکار و عقد قرارداد
۶ زمان بندی و برنامه نگهداشت
۱۲ نگهداشت
۱۳ گزارش نگهداشت
۱۴ نیاز تعمیرات؟
خیر
۱۵ صورت جلسه دفتر فنی
بلی
۱۶ تعمیرات
۱۷ گزارش نگهداشت
۱۸ زمان بندی و برنامه نگهداشت
۱۹ نظارت بر عملکرد
۲۰ گزارش عملیات
(ج) (و) دفتر فنی، مدیریت
(ط) ا مور مالی،
(ه) معاونت پشتیبانی کاربر
۴
(ج) (و) دفتر فنی، مدیریت
(ه) ا مور مالی،
(ط) معاونت پشتیبانی
۸
(ج)(و) مدیریت ، دفتر فنی
(ه) ا مور مالی، تدارکات پیمانکار
(ط) معاونت پشتیبانی
۱۰
(ج)(و) شرکت راهبر، پیمانکار
(ه) مدیریت
(ط)دفترفنی معاونت پشتیبانی
۱۲
(و) پیمانکار شرکت راهبر،دفتر فنی
(ه) مدیریت
(ط) معاونت پشتیبانی ، امور مالی
۱۴
(و) مدیریت ، دفترفنی
(ه) شرکت راه انداز پیمانکار
(ط) تدارکات،امور مالی
۱۶
(ج)شرکت راهبر، پیمانکار
(ه) مدیریت
(و) دفتر فنی
۱۹

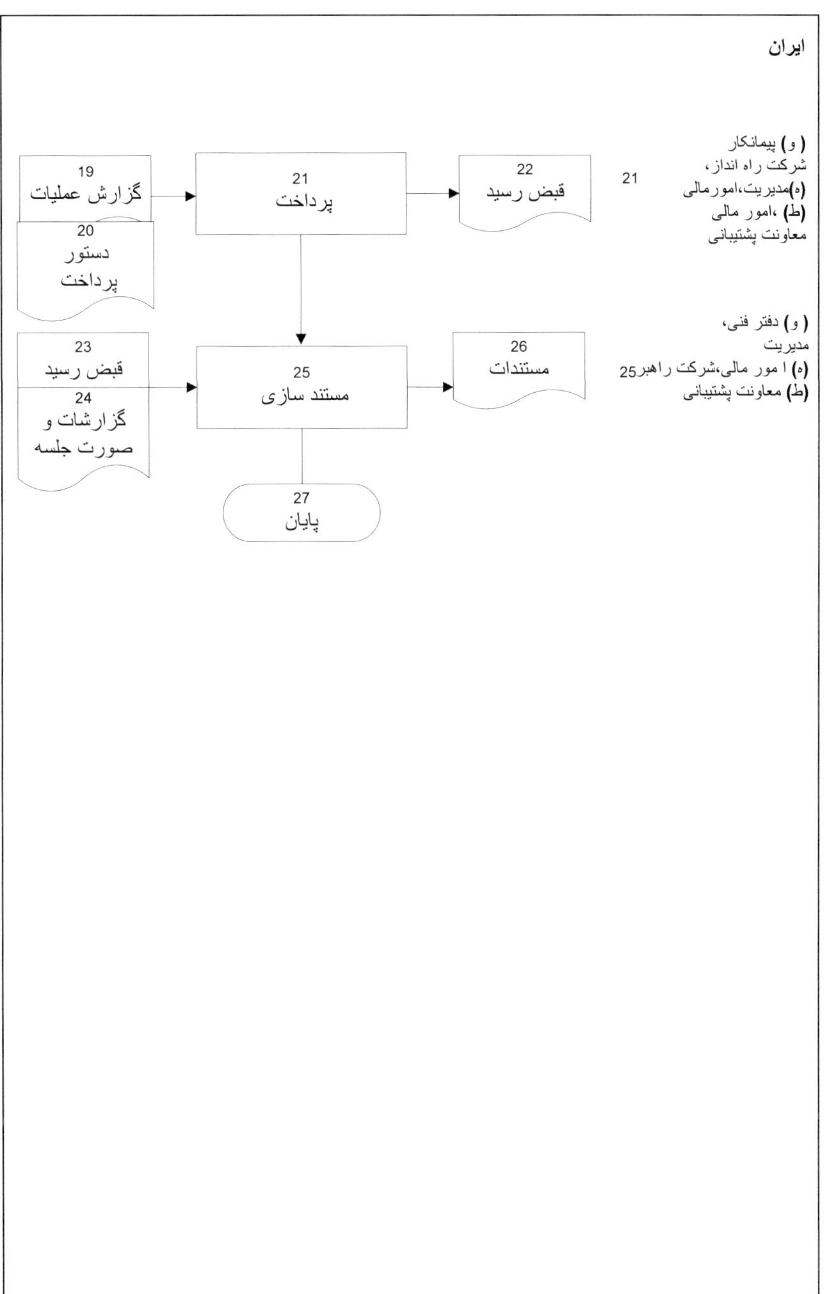
ایران

19
گزارش عملیات

20
دستور
پرداخت

21
پرداخت

22
قبض رسید

23
قبض رسید

24
گزارشات و
صورت جلسه

25
مستند سازی

26
مستندات

27
پایان

21

(و) پیمانکار
شرکت راه انداز،
(ه)مدیریت،امورمالی
(ط) ،امور مالی
معاونت پشتیبانی

(و) دفتر فنی،
مدیریت
(ه) ا مور مالی،شرکت راهبر25
(ط) معاونت پشتیبانی

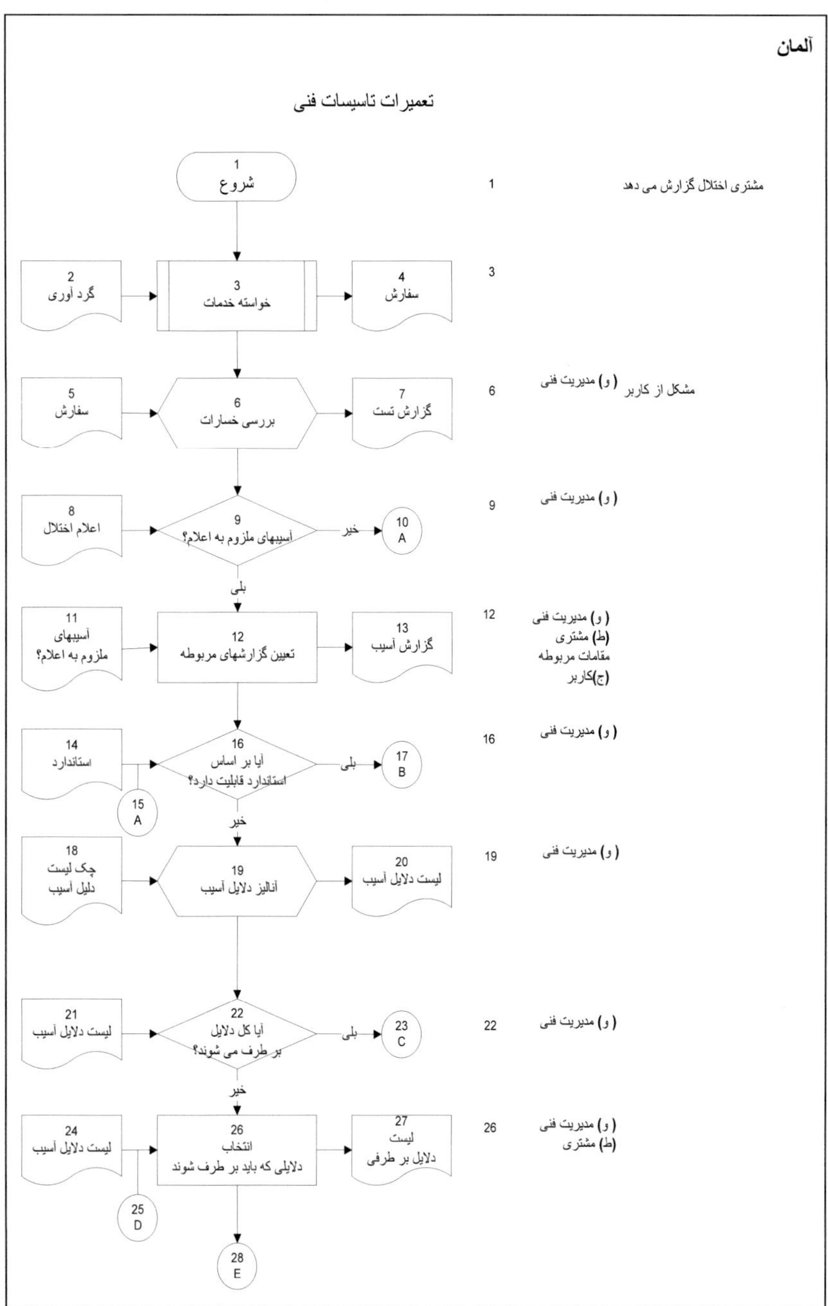
آلمان

تعمیرات تاسیسات فنی

1
شروع

1
مشتری اختلال گزارش می دهد

2
گرد آوری

3
خواسته خدمات

4
سفارش

3

5
سفارش

6
بررسی خسارات

7
گزارش تست

6
(و) مدیریت فنی   مشکل از کاربر

8
اعلام اختلال

9
آسیبهای ملزوم به اعلام؟

خیر

10
A

9
(و) مدیریت فنی

بلی

11
آسیبهای ملزوم به اعلام؟

12
تعیین گزارشهای مربوطه

13
گزارش آسیب

12
(و) مدیریت فنی
(ط) مشتری
مقامات مربوطه
(ج)کاربر

14
استاندارد

16
آیا بر اساس استاندارد قابلیت دارد؟

بلی

17
B

16
(و) مدیریت فنی

15
A

خیر

18
چک لیست دلیل آسیب

19
آنالیز دلایل آسیب

20
لیست دلایل آسیب

19
(و) مدیریت فنی

21
لیست دلایل آسیب

22
آیا کل دلایل بر طرف می شوند؟

بلی

23
C

22
(و) مدیریت فنی

خیر

24
لیست دلایل آسیب

26
انتخاب دلایلی که باید بر طرف شوند

27
لیست دلایل بر طرفی

26
(و) مدیریت فنی
(ط) مشتری

25
D

28
E

آلمان

تعمیرات تاسیسات فنی

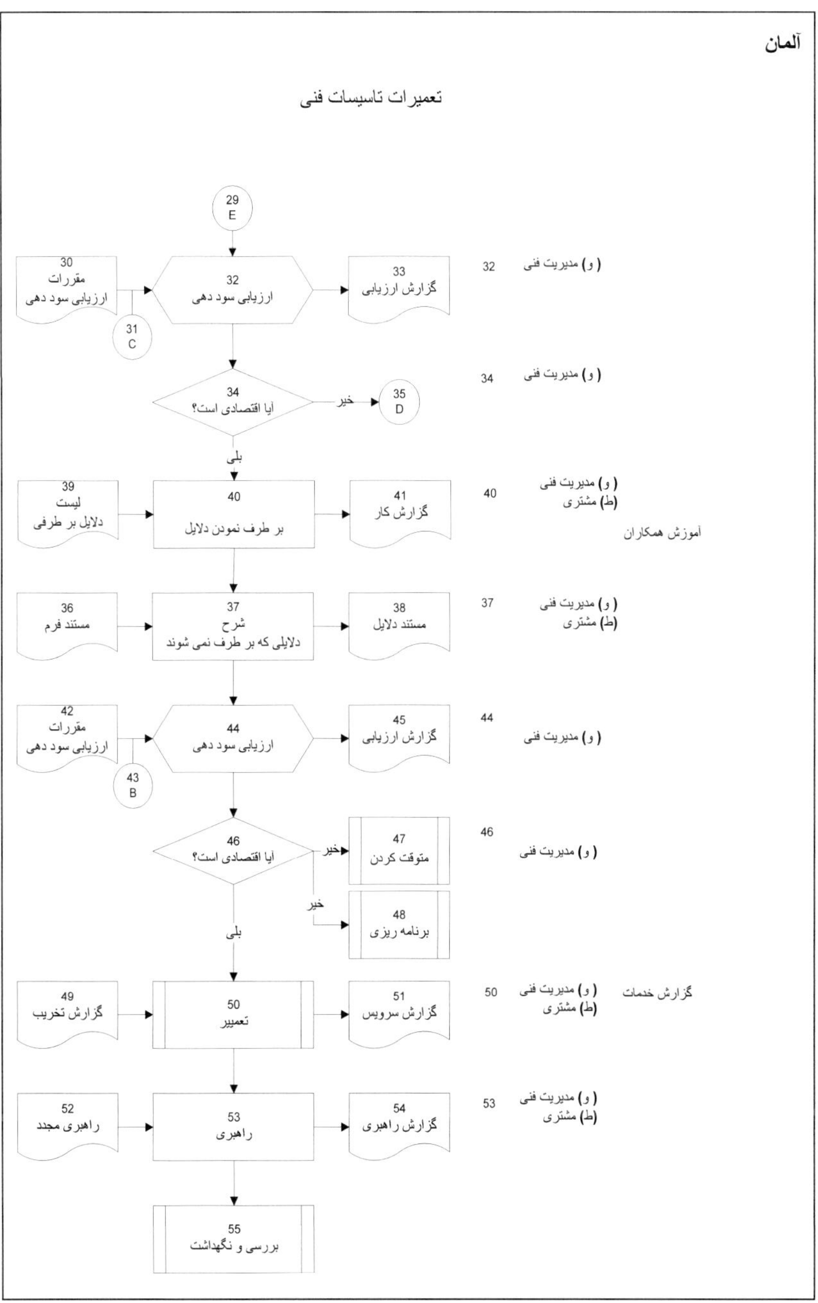

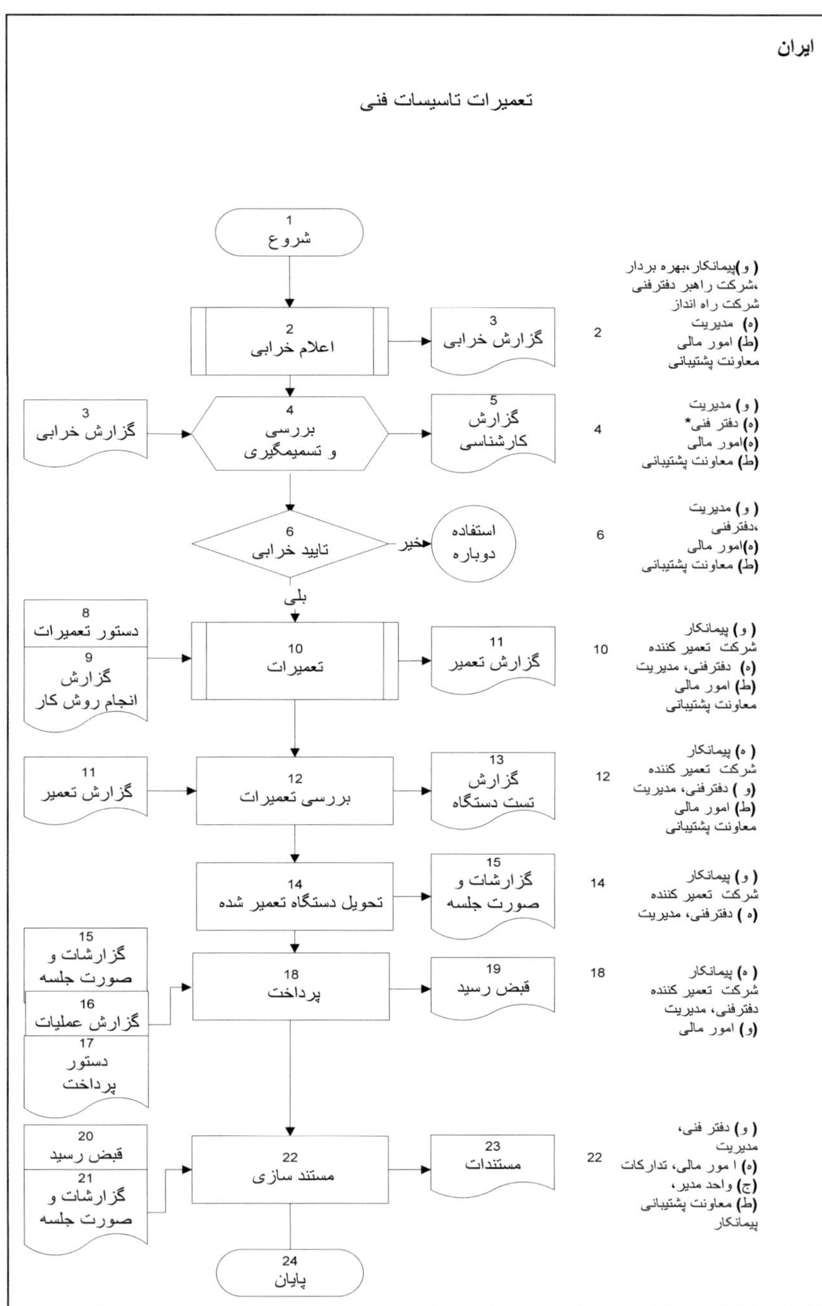
ایران
تعمیرات تاسیسات فنی
1 شروع
2 اعلام خرابی
3 گزارش خرابی
4 بررسی و تسمیمگیری
5 گزارش کارشناسی
3 گزارش خرابی
6 تایید خرابی
خیر
استفاده دوباره
بلی
8 دستور تعمیرات
9 گزارش انجام روش کار
10 تعمیرات
11 گزارش تعمیر
11 گزارش تعمیر
12 بررسی تعمیرات
13 گزارش تست دستگاه
14 تحویل دستگاه تعمیر شده
15 گزارشات و صورت جلسه
15 گزارشات و صورت جلسه
16 گزارش عملیات
17 دستور پرداخت
18 پرداخت
19 قبض رسید
20 قبض رسید
21 گزارشات و صورت جلسه
22 مستند سازی
23 مستندات
24 پایان
(و)پیمانکار ،بهره بردار، شرکت راهبر دفترفنی شرکت راه انداز
(ه) مدیریت
(ط) امور مالی
معاونت پشتیبانی
2
(و) مدیریت
(ه) دفتر فنی*
(ه)امور مالی
(ط) معاونت پشتیبانی
4
(و) مدیریت
،دفترفنی
(ه)امور مالی
(ط) معاونت پشتیبانی
6
(و) پیمانکار
شرکت تعمیر کننده
(ه) دفترفنی، مدیریت
(ط) امور مالی
معاونت پشتیبانی
10
(ه) پیمانکار
شرکت تعمیر کننده
(و) دفترفنی، مدیریت
(ط) امور مالی
معاونت پشتیبانی
12
(و) پیمانکار
شرکت تعمیر کننده
(ه) دفترفنی، مدیریت
14
(ه) پیمانکار
شرکت تعمیر کننده
دفترفنی، مدیریت
(و) امور مالی
18
(و) دفتر فنی،
مدیریت
(ه ا امور مالی، تدارکات
(ج) واحد مدیر،
(ط) معاونت پشتیبانی
پیمانکار
22

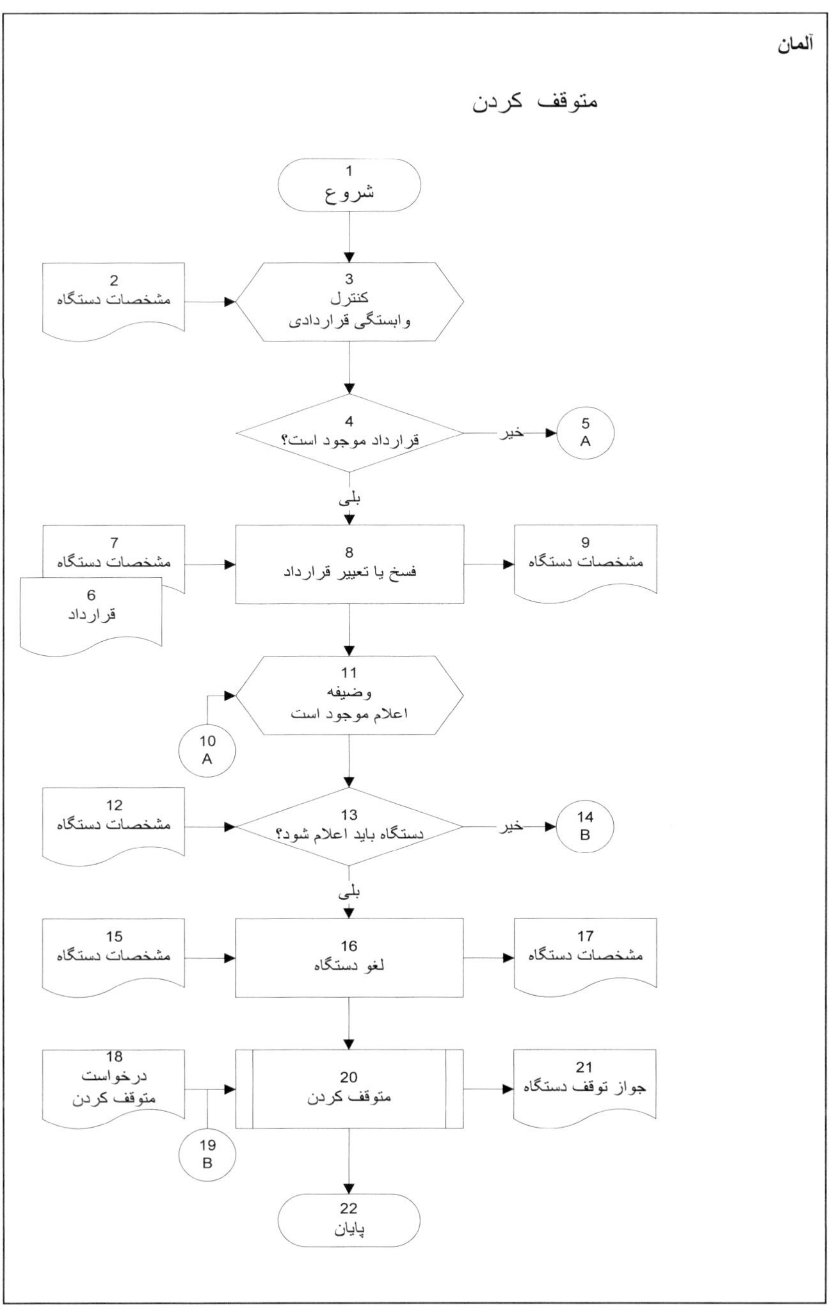
آلمان

متوقف کردن

1
شروع

2
مشخصات دستگاه

3
کنترل
وابستگی قراردادی

4
قرارداد موجود است؟

خیر

5
A

بلی

7
مشخصات دستگاه

6
قرارداد

8
فسخ یا تغییر قرارداد

9
مشخصات دستگاه

11
وظیفه
اعلام موجود است

10
A

12
مشخصات دستگاه

13
دستگاه باید اعلام شود؟

خیر

14
B

بلی

15
مشخصات دستگاه

16
لغو دستگاه

17
مشخصات دستگاه

18
درخواست
متوقف کردن

19
B

20
متوقف کردن

21
جواز توقف دستگاه

22
پایان

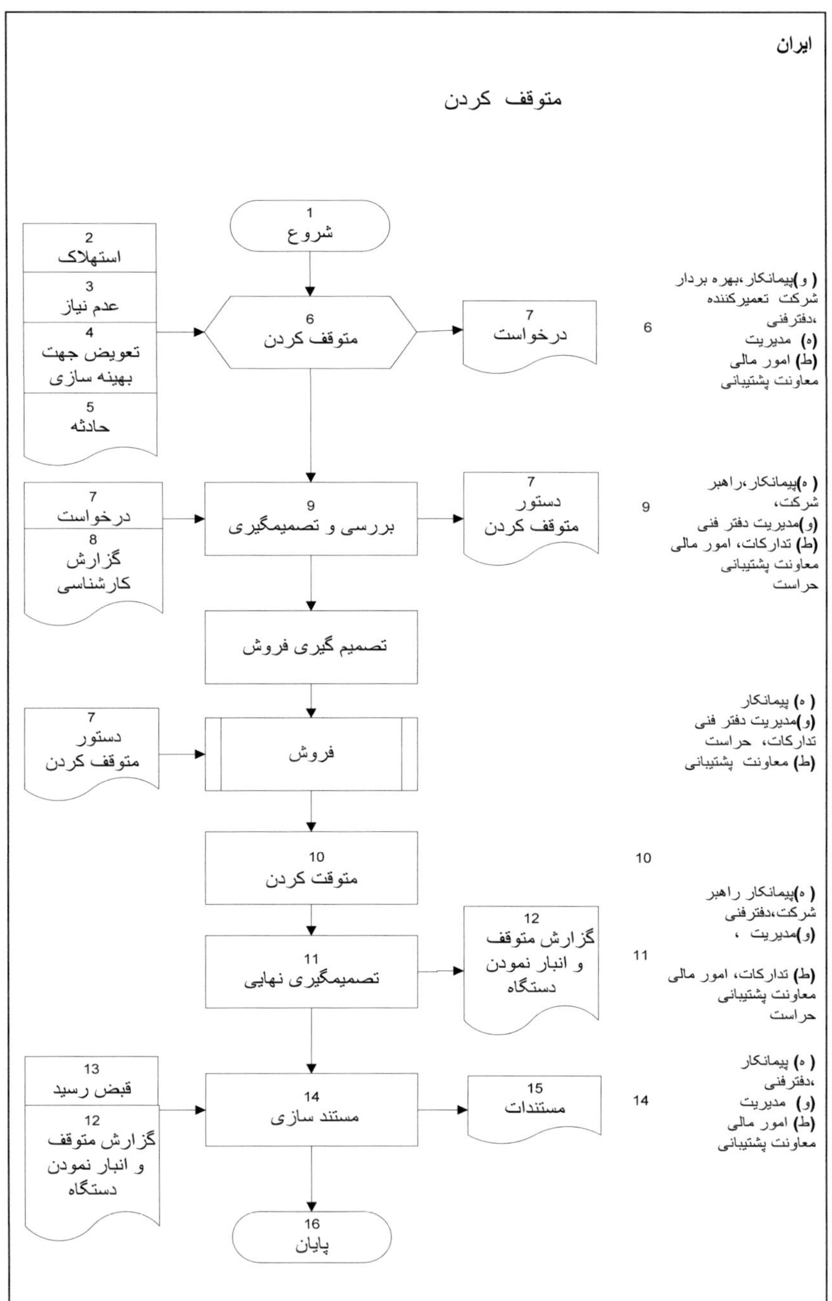
ایران
متوقف کردن
1
شروع
2
استهلاک
3
عدم نیاز
4
تعویض جهت بهینه سازی
5
حادثه
6
متوقف کردن
7
درخواست
6
(و)پیمانکار ،بهره بردار
شرکت تعمیرکننده ،دفترفنی
(ه) مدیریت
(ط) امور مالی
معاونت پشتیبانی
7
درخواست
8
گزارش کارشناسی
9
بررسی و تصمیمگیری
7
دستور متوقف کردن
9
(ه)پیمانکار،راهبر شرکت،
(و)مدیریت دفتر فنی
(ط) تدارکات، امور مالی
معاونت پشتیبانی
حراست
تصمیم گیری فروش
7
دستور متوقف کردن
فروش
(ه) پیمانکار
(و)مدیریت دفتر فنی
تدارکات، حراست
(ط) معاونت پشتیبانی
10
متوقت کردن
10
11
تصمیمگیری نهایی
12
گزارش متوقف و انبار نمودن دستگاه
11
(ه)پیمانکار راهبر شرکت،دفترفنی
(و)مدیریت ،
(ط) تدارکات، امور مالی
معاونت پشتیبانی
حراست
13
قبض رسید
12
گزارش متوقف و انبار نمودن دستگاه
14
مستند سازی
15
مستندات
14
(ه) پیمانکار ،دفترفنی
(و) مدیریت
(ط) امور مالی
معاونت پشتیبانی
16
پایان

APPENDIX - A14

Interfaces of the Process - Technical Facilities

Deutschland

Prozess	Prozessschritt	Kostenfaktor[1]	Wertung*	Wertung Zahlenwert	Qualitätsfaktor[2]	Wertung*	Wertung Zahlenwert	Zeitfaktor[3]	Wertung*	Wertung Zahlenwert
Instandhaltung allgem.		Instandhaltungskosten bezogen auf Wiederbeschaffungswert Anschaffungskosten bezogen auf Anzahl der Anwendungen								
	Planung	Planungskosten bezogen auf Investitionskosten	B	1,9	Anzahl notwendiger Nachplanungen Verhältnis geplante Kosten zu tatsächliche Kosten Verhältnis geplante Ausführungszeit/tatsächliche Zeit	A	1,1	Dauer der Planung	B	2,2
	Beschaffung	Anschaffungskosten bezogen auf Pflegetage Preis pro Gerät (gerätespezifisch)	B	1,6	Anzahl der beanstandeten Beschaffungen	B	1,9	Dauer der Beschaffung	C	3,0
	Inbetriebnahme	Schulungsaufwand pro Gerät (gerätespezifisch)	C	2,6	Anzahl der fehlgeschlagenen Testläufe	B	2,1	Dauer der Inbetriebnahme	C	3,0
	Betrieb	Energieverbrauch pro Anwendung Materialkosten pro Anwendung	A	1,3	Auslastung der Geräte	A	1,3	Dauer für den Betrieb	B	1,5
	Wartung und Inspektion	Wartungskosten bezogen auf Wiederbeschaffungswert	A	1,4	Anzahl der meldepflichtigen Ereignisse	B	1,6	Dauer der Wartung	B	1,5
	Instandsetzung	Instandsetzungskosten bezogen auf Wiederbeschaffungswert	A	1,3	Dauer der Instandsetzung	B	1,6	Dauer der Instandsetzung	B	1,5
	Stilllegung und Entsorgung	Entsorgungskosten bezogen auf Anschaffungswert	C	2,7	Anzahl der nicht dokumentierten Entsorgungen Dauer des Prozesses	C	2,9	Dauer	C	2,8
Planung										
	Festlegung Nutzungskonzept	Arbeitszeit	B	1,9	Anzahl der Nachbesserungen nach der Inbetriebnahme oder während der Bauphase aufgrund fehlerhaftem Konzept	B	1,6	Dauer	C	2,7
	Vorplanung	Arbeitszeit Honorarkosten	B	2,3	Anzahl der Nachbesserungen nach der Inbetriebnahme oder während der Bauphase aufgrund fehlerhaftem Konzept Grad der Kostengenauigkeit im Vergleich tatsächlicher Kosten	B	1,7	Dauer	B	2,5
	Entwurfsplanung	Arbeitszeit Honorarkosten	B	2,1	Anzahl der Nachbesserungen nach der Inbetriebnahme oder während der Bauphase aufgrund fehlerhaftem Konzept	B	1,7	Dauer	B	2,0
	Wirtschaftlichkeitsbetrachtung	Arbeitszeit Honorarkosten	C	2,6	Vergleich Soll/Istkosten	B	1,6	Dauer	C	2,8
	Genehmigungsplanung	Arbeitszeit Honorarkosten	B	2,3	Anzahl der Beanstandungen oder Nachträge Kosten für Nachbesserungen	B	2,3	Dauer	B	2,5
	Ausführungsplanung	Arbeitszeit Honorarkosten	B	1,9	Anzahl der fehlenden Unterlagen Anzahl der Nachfragen der ausführenden Unternehmen	B	2,0	Dauer	B	2,2
	Vorbereitung der Vergabe	Arbeitszeit Honorarkosten	B	2,3	Anzahl der Einsprüche aufgrund fehlerhafter Vergabe	B	2,4	Dauer	B	2,2
	Erstellung Kostenrahmen	Arbeitszeit Honorarkosten	B	2,3	Abweichung der Kosten vom Soll	B	2,0	Dauer	B	2,5
	Festlegung Instandhaltungsstrategien	Arbeitszeit	B	1,7	Instandhaltungskosten/Wiederbeschaffungswert Anzahl der Ausfälle Erreichung der geplanten Nutzungsdauer	A	1,4	Dauer	B	2,2

Deutschland

Prozess	Prozessschritt	Kostenfaktor[1]	Wertung*	Wertung Zahlenwert	Qualitätsfaktor[2]	Wertung*	Wertung Zahlenwert	Zeitfaktor[3]	Wertung*	Wertung Zahlenwert
Beschaffung										
	Beschaffungs-entscheidung	Abstimmungsaufwand (Zeit) pro Beschaffung	B	2,0	Anzahl der Kundenbeschwerden Vergleich der Folgekosten (Weiterbetrieb, Wiederbeschaffung, Neubeschaffung)	B	1,6	Dauer	B	2,5
	Ausschreibung	Arbeitszeit pro Ausschreibung	B	2,3	Anzahl der Rückfragen der Bieter Anzahl der Nachträge	B	1,7	Dauer	B	2,0
	Bestellung	Arbeitszeit pro Bestellung	B	2,4	Anzahl der Stornos Anzahl der Reklamationen	B	2,3	Dauer	C	2,7
	Lieferung	Arbeitszeit für Empfang der Lieferung bezogen auf Bestellwert	C	2,9	Anzahl der Nachlieferung Termintreue	B	1,9	Dauer	C	2,7
	Lieferantenbewertung	Arbeitszeit pro Analyse	B	2,4	Anzahl der Gewährleistungsmängel	B	1,9	Dauer	C	2,7
	Dokumentation der gerätebezogenen Vorgänge	Arbeitszeit pro Dokumentation	B	2,4	Anzahl der unvollständigen Datensätze nach Beschaffung	B	2,1	Dauer	B	2,3
Inbetriebnahme										
	Planung und Festlegung der Anlieferung und Installation	Planungskosten bezogen auf Investitionskosten	B	2,3	Anzahl der Terminüberschreitungen Kundenzufriedenheit Anzahl der Nachbesserungen	B	1,6	Dauer	B	2,3
	Installation	Arbeitszeit pro Installation bezogen auf Gesamtkosten Kosten pro Beschaffung	B	2,1	Dauer der Montage, Terminabweichungen Anzahl der Abnahmemängel Grad der Flexibilität	A	1,4	Dauer	B	2,0
	Funktionsprüfung	Arbeitszeit pro Funktionsprüfung bezogen auf Beschaffungswert	C	2,7	Anzahl der Wiederholungen Zielerreichungsgrad	B	2,0	Dauer	B	2,3
	Abnahme	Arbeitszeit pro Abnahme bezogen auf Gesamtkosten der Beschaffung	C	3,0	Anzahl der bei der Abnahme übersehenen Mängel	B	1,7	Dauer	B	2,5
	Durchführung von Einweisungen, Schulungen	Arbeitszeit pro Beschaffung, Schulungskosten pro Beschaffung	B	2,3	Anzahl Bedienfehler (evtl. Resultat Störung)	B	1,9	Dauer	B	2,0
	Übergabe	Arbeitszeit pro Beschaffung	C	2,7	Anzahl der nicht dokumentierten Übergaben und Einweisungen	B	2,3	Dauer	B	2,5
	Aktualisierung der Dokumentation	Arbeitszeit pro Beschaffung	C	2,6	Anzahl der Fehlaufträge Anzahl der Störungen Höhe der Verlängerung der Störungsbehebungszeit	B	1,6	Dauer	B	2,5
	Meldung an zuständige Behörden	Genehmigungsgebühren pro Beschaffung Arbeitszeit pro Meldung	C	2,6	Höhe der Bußgelder Anzahl der Rückfragen	B	2,4	Dauer	B	2,5

Deutschland

Prozess	Prozessschritt	Kostenfaktor[1]	Wertung*	Wertung Zahlenwert	Qualitätsfaktor[2]	Wertung*	Wertung Zahlenwert	Zeitfaktor[3]	Wertung*	Wertung Zahlenwert
Wartung und Inspektion										
	Instandhaltungsplanung / Strategieplanung	Arbeitszeit pro Ereignis/Auftrag Hard-/Softwarekosten je Ereignis	A	1,4	Verhältnis zwischen geplanten und ungeplanten Aufträgen	A	1,3	Dauer	B	2,2
	Arbeitsplanerstellung	Arbeitszeit pro Ereignis/Auftrag	B	1,6	Anzahl der Korrekturen pro Arbeitsplan Anzahl der Rückfragen	A	1,4	Dauer	B	2,2
	Terminplanung	Arbeitszeit pro Ereignis/Auftrag	B	1,9	Anzahl der Terminüberschreitungen Anzahl der Nachbesserungen Grad der Kapazitätsauslastung	A	1,1	Dauer	B	1,8
	Steuerung, Durchführung, Prüfung, Überwachung	Arbeitszeit pro Ereignis/Auftrag Materialkosten pro Auftrag Telefonkosten pro Auftrag Ausfallkosten pro Auftrag	B	1,6	Anzahl der Nachbesserungen Höhe der Ausfallzeiten Anzahl der Überschreitung der Vorgabezeiten	B	1,7	Dauer	B	1,7
	Dokumentation von eigen/fremd - Anteil	Arbeitszeit für Dokumentation pro Ereignis	B	2,4	Anzahl der nicht vollständig ausgefüllten Dokumentationen	B	1,9	Dauer	B	2,3
	Kostenübermittlung	Arbeitszeit pro Auftrag	C	2,7	Anzahl der unvollständigen Rechnungen Anzahl der Nachfragen	B	2,1	Dauer	B	2,5
	Auftragsabweichungsanalyse	Arbeitszeit pro Auftrag	C	2,6	Anzahl der Abweichungen je Auftrag (Gradient)	B	2,4	Dauer	C	2,7
	Inspektion analysieren	Arbeitszeit pro Auftrag	C	2,6	Anzahl der nicht rechtzeitig erkannten Instandsetzungen Anzahl der unvorhergesehenen Ausfälle und Störungen	B	2,0	Dauer	B	2,5
Instandsetzung										
	Schaden prüfen	Arbeitszeit pro Auftrag	A	1,4	Anzahl der eingeleiteten Fehlmaßnahmen	A	1,3	Dauer	B	1,8
	Zugehörige Berichte erstellen Durchführung entsp. Maßnahmen	Arbeitszeit pro Auftrag	B	2,4	Anzahl Beanstandungen durch zuständige Behörden	B	2,1	Dauer	B	2,5
	Ursachen analysieren	Arbeitszeit pro Auftrag Hard-/Softwarekosten je Auftrag	B	1,7	Anzahl der Fehldiagnosen	A	1,3	Dauer	B	1,7
	Auswahl der zu beseitigenden Ursachen	Arbeitszeit pro Auftrag	B	2,4	Anzahl der fehlgeschlagenen Behebungen	B	1,9	Dauer	B	2,5
	Wirtschaftlichkeit prüfen	Arbeitszeit für Prüfung pro Auftrag	B	2,0	Anzahl und relative Höhe der negativen Nachkalkulationen	B	1,7	Dauer	B	2,5
	Ursachen beseitigen	Arbeitszeit pro Auftrag	B	1,6	Anzahl der Wiederholungen gleicher Ursachen	A	1,4	Dauer	B	1,7
	zunächst nicht zu beseitigende Ursachen dokumentieren	Arbeitszeit pro Auftrag	C	2,6	Anzahl nicht dokumentierter Ursachen	B	2,3	Dauer	B	2,3
	Instandsetzen	Arbeitszeit pro Auftrag Materialkosten pro Auftrag Hard-/Softwarekosten je Auftrag	A	1,0	Anzahl der Nachbesserungen Anzahl der Reklamationen durch Kunden (Kundenzufriedenheit)	A	1,0	Dauer	B	1,5
	Wiederinbetriebnahme	Arbeitszeit pro Auftrag	B	2,3	Höhe der Stillstands- oder Ausfallzeit	A	1,1	Dauer	B	2,0
	Schaden prüfen	Arbeitszeit pro Auftrag	A	1,4	Anzahl der eingeleiteten Fehlmaßnahmen	A	1,3	Dauer	B	1,8
	Zugehörige Berichte erstellen Durchführung entsp. Maßnahmen	Arbeitszeit pro Auftrag	B	2,4	Anzahl Beanstandungen durch zuständige Behörden	B	2,1	Dauer	B	2,5
	Ursachen analysieren	Arbeitszeit pro Auftrag Hard-/Softwarekosten je Auftrag	B	1,7	Anzahl der Fehldiagnosen	A	1,3	Dauer	B	1,7
	Auswahl der zu beseitigenden Ursachen	Arbeitszeit pro Auftrag	B	2,4	Anzahl der fehlgeschlagenen Behebungen	B	1,9	Dauer	B	2,5
	Wirtschaftlichkeit prüfen	Arbeitszeit für Prüfung pro Auftrag	B	2,0	Anzahl und relative Höhe der negativen Nachkalkulationen	B	1,7	Dauer	B	2,5
	Ursachen beseitigen	Arbeitszeit pro Auftrag	B	1,6	Anzahl der Wiederholungen gleicher Ursachen	A	1,4	Dauer	B	1,7
	zunächst nicht zu beseitigende Ursachen dokumentieren	Arbeitszeit pro Auftrag	C	2,6	Anzahl nicht dokumentierter Ursachen	B	2,3	Dauer	B	2,3
	Instandsetzen	Arbeitszeit pro Auftrag Materialkosten pro Auftrag Hard-/Softwarekosten je Auftrag	A	1,0	Anzahl der Nachbesserungen Anzahl der Reklamationen durch Kunden (Kundenzufriedenheit)	A	1,0	Dauer	B	1,5
	Wiederinbetriebnahme	Arbeitszeit pro Auftrag	B	2,3	Höhe der Stillstands- oder Ausfallzeit	A	1,1	Dauer	B	2,0

Deutschland

Prozess	Prozessschritt	Kostenfaktor[1]	Wertung*	Wertung Zahlenwert	Qualitätsfaktor[2]	Wertung*	Wertung Zahlenwert	Zeitfaktor[3]	Wertung*	Wertung Zahlenwert
Stilllegung/ Entsorgung										
	Vertragliche Bindungen prüfen	Arbeitszeit	B	2,1	Anzahl der Rechtsfehler Höhe der Blindkosten	B	2,0	Dauer	C	2,7
	Verträge kündigen / umschreiben	Arbeitszeit Gebührenkosten (Anwalt)	B	2,4	Anzahl der Rechtsauseinandersetzungen	B	2,0	Dauer	C	2,7
	Meldepflicht prüfen	Arbeitszeit	C	2,6	Anzahl der Bußgeldbescheide Anzahl der Beanstandungen	B	2,0	Dauer	C	2,7
	Gerät bei Behörden abmelden	Arbeitszeit	C	2,9	Anzahl der Bußgeldbescheide Anzahl der Beanstandungen	B	2,3	Dauer	C	3,0
	Entsorgung	Entsorgungskosten Arbeitszeit	B	1,9	Anzahl der Bußgeldbescheide Anzahl der Beanstandungen	B	2,1	Dauer	B	2,2

Iran

Prozess	Prozessschritt	Kostenfaktor	Wertung	Qualitätsfaktor	Wertung	Zeitfaktor	Wertung
Planung und Beschaffung	Anfrage / Produktwahl	Ermittlung der Arbeitsstunden, Abstimmungszeit	C	Anzahl und Genauigkeit der Geräteinformationen	C	Arbeitszeit	B
	Analyse und Entscheidung	Arbeitszeit diverser Abteilungen	B	Anzahl der Reklamationen	B	Arbeitszeit	A
	Ausschreibung	Arbeitszeit	B	Anzahl und Genauigkeit der Ausschreibungs-information	A	Dauer	A
	Bestellung	Bestellungszeit	B	Anzahl der fehlerhaften Bestellungen, Anzahl der Reklamationen	B	Dauer	A
	Gerätelieferung	Arbeitszeit der Annahme der Lieferung bezogen auf Bestellwert	B	Anzahl der Nachlieferungen	B	Dauer	B
	Zahlung	Termintreue, Arbeitszeit	B	Anzahl der Geräteunterlagen und Anzahl der notwendigen Genehmigungen	B	Dauer	A
	Dokumentation	Arbeitszeit pro Dokumentation	B	Anzahl unvollständiger Dokumente	B	Arbeitszeit	B
Inbetriebnahme	Planung	Arbeitszeit	B	Anzahl und Genauigkeit des Inbetriebzu-nehmenden Geräts	B	Arbeitszeit	B
	Analyse und Entscheidung	Koordinierung unterschiedlicher Einheiten, Planungskosten bezogen auf Investitionskosten	B	Anzahl der Reklamationen	B	Arbeitszeit	A
	Ausschreibung	Arbeitszeit	B	Anzahl und Genauigkeit der Ausschreibungs-information	B	Dauer	A
	Anzahlung	Arbeitszeit	B			Dauer	A
	Inbetriebnahme	Arbeitszeit pro Installation bezogen auf Gesamtbeschaffungskosten	A	Installationszeit, Terminabweichungen , Anzahl der nicht behobenen Mängel	A	Dauer	A
	Testbetrieb, Prüfung	Arbeitszeit pro Test bezogen auf Investitionswert	B	Anzahl der Test Wiederholungen, Anpassungsgrad an Normen	B	Dauer	B
	Übergabe an Abteilung	Arbeitszeit pro Übergabe bezogen auf Gesamtkosten der Beschaffung	B	Anzahl der fehlerhaften Tests	B	Dauer	B
	Schulungen	Schulungskosten bezogen auf Gerätekosten	A	Anzahl der anwenderbedingten Fehler oder Störungen	B	Dauer	B
	Endübergabe	Arbeitszeit bezogen auf Gesamtkosten der Beschaffung	B	Anzahl der unvollständigen Unterlagen	B	Dauer	B
	Abrechnung	Arbeitszeit	B			Dauer	B
	Dokumentation	Arbeitszeit pro Dokumentation	B	Anzahl der unvollständigen, fehlerhaften Dokumente	B	Dauer	B
Wartung und Inspektion	Planung der Wartung und Inspektion	Arbeitszeit	B	Anzahl und Genauigkeit des zu wartenden Geräts	B	Arbeitszeit	B
	Analyse und Entscheidung	Koordination verschiedener Einheiten, Planungskosten bezogen auf Investitionskosten	A	Anzahl der Reklamationen	B	Arbeitszeit	A
	Ausschreibung	Arbeitszeit	B	Anzahl und Genauigkeit der Ausschreibungs-information, Anzahl der an NU vergebenen Aufträge	B	Dauer	B
	Wartung	Wartungszeit bezogen auf Gesamtkosten der Beschaffung	C	Arbeitszeit pro Inspektion, Anzahl der zu behebenden Mängel	B	Dauer	B

Iran

Prozess	Prozessschritt	Kostenfaktor	Wertung	Qualitätsfaktor	Wertung	Zeitfaktor	Wertung
	Überwachung, Steuerung	Arbeitszeit pro Wartung bezogen auf Beschaffungswert	B	Anzahl der Wartung pro Vertragsauer, Anpassungsgrad an Normbedingungen	B	Dauer	B
	Dokumentation	Arbeitszeit pro Dokumentation	B	Anzahl der unvollständigen, fehlerhaften Dokumente	B	Dauer	B
Instandsetzung	Schadensanalyse und -bericht	Arbeitszeit	C	Anzahl der Schäden, Anzahl der Reklamationen bezogen auf den Nutzer	B	Arbeitszeit	B
	Instandsetzung	Arbeitszeit, Materialkosten	B	Anzahl der Nachbesserungen, Kundenzufriedenheit	A	Dauer	B
	Geräteübergabe	Arbeitszeit, Kosten der Ausfallzeit bezogen auf die Einnahmen des Krankenhauses	B	Höhe der Stillstands- oder Ausfallzeiten	B	Dauer	B
Entsorgung	Entsorgungs-antrag	Arbeitszeit	A	Anzahl und Genauigkeit der Geräteinformationen	B	Dauer	A
	Analyse und Entscheidung	Arbeitszeit	B	Anzahl und Genauigkeit der Geräteinformationen	B	Dauer	A
	Entsorgung	Kosten der Ausfallzeit bezogen auf die Einnahmen die dem Krankenhaus dadurch entgehen	A	Beschleunigung der Stilllegung des Geräts	B	Dauer	B

آلمان

ارزش های مدتی	ارزش	فاکتورهای زمان	ارزش های مدتی	ارزش	فاکتورهای کیفیت	ارزش های مدتی	ارزش	فاکتورهای هزینه	نام فرایند	فرایند
								هزینه های نگهداری بر حسب ارزش های تهیه مجدد، حسب تعداد کاربرد، هزینه های تهیه بر حسب تعداد کاربرد		نگهداری تجهیزات پزشکی
2,2	B	مدت برنامه ریزی	1,1	A	تعداد برنامه ریزی های مجدد لازم بندی، نسبت هزینه های پیش بینی شده به هزینه های واقعی، نسبت زمان اجرایی پیش بینی شده به واقعی	1,9	B	هزینه های برنامه ریزی بازا هزینه های سرمایه گذاری	برنامه ریزی	
3,0	C	مدت خرید	1,9	B	تعداد خرید های ایراد دار	1,5	B	هزینه اکتساب بر حسب روز های مراقبت، قیمت بازا دستگاه	خرید	
3,0	C	مدت راه اندازی	2,1	B	تعداد نصبهای نا موفق	2,5	C	هزینه آموزش بازا تجهیزات	راه اندازی	
1,5	B	مدت بهره برداری	1,3	A	بهره برداری از ظرفیت دستگاه	1,3	A	مصرف انرژی و مواد بازا کاربرد	بهره برداری	
1,5	B	مدت نگهداری و بازرسی	1,6	B	تعداد اقلات لازم به اعلام	1,4	A	هزینه های نگهداری بر حسب ارزش های تهیه مجدد	نگهداری و بازرسی	
1,5	B	مدت تعمیر	1,6	B	مدت تعمیر	1,3	A	هزینه های تعمیر بر حسب ارزش های تهیه مجدد	تعمیر	
2,8	C	مدت	2,9	C	تعداد اسقاط ثبت نشد، مدت فرایند	2,7	C	هزینه های اسقاط بر حسب قیمت خرید	خاتمه بهره برداری و اسقاط	

ارزش های مدتی	ارزش	فاکتورهای زمان	ارزش های مدتی	ارزش	فاکتورهای کیفیت	ارزش های مدتی	ارزش	فاکتورهای هزینه	نام فرایند	فرایند
										برنامه ریزی
2,7	C	مدت	1,6	B	تعداد تصحیحات بعد از راه اندازی با در طول مدت بلت مطلوب بودن طرح	1,9	B	زمان کار	تعیین اصول استفاده	
2,5	B	مدت	1,7	B	تعداد تصحیحات بعد از راه اندازی با در طول مدت بلت مطلوب بودن طرح، درجه دقت هزینه در مقایسه هزینه های واقعی	2,3	B	زمان کار، حق النفع	پهنای برنامه	
2,0	B	مدت	1,7	B	تعداد تصحیحات بعد از راه اندازی با در طول مدت بلت مطلوب بودن طرح	2,1	B	زمان کار، حق النفع	طرح ریزی	
2,8	C	مدت	1,6	B	مقایسه هزینه های پیش بینی شده به هزینه های واقعی	2,4	C	زمان کار، حق النفع	محاسبه بازدهی اقتصادی	
2,5	B	مدت	2,3	B	تعداد ایرادات با ضمیمه ها، هزینه تصحیحات	2,3	B	زمان کار، حق النفع	برنامه ریزی اخذ مجوز	
2,2	B	مدت	2,0	B	تعداد مشارک تشکیل، تعداد سوالات بعدی شرکتهای اجرایی	1,9	B	زمان کار، حق النفع	برنامه ریزی اجرا	
2,2	B	مدت	2,4	B	تعداد اجرا صفت بلت مطلوب بودن واگذاری	2,3	B	زمان کار، حق النفع	آماده سازی واگذاری	
2,5	B	مدت	2,0	B	تفاوت هزینه های واقعی از هزینه های پیش بینی شده	2,3	B	زمان کار، حق النفع	ابلاغ چارچوب هزینه	
2,2	B	مدت	1,4	A	هزینه های نگهداری بر حسب ارزش های تهیه مجدد، تعداد خرابی، رسیدن به عمر استفاده برنامه ریزی شد	1,7	B	زمان کار	تعیین اجرائیات نگهداری	

آلمان

فرايند	نام فرايند	فاكتورهاى هزينه	شاخص	شاخص هدفى	فاكتورهاى كيفيت	شاخص	شاخص هدفى	فاكتورهاى زمان	شاخص	شاخص هدفى
خريد										
	تصميم خريد	مدت شناسكى بازار خريد	B	2,0	تعداد شكايات مشتريها، مقايسه هزينه هاى مختلف	B	1,6	مدت	B	2,5
	مناقصه	مدت كل بازار مناقصه	B	2,3	تعداد سوالات بندى فروشندگان، تعداد منسجم	B	1,7	مدت	B	2,0
	سفارش	مدت كل بازار سفارش	B	2,4	تعداد لغو، تعداد شكايات	B	2,3	مدت	C	2,7
	ارسال	مدت كل براى دريافت بازار ارزش سفارش	C	2,9	تناسبين ارسالات، رعايت موعدها	B	1,9	مدت	C	2,7
	ارزيابى فروشندگان	مدت كل بازار ارزيابى	B	2,4	تعداد كمبود هاى كارانتى	B	1,9	مدت	C	2,7
	مستند سازى مراحل مربوط به نمدگاه	مدت كل بازار سند	B	2,4	تعداد اطلاعات ناكامل بد از خريد	B	2,1	مدت	B	2,3

فرايند	نام فرايند	فاكتورهاى هزينه	شاخص	شاخص هدفى	فاكتورهاى كيفيت	شاخص	شاخص هدفى	فاكتورهاى زمان	شاخص	شاخص هدفى
راه اندازى										
	برنامه ريزى و تعيين ارسال و نصب	هزينه هاى برنامه ريزى بازا هزينه هاى سرمايه گذارى	B	2,3	تعداد عدم رعايت موعد، رضايت بخش، تعداد تصحيحات	B	1,6	مدت	B	2,3
	نصب	زمان كار بازا نصب هزينه بر حسب كل هزينه بازا سفارش	B	2,1	مدت موتلا، انحراف از موعد، تعداد كسونها در زمان تحويل، درجه انبطاف	A	1,4	مدت	B	2,0
	آزمون كاركرد	زمان كار بازا آزمون كاركرد بر حسب ارزش خريد	C	2,7	تعداد تكرارات، درجه رسيدن به هدف	B	2,0	مدت	B	2,3
	تحويل	زمان كار بازا تحويل بر حسب ارزش خريد	C	3,0	تعداد كسود هاى ده نشد در زمان تحويل	B	1,7	مدت	B	2,5
	تهيه را هنماى كاربردى، آموزش	زمان كار بازا خريد، هزينه آموزش بازا خريد	B	2,3	تعداد اشتباهات ناشى از اپراتور	B	1,9	مدت	B	2,0
	واگذارى	زمان كار بازا خريد	C	2,7	تعداد تحويلات و تصورات ثبت نشد	B	2,3	مدت	B	2,5
	تجديد مدارك	زمان كار بازا خريد	C	2,4	تعداد سفار شك اشتباه، تعداد اختلالات، ميزان افزايش زمان رفع اختلال	B	1,6	مدت	B	2,5
	ثبت در ادارات مربوطه	هزينه اخذ جواز بازا خريد، زمان كار بازا ثبت	C	2,4	ميزان جريمه پرداختى، تعداد سوالات بندى	B	2,4	مدت	B	2,5

آلمان

فرایند	نام فرایند	فاکتورهای هزینه	شاخص	شاخص هزینه	فاکتورهای کیفیت	شاخص	شاخص هزینه	فاکتور زمان	شاخص	شاخص هزینه
نگهداری و بازرسی										
	برنامه ریزی نگهداری، برنامه ریزی اضطراری	زمان کار بازا مظاهری، هزینه نرم و سخت الزار کار بازا مظاهری	A	1,4	رابطه بین مظاهرشدت برنامه ریزی شده و غیر برنامه ریزی شده	A	1,3	مدت	B	2,2
	ابدا برنامه ریزی کار	زمان کار بازا مظاهری	B	1,6	تعداد تصمیمات بازا ... بندی	A	1,4	مدت	B	2,2
	برنامه ریزی زمان	زمان کار بازا مظاهری	B	1,9	تعداد عدم رعایت موعد، تعداد تصمیمات، میزان استفاده از ظرفیت کار	A	1,1	مدت	B	1,8
	کنترل، اجرا، تست، نظارت	زمان کار بازا مظاهری، هزینه مواد بازا مظاهری، هزینه تلف بازا مظاهری، هزینه خرابی بازا مظاهری	B	1,6	تعداد تصمیمات، مدت زمان خوابیدن دستگاه، میزان تخطی از زمان برنامه ریزی شده	B	1,7	مدت	B	1,7
	مستند سازی قسمت داخلی و خارجی	زمان مستند سازی بازا مظاهری	B	2,4	تعداد مدارک ناکامل	B	1,9	مدت	B	2,3
	برآورد هزینه	زمان کار بازا مظاهری	C	2,7	تعداد صورتحسابهای ناکامل، تعداد سوالات بندی	B	2,1	مدت	B	2,5
	بررسی انحرافات از قرارداد	زمان کار بازا مظاهری	C	2,6	تعداد انحرافات از مظاهری	B	2,4	مدت	C	2,7
	بررسی نتایج بازرسی	زمان کار بازا مظاهری	C	2,6	تعداد تعمیراتی که سر موقع مشخص نشده اند، تعداد خرابیها و اختلالات پیشبینی نشده	B	2,0	مدت	B	2,5
تعمیر										
	بررسی خسارات	زمان کار بازا مظاهری	A	1,4	تعداد اقدامات اشتباه	A	1,3	مدت	B	1,8
	تهیه گزارشات مربوطه، اجرای اقدامات لازم	زمان کار بازا مظاهری	B	2,4	تعداد ایرادات از طریق اقدامات مربوطه	B	2,1	مدت	B	2,5
	بررسی علل	زمان کار بازا مظاهری، هزینه نرم و سخت الزار کار بازا مظاهری	B	1,7	تعداد تشخیصات اشتباه	A	1,3	مدت	B	1,7
	انتخاب علل لازم به رفع	زمان کار بازا مظاهری	B	2,4	تعداد اشکال زدایی های اشتباه	B	1,9	مدت	B	2,5
	بررسی موردی	زمان کار بازا بررسی مظاهری	B	2,0	تعداد و میزان نسبی محاسبات منفی بندی	B	1,7	مدت	B	2,5
	رفع علل	زمان کار بازا مظاهری	B	1,6	تعداد تکرار علل مشابه	A	1,4	مدت	B	1,7
	ابتدا درج علل غیرلازم به رفع	زمان کار بازا مظاهری	C	2,6	تعداد نقطه ضعفهای ثبت نشد	B	2,3	مدت	B	2,3
	تعمیر	زمان کار بازا مظاهری، هزینه مواد بازا مظاهری، هزینه نرم و سخت الزار کار بازا مظاهری	A	1,0	تعداد تصمیمات، تعداد ایرادات بخش، رضایت بخش	A	1,0	مدت	B	1,0
	راه اندازی مجدد	زمان کار بازا مظاهری	B	2,3	میزان خوابیدن با خرابی	A	1,1	مدت	B	1,1
وظایف، اسقاط										
	بررسی تعهدات قرارداد	زمان کار	B	2,1	تعداد اشتباهات حقوقی، میزان هزینههای منفی	B	2,0	مدت	C	2,7
	فسخ قرارداد، تفسیر قرارداد	زمان کار، هزینه مبلغ	B	2,4	تعداد دعاوی حقوقی	B	2,0	مدت	C	2,7
	بررسی تعهدات ثبت	زمان کار	C	2,6	تعداد جریمه ها و شکایات	B	2,0	مدت	C	2,7
	لغو ثبت دستگاه نزد ادارات	زمان کار	C	2,9	تعداد جریمه ها و شکایات	B	2,3	مدت	B	3,0
	اسقاط	هزینه اسقاط، زمان کار	B	1,9	تعداد جریمه ها و شکایات	B	2,1	مدت	B	2,2

فرایند	گام فرایند		فاکترهای هزینه ای		فاکترهای کیفیت		فاکترهای زمان
ایران							
برنامه ریزی و خرید							
	درخواست		تناسب نیاز ساعت کاری		تعداد و دقیق بودن اطلاعات دستگاه مورد نیاز		زمان اداری
	بررسی و تصمیم گیری		ساعت اداری متخصصین و بخش های مختلف		تعداد اعتراض مشتری آینده نگری		زمان اداری
	مناقصه		ساعت اداری برای مناقصه		تعداد دقیق بودن اطلاعات شرایط مناقصه		زمان
	سفارش		زمان سفارش		تعداد مرجوعات تعداد اعتراضات		زمان
	دریافت دستگاه		زمان کاری دریافت کالا در مقایسه با ارزش کالا		تعداد ارسال کالا		زمان
	پرداخت هزینه		انجام به موقع موضوع قرارداد ساعت اداری		تعداد مدارک دستگاه خریداری شده و تعداد گواهی های لازم		زمان
	مستندسازی		زمان کاری برای مستندسازی		تعداد ناقص بودن مستنداد		زمان اداری
راه اندازی							
	برنامه ریزی		ساعت اداری متخصصین بخش های مختلف		تعداد و دقیق بودن اطلاعات دستگاهی که راه اندازی می شود		زمان اداری متخصصی
	بررسی و تصمیم گیری		هماهنگی بخش های مختلف مقایسه هزینه برنامه‌ریزی با هزینه سرمایه گزاری		تعداد اعتراض استفاده کننده/ کاربر		زمان اداری
	مناقصه		ساعت اداری برای مناقصه		تعداد و دقیق بودن اطلاعات شرایط مناقصه		زمان
	پیش پرداخت		به موقع موضوع قرارداد ساعت کار داری				بازه زمان
	راه اندازی		صرف زمان راه اندازی به کل هزینه خرید		زمان نصب دستگاه تعداد نواقص برطرف نبرده تاخیرو تسریع در انجام کار		بازه زمان
	تست دستگاه		ساعت کار هر تست در مقایسه با ارزش سرمایه گزاری		تعداد تکرار تست درجه انتباغ با شرایط استاندارد		بازه زمان
	تحویل دستگاه راه اندازی شده		ساعت کار هرتحویل دستگاه در مقایسه با هزینه خرید		تعداد اشکالات در تست دیده نشده		بازه زمان
	آموزش / کلاس توجیهی		هزینه آموزش در در مقایسه با هزینه دستگاه		تعداد اشتباهات کاربر		بازه زمان
	تحویل نهایی		زمان کار در مقایسه با خرید		تعداد نواقص مدارک		بازه زمان
	تصویه حساب		به موقع موضوع قرارداد ساعت کار داری				بازه زمان
	مستندسازی در سوابق		زمان کار در مقایسه با خرید		تعداد گزارشهای اشکالات زمان برطرف نمودن خرابی/ اشکالات		بازه زمان

فاکترهای زمان		فاکترهای کیفیت		فاکترهای هزینه ای	گام فرایند	فرایند
						نگهداشت
زمان اداری متخصصی		تعداد و دقیق بودن اطلاعات دستگاه		ساعت اداری متخصصین بخش های مختلف	برنامه ریزی	
زمان اداری		تعداد اعتراض استفاده کننده/ کاربر		هماهنگی بخش های مختلف مقایسه هزینه برنامه ریزی با هزینه سرمایه گذاری	بررسی و تصمیم گیری	
زمان		تعداد و دقیق بودن اطلاعات شرایط مناقصه تعداد ارجاع کار به شرکت/ مجری		ساعت اداری برای مناقصه	مناقصه	
بازه زمان		زمان لازم برای سرویس تعداد برطرف نمودن نواقص متابقت اجرا با برنامه ریزی		صرف زمان نگهداشت به کل هزینه خرید	نگهداشت	
بازه زمان		تعداد نظارت به طول زمان قرارداد درجه انتباغ با شرایط استاندارد		ساعت کار هر نظارت در مقایسه با ارزش خرید	نظارت بر عملکرد	
بازه زمان		تعداد گزارشهای/ مدارک ناقص		زمان کار هر مستند	مستندسازی در سوابق	
						تعمیرات
زمان اداری متخصصی		تعداد اعلان خرابی بودن توجه به صحت و سقم اعلام تعداد شکایات هم از کاربر و هم از دفتر فنی		ساعت اداری متخصصین بخش های مختلف	بررسی و گزارش خرابی	
بازه زمان		تعداد تعمیرات ثانویه تعداد اعتراض مشتری/ کاربر و هم از دفتر فنی		ساعت کار تعمیر کننده هزینه لوازم جاری هر تعمیر	تعمیرات	
بازه زمان		بالا بودن خواب دستگاه		ساعت کار اداری هزینه زمان خواب دستگاه در مقایسه با درآمد بیمارستان	تحویل دستگاه	
						متوقف کردن/ اسقاط
بازه زمان		تعداد و دقیق بودن اطلاعات		ساعت کار اداری متخصصین	درخواست	
بازه زمان		تعداد و دقیق بودن اطلاعات		ساعت کار اداری متخصصین	بررسی و تصمیم گیری	
بازه زمان		تسریع عمل متوقف کردن		هزینه زمان خواب دستگاه در مقایسه با هزینه پرسنل که نمی تواند کار کند	متوقف کردن/ اسقاط	

ایران

APPENDIX - A15

Identity Values -Technical Facilities

Deutschland

Prozess	Prozessschritt	Sender	Information/Inhalt	Empfänger
Instandhaltung geb.technische Anlagen allgem.				
Planung				
	Festlegung Nutzungskonzept	Kunde	Nutzungskonzept	TD
	Festlegung Nutzungskonzept	TD	Nutzungskonzept	TD
	Vorplanung	Kunde	Vorplanung	TD
	Vorplanung	Externer Planer	Vorplanung	TD
	Vorplanung	TD	Vorplanung	Kunde
	Entwurfsplanung	Kunde	Entwurfsplanung	TD
	Entwurfsplanung	Externer Planer	Entwurfsplanung	TD
	Entwurfsplanung	TD	Entwurfsplanung	Kunde
	Wirtschaftlichkeitsbetrachtung	Kunde	Wirtschaftlichkeitsbetrachtung	TD
	Wirtschaftlichkeitsbetrachtung	Externer Planer	Wirtschaftlichkeitsbetrachtung	TD
	Wirtschaftlichkeitsbetrachtung	TD	Wirtschaftlichkeitsbetrachtung	Kunde
	Genehmigungsplanung	Kunde	Genehmigungsplanung	TD
	Genehmigungsplanung	Externer Planer	Genehmigungsplanung	TD
	Genehmigungsplanung	TD	Genehmigungsplanung	Kunde
	Ausführungsplanung	Kunde	Ausführungsplanung	TD
	Ausführungsplanung	Externer Planer	Ausführungsplanung	TD
	Ausführungsplanung	TD	Ausführungsplanung	Kunde
	Vorbereitung der Vergabe	Kunde	LV und Mengenermittlung	TD
	Vorbereitung der Vergabe	Externer Planer	LV und Mengenermittlung	TD
	Vorbereitung der Vergabe	TD	LV und Mengenermittlung	Kunde
	Erstellung Kostenrahmen	Kunde	Kostenrahmen	TD
	Erstellung Kostenrahmen	TD	Kostenrahmen	Externer Planer
	Erstellung Kostenrahmen	TD	Kostenrahmen	Kunde
	Festlegung Instandhaltungsstrategien	Kunde	Instandhaltungsstrategie	TD
	Festlegung Instandhaltungsstrategien	TD	Instandhaltungsstrategie	Externer Planer
	Festlegung Instandhaltungsstrategien	TD	Instandhaltungsstrategie	Kunde
Beschaffung gebtechnische Anlagen				
	Beschaffungsentscheidung	Kunde (Besteller)	Gerätespezifikationen, Termine	Einkauf / TD
	Beschaffungsentscheidung	Technischer Dienst	Zustimmung Gerät	Einkauf
	Beschaffungsentscheidung	Einkauf	Kosten	Technischer Dienst
	Beschaffungsentscheidung	Technischer Dienst	Entscheidungsvorlage	GF
	Ausschreibung	TD		Einkauf
	Bestellung	Einkauf, TD	Kosten, Gerätespezifikationen, Aufstellungsort, begleitende Maßnahmen, Termine	Lieferant
	Lieferung	Lieferant	Lieferschein	TD
	Lieferantenbewertung	TD	Bewertung	Einkauf
	Dokumentation der gerätebezogenen Vorgänge	Einkauf	Gerätedaten	Technischer Dienst

Deutschland

Prozess	Prozessschritt	Sender	Information/Inhalt	Empfänger
Inbetriebnahme gebtechnischer Anlagen				
	Planung und Festlegung der Anlieferung und Installation	Kunde (Besteller)	Ort, Zeitpunkt, Art der Montage	Technischer Dienst
	Planung und Festlegung der Anlieferung und Installation	Technischer Dienst	Ort, Zeitpunkt, Art der Montage	beteiligte Organisationseinheiten
	Planung und Festlegung der Anlieferung und Installation	Technischer Dienst	Ort, Zeitpunkt, Art der Montage	Lieferant
	Installation	Technischer Dienst	Ort, Zeitpunkt, Art der Montage	Kunde (Nutzer)
	Funktionsprüfung	Lieferant	Testbericht, Prüfbericht, Fehlerdokumentation	Technischer Dienst
	Funktionsprüfung	Technischer Dienst	Terminbestätigung oder Terminänderung	Kunde (Nutzer)
	Abnahme	TD / Nutzer	Abnahmeprotokoll	Lieferant / ausf. Firma
	Abnahme	Sachverständiger	Abnahmeprotokoll	TD dann weiter ggf. an Behörden etc.
	Abnahme	Sachverständiger	Abnahmeprotokoll	TD
	Durchführung von Einweisungen, Schulungen	Lieferant	Bedienungshinweise	Kunde (Nutzer)
	Durchführung von Einweisungen, Schulungen	Lieferant	Servicehinweise, Bedienungshinweise	Technischer Dienst
	Übergabe	Technischer Dienst	Gerät, Bedienungsanleitung, Übergabeprotokoll	Kunde (Verantwortlicher)
	Aktualisierung der Dokumentation	Technischer Dienst		TD
	Meldung an zuständige Behörden	Betreiber	Anmeldung, Anzeige Geräteinformationen	zuständige Behörden
Wartung und Inspektion geb.technischer Anlagen				
	Instandhaltungs-planung	Technischer Dienst	Wartungsplan, wer, wann, wo , was	Kunde (Verantwortlicher)
	Instandhaltungs-planung	Dienstleister / Lieferant	Wartungsplan, wer, wann, wo , was	TD
	Arbeitsplanerstellung	Technischer Dienst	Arbeitsplan	Technischer Dienst
	Terminplanung	Technischer Dienst	Termine	Kunde (Nutzer)
	Steuerung, Durchführung, Überwachung	Technischer Dienst	Termine, Qualitäten	Kunde (Besteller)
	Dokumentation von eigen/fremd Anteil	Technischer Dienst	Dokumentation	TD
	Kostenübermittlung	Technischer Dienst	Kosten (Rechnung)	Kunde (Besteller)
	Auftragsabweichungsanalyse	Technischer Dienst	alle Auftragsdaten	Technischer Dienst
	Inspektion analysieren	Technischer Dienst	Inspektionsdaten	Technischer Dienst

Deutschland

Prozess	Prozessschritt	Sender	Information/Inhalt	Empfänger
Instandsetzung geb.technischer Anlagen				
	Schaden prüfen	ggf. Dienstleister / Lieferant	Prüfbericht	TD
	Schaden prüfen	Technischer Dienst	Prüfbericht	Kunde (Verantwortlicher)
	Zugehörige Berichte erstellen Durchführung entsp. Maßnahmen	Technischer Dienst	Meldung über Vorkommnis	Betreiber
	Zugehörige Berichte erstellen Durchführung entsp. Maßnahmen	Betreiber	Meldung über Vorkommnis	Behörden etc.
	Ursachen analysieren	ggf. Gutachter	Schadensbild	Technischer Dienst
	Auswahl der zu beseitigenden Ursachen	Technischer Dienst	Ursachen mit Behebungsvorschlägen	Kunde (Verantwortlicher)
	Auswahl der zu beseitigenden Ursachen	Kunde (Verantwortlicher)	zu beseitigende Ursachen	Technischer Dienst
	Wirtschaftlichkeit prüfen	TD	Wirtschaftlichkeitsprüfung	GF oder Kunde oder TD
	Ursachen beseitigen	Technischer Dienst	Schulung, Änderungsinformation	Kunde (Nutzer)
	zunächst nicht zu beseitigende Ursachen dokumentieren	Technischer Dienst	Auflistung nicht zu beseitigender Ursachen	Technischer Dienst
	Wirtschaftlichkeit prüfen	TD	Wirtschaftlichkeitsprüfung	GF oder Kunde oder TD
	Instandsetzen	Technischer Dienst	Servicebericht	Kunde (Besteller)
	Wiederinbetriebnahme	Technischer Dienst	Protokoll, Testbericht	Betreiber
	Wiederinbetriebnahme	TD	Protokoll, Testbericht	ggf. Hygiene oder Arbeitssicherheit oder Strahlenschutzbeauftragter
	Wiederinbetriebnahme	Betreiber	Anmeldung, Anzeige Geräteinformationen	zuständige Behörden
Stillegung/Entsorgung				
	Vertragliche Bindungen prüfen	Technischer Dienst		TD
	Verträge kündigen / umschreiben	Technischer Dienst	Kündigung	Vertragspartner
	Meldepflicht prüfen	TD	Prüfungsergebnis	TD
	Gerät bei Behörden abmelden	TD	Stilllegungsanzeige	Behörden etc.
	Entsorgung	TD	ggf. Vollzugsmeldung	Behörden etc.
	Entsorgung	Technischer Dienst	Vollzugsmeldung	Kunde (Verantwortlicher)

Iran

Prozess	Prozessschritt	Sender	Information/Inhalt	Empfänger
Beschaffung	Bedarfserklärung	Kunde/ Abteilungsleitung	Bedarfsanfrage / Firmenangebote	TB (Technisches Büro)
	Analyse und Entscheidung	TB	Bedarfsmeldung, Bericht des Technisches Büro, Berichte der Anbieterfirmen	Management
	Analyse und Entscheidung	Management	Bedarfsmeldung der zu Beschaffenden Geräte, Kosten, Gerätedaten	Service Center, Finanzabtl.
	Analyse und Entscheidung	Management	Bericht des Technischen Büros, Kalkulation	Unterstützungseinheit(falls Geräteanschaffung sehr teuer)
	Analyse und Entscheidung	Unterstützungseinheit	Komplette Informationen oder Berichte, Genehmigungsausfertigung	Management oder TB
	Analyse und Entscheidung / Anfrage des Management	Management	Bericht der technischen Anlagen oder des Wartungsunternehmens, Genehmigung	TB
	Ausschreibung	TB	Ausschreibungsbedingungen	Management
	Ausschreibung	Management	Kalkulation des TB und Ausführungsanweisung	Finanzabtl.
	Ausschreibung	Management und Ausschreibungskommission	Auswahl des Nachunternehmers und Vertragsschließung	Unterstützungseinheit
	Ausschreibung	Management	Vertrag	Nachunternehmer
	Ausschreibung	Management	Kopie des Vertrags	Büro der Kommissionsmitglieder
	Ausschreibung	Management	Anweisung für die Vorkasse	Finanzabteilung
	Vorkasse	Finanzabteilung	Überweisungs-Durchführung	Service Center
	Bestellung	Finanzabteilung (mit Kontrolle des TB)	Bestimmungen des Vertrags und ergänzende Bestimmungen	Anbietende Firmen
	Warenannahme	Lieferant/ Nachunternehmen	Gesamte Geräteinformationen	Abteilungsleiter, TB
	Warenannahme	Service-Center	Gesamte Geräteinformationen	Lager
	Warenannahme	Lieferant/ Nachunternehmen	Arbeitsbericht	Management (TB)
	Warenuntersuchung	TB	Untersuchungsbericht	Management
	Warenübergabe	Lieferant	Informationen, Gerätedaten	Service-Center, TB
	Warenübergabe	Lager	Lieferschein	Management, Finanzabteilung
	Warenübergabe	Management	Geräteankunft	TB
	Abrechnung	Management	Zahlungsanweisung	Finanzabteilung
	Abrechnung	Finanzabteilung	Ausführung der Vertragsbestimmungen	Lieferant, Nachunternehmen
	Dokumentation	Finanzabteilung, TB	Gesamte Daten	Abteilungsleiter, Management, Abtl. Sicherheit, Finanzabteilung der Universität
Inbetrieb-nahme	Planung	Management	Anfrage	TB
	Analyse und Entscheidung	TB	Beschaffungsantrag, Gutachten, Bericht, Lieferantenbericht	Management
	Analyse und Entscheidung	Management	Antrag auf Inbetriebnahme, Kosten, Bestimmungen der Inbetriebnahme	Service Center, Finanzabteilung
	Analyse und Entscheidung	Management	Bericht des TB, Kalkulation	Unterstützende Einheit (Falls der Preis sehr hoch ist)
	Analyse und Entscheidung	Unterstützende Einheit	Gesamte Informationen und Berichte, Genehmigungen	Management
	Analyse und Entscheidung/ Anfrage des Management	Management	Bericht TB oder Bericht des Inbetriebnehmenden Unternehmens, Genehmigung	TB
	Ausschreibung	TB	Arbeitsmethode, Ausschreibungsbedingungen	Management
	Ausschreibung	Management	Anweisung zur Durchführung der Arbeitsmethode, Ausschreibungsbedingungen	Finanzabteilung
	Ausschreibung	Management und Ausschreibungskommission	Auswahl der Nachunternehmens und Vertragsschließung	Unterstützende Einheit
	Ausschreibung	Management	Weiterleitung eines Exemplars des Vertrags	Nachunternehmen
	Ausschreibung	Management	Eine Kopie des Vertrags	Büro der Kommissionsmitglieder
	Inbetriebnahme	Inbetriebnehmendes Unternehmen	Bericht der operativen Inbetriebnahme	Management, TB
	Abschluss der Inbetriebnahme	Inbetriebnehmendes Unternehmen	Berichte und Tests	Management, TB
	Untersuchung des Berichte	TB	Untersuchung der Testberichte und der operativen Inbetriebnahme	Management, Inbetriebnehmendes Unternehmen
	Geräte -übergabe	TB, Inbetriebnehendes Unternehmen	Sitzungsbericht	Management

Iran

Prozess	Prozessschritt	Sender	Information/Inhalt	Empfänger
	Schulungen	Inbetriebnehendes Unternehmen	Wartungsbestimmungen, Schulungsinformationen	Kunde, Nutzer (Abteilung)
	Vorkasse	Management	Zahlungsanweisung	Finanzabteilung
	Vorkasse	Finanzabteilung	Zahlung	Inbetriebnehmendes Unternehmen, Nachunternehmen
	Endübergabe	Inbetriebnehmendes Unternehmen, TB, Abteilungsleiter	Gesamte Informationen	Management
	Rechnungsabschluss	Management	Endabrechnungsanweisung	Finanzabteilung
	Rechnungsabschluss	Finanzabteilung	Endzahlung	Inbetriebnehmendes Unternehmen, Nachunternehmen
	Dokumentation	Finanzabteilung, TB	Gesamte Informationen	Abteilungsleiter, Management, Abtl. Sicherheit
Wartung und Inspektion	Planung	TB	Wartungsplan und Antrag	Management
	Planung	Management	Antragsanalyse, Wartungsplanung	TB, Unterstützende Einheit
	Analyse und Entscheidung	Management	Wartungsantrag, Kosten, Wartungsbestimmungen	Finanzabteilung
	Analyse und Entscheidung	Unterstützende Einheit	Gesamte Informationen und Berichte, Genehmigung	Management
	Analyse und Entscheidung	Management	Gesamte Informationen und Berichte, Genehmigung	TB, Unterstützende Einheit
	Ausschreibung	TB	Arbeitsmethode, Ausschreibungsbedingungen	Management
	Ausschreibung	Management	Ausführungsanweisung	Finanzabteilung
	Ausschreibung	Management und Ausschreibungskommission	Auswahl der Nachunternehmens und Vertragsschließung	Unterstützende Einheit
	Ausschreibung	Management	Weiterleitung einer Exemplars des Vertrags	Nachunternehmen
	Zahlung	Finanzabteilung	Zahlung	Wartungs-unternehmen, NU
	Dokumentation	TB, Finanzabteilung, Wartungs-unternehmen	Gesamte Information	Management
Instandsetzung	Anfrage	NU, Nutzer, Verantwortlicher TB	Schadensmeldung	Management
	Analyse und Entscheidung	Management	Instandsetzungsantrag	NU (kleine Instandsetzungsmaßnahmen), TB
	Analyse	TB	Gutachten	Management
	Entscheidung	Management	Instandsetzungsanweisung	Service-Center, Finanzabteilung
	Vertragsangebot	Management, TB	Vertragsschließung	NU, Instandsetzungsunternehmen
	Instandsetzung	NU, Instandsetzendes unternehmen	Instandsetzungsbericht	Management, TB
	Instandsetzungsanalyse	TB	Gutachten und Gerätetest	Management
	Instandsetzungsanalyse	TB, NU	Instandsetzungsbericht	Management
	Übergabe des reparierten Geräts	NU, Instandsetzendes. unternehmen, TB	Gesamte Berichte	NU, Nutzer
	Zahlung	Management	Zahlungsanweisung	Finanzabteilung
	Zahlung	Finanzabtl.	Zahlung	Instandsetzungsunternehmen, NU
	Dokumentation	TB, Finanzabtl.., Instandsetzungs-unternehmen	Gesamte Information	Management
Stilllegung	Entsorgungs-antrag	TB, NU, Abteilungsleiter	Abnutzung, nicht reparierfähiges Gerät, Unfall, Optimierung	Management
	Analyse und Entscheidung	Management	Entsorgungsantrag	TB, Unterstützende Einheit
	Analyse und Entscheidung	TB	Gutachten	Management, Unterstützende Einheit
	Analyse und Entscheidung	Management	Entsorgungsanweisung mit Bericht	Finanzabteilung, TB, NU, Nutzer
	Entsorgung	Entsorgungslager	Bericht	Finanzabteilung, TB, Nutzer
	Analyse und Entscheidung	Finanzabtl.	Bericht	Management

آلمان

فرایند	نام فرایند	فرستنده	اطلاعات، محتوی	گیرنده
راه اندازی تأسیسات ساختمانی				
	برنامه ریزی و تعیین ارسال و نصب	مشتری	اصول اعتقاد	خدمات فنی
	برنامه ریزی و تعیین ارسال و نصب	خدمات فنی	اصول اعتقاد	خدمات فنی
	برنامه ریزی و تعیین ارسال و نصب	مشتری	برنامه ریزی اولیه	خدمات فنی
	نصب	برنامه ریز خارجی	برنامه ریزی اولیه	خدمات فنی
	آزمون کارکرد	خدمات فنی	برنامه ریزی اولیه	مشتری
	آزمون کارکرد	مشتری	برنامه ریزی طرح	خدمات فنی
	تحویل	برنامه ریز خارجی	برنامه ریزی طرح	خدمات فنی
	تحویل	خدمات فنی	برنامه ریزی طرح	مشتری
	تحویل	مشتری	بررسی المصاله	خدمات فنی
	بررسی المصاله	برنامه ریز خارجی	بررسی المصاله	خدمات فنی
	بررسی المصاله	خدمات فنی	بررسی المصاله	مشتری
	برنامه ریزی اخذ مجوز	مشتری	برنامه ریزی اخذ مجوز	خدمات فنی
	برنامه ریزی اخذ مجوز	برنامه ریز خارجی	برنامه ریزی اخذ مجوز	خدمات فنی
	برنامه ریزی اخذ مجوز	خدمات فنی	برنامه ریزی اخذ مجوز	مشتری
برنامه ریزی اجرا	برنامه ریز خارجی	برنامه ریزی اجرا	خدمات فنی	
	برنامه ریزی اجرا	خدمات فنی	برنامه ریزی اجرا	مشتری
	آماده سازی واگذاری	مشتری	لیست خدمات و مقادیر	خدمات فنی
	آماده سازی واگذاری	برنامه ریز خارجی	لیست خدمات و مقادیر	خدمات فنی
	آماده سازی واگذاری	خدمات فنی	لیست خدمات و مقادیر	مشتری
	ابلاغ چارچوب هزینه	مشتری	چارچوب هزینه	خدمات فنی
	ابلاغ چارچوب هزینه	خدمات فنی	چارچوب هزینه	برنامه ریز خارجی
	ابلاغ چارچوب هزینه	خدمات فنی	چارچوب هزینه	مشتری
	تعیین استراتژی نگهداری	مشتری	استراتژی نگهداری	خدمات فنی

آلمان

فرایند	نام فرایند	فرستنده	اطلاعات، محتوی	گیرنده
راه اندازی تسهیلات ساختمانی				
	بر نامه ریزی و تسین ارسال و نصب	مشتری (سفارش دهنده)	محل زمان و روش مونتاژ	خدمات فنی
	بر نامه ریزی و تسین ارسال و نصب	خدمات فنی	محل زمان و روش مونتاژ	سازمانهای مربوطه
	بر نامه ریزی و تسین ارسال و نصب	خدمات فنی	محل زمان و روش مونتاژ	فروشنده
	نصب	خدمات فنی	محل زمان و روش مونتاژ	مشتری
	آزمون کارکرد	فروشنده	گزارش تست، مدارک ضمائم	خدمات فنی
	آزمون کارکرد	خدمات فنی	تایید با انبیر موعد	مشتری
	تحویل	خدمات فنی، استفاد کنند	گزارش تحویل	فروشنده
	تحویل	کارشناس	گزارش تحویل	خدمات فنی
	تحویل	کارشناس	گزارش تحویل	خدمات فنی
	تهیه راهنمای کاربردی، آموزش	فروشنده	اطلاعات کاربرد	مشتری
	تهیه راهنمای کاربردی، آموزش	فروشنده	اطلاعات سرویس، کاربرد	خدمات فنی
	واگذاری	خدمات فنی	دستگاه، برشور کار، گزارش	مشتری مسئول
	تجدید مدارک	خدمات فنی		خدمات فنی
	ثبت در ادارات مربوطه	اپراتور	ثبت،اعلام اطلاعات دستگاه	ادارات مسئول
نگهداری و بازرسی				
	برنامه ریزی نگهداری	خدمات فنی	برنامه ریزی نگهداری، کی،	مشتری مسئول
	برنامه ریزی نگهداری	خدمات خارجی، فروشنده	برنامه ریزی نگهداری، کی،	خدمات فنی
	برنامه ریزی نگهداری، برنامه ریزی اضطراری	خدمات فنی	برنامه کار	خدمات فنی
	ابجاد برنامه ریزی کار	خدمات فنی	موعد ها	مشتری (مصرف کننده)
	کنترل، اجرا، تست، نظارت	خدمات فنی	موعد ها، کیفیتها	مشتری سفارشگر
	مستند سازی تست داخلی و خارجی	خدمات فنی	مستند سازی	خدمات فنی
	برآورد هزینه	خدمات فنی	هزینه	مشتری سفارشگر
	بررسی انحرافات از قرارداد	خدمات فنی	کل اطلاعات مربوط به قرارداد	خدمات فنی
	بررسی نتایج بازرسی	خدمات فنی	اطلاعات مربوط به بازرسی	خدمات فنی

آلمان

فرایند	نام فرایند	فرستنده	اطلاعات، محتوی	گیرنده
تعمیر دستگاه مانیتور				
	بررسی خسارات	خدمات خارجی، فروشنده	گزارش تست	خدمات فنی
	بررسی خسارات	خدمات فنی	گزارش تست	مشتری
	تهیه گزارشات مربوطه، اجرای اقدامات لازم	خدمات فنی	معرفی حوادث	اپراتور
	تهیه گزارشات مربوطه، اجرای اقدامات لازم	اپراتور	معرفی حوادث	ادارات دولتی
	بررسی علل	کارشناس	تشریح خسارت	خدمات فنی
	انتخاب علل لازم به رفع	خدمات فنی	علتها و پیشنهادات حل آنها	مشتری
	انتخاب علل لازم به رفع	مشتری	علل لازم به رفع	خدمات فنی
	بررسی موندنی	خدمات فنی	بررسی موندنی گزارش	خدمات فنی، مدیریت
	رفع علل	مشتری	آموزش، اطلاعات تنیرات	مشتری
	ابتدانر ج علل عیرلازم به رفع	خدمات فنی	لیست علل عیرلازم به رفع	خدمات فنی
	بررسی موندنی	خدمات فنی	بررسی موندنی گزارش	خدمات فنی، مدیریت
	تعمیر	خدمات فنی	گزارش سرویس	مشتری
	راه اندازی مجدد	خدمات فنی	گزارش تست	اپراتور
	راه اندازی مجدد	خدمات فنی	گزارش تست	مسئول بهداشت
	راه اندازی مجدد	اپراتور	ثبت اطلاعات دستگاه	ادارات مسئول

ایران

گیرنده	اطلاعات	فرستنده	گام فرایند	فرایند
خرید				
دفتر فنی	اعلام درخواست و گزارش شرکت فروشنده	بهره بردار / مدیر واحد	درخواست	
مدیریت	برگه درخواست، گزارش دفتر فنی، گزارش شرکت فروشنده	دفتر فنی	بررسی و تصمیم گیری	
تدارکات / امور مالی	برگه درخواست لوازم خرید، هزینه، مشخصات دستگاه	مدیریت	بررسی و تصمیم گیری	
معاونت پشتیبانی (اگرهزینه خیلی بالا باشد)	گزارش دفتر فنی، براورد هزینه	مدیریت	بررسی و تصمیم گیری	
مدیریت یا دفتر فنی	کل اطلاعات و گزارش و صدور مجوز	معاونت پشتیبانی	بررسی و تصمیم گیری	
دفتر فنی	گزارش تاسیسات یا گزارش شرکت راهبر، مجوز	مدیریت	بررسی و تصمیم گیری / درخواست مدیریت	
مدیریت	نحوه کار، شرایط مناقصه	دفتر فنی	مناقصه	
امور مالی	براورد مالی دفتر فنی و دستور اجرا	مدیریت	مناقصه	
معاونت پشتیبانی	انتخاب پیمانکارو عقد قرارداد	مدیریت و کمیسیون مناقصه	مناقصه	
پیمانکار	ابلاغ یک نسخه از قرارداد	مدیریت	مناقصه	
دفاتر اعضای کمیسیون	یک نسخه از قرارداد	مدیریت	مناقصه	
امور مالی	دستور پیش پرداخت	مدیریت	مناقصه	
تدارکات	انجام پیش پرداخت	امور مالی	پیش فاکتور	
شرکت فروشنده	مشخصات ذکر شده در قرارداد و مشخصات تکمیلی	مدیریت (با نظارت دفتر فنی)	سفارش (انجام موضوع قرارداد)	
نبار	کل اطلاعات	تدارکات	دریافت دستگاه	
مدیریت (دفتر فنی)	گزارش انجام کار	شرکت فروشنده، پیمانکار	دریافت دستگاه	
مدیریت	گزارش بررسی	دفتر فنی	بررسی دستگاه	
تدارکات، دفتر فنی	اطلاعات، فاکتور دستگاه	شرکت فروشنده	تحویل دستگاه	
مدیریت، امور مالی	قبض انبار	انبار	تحویل دستگاه	
دفتر فنی	رسیدن دستگاه	مدیریت	تحویل دستگاه	
امور مالی	دستور پرداخت	مدیریت	پرداخت هزینه	
شرکت فروشنده، پیمانکار	انجام تعهدات مالی موضوع قرارداد	امور مالی	پرداخت هزینه	
مدیر واحد، مدیریت، حراست، امور مالی دانشگاه	کل سوابق	امور مالی، دفتر فنی	مستندسازی	
راه اندازی				
دفتر فنی	درخواست	مدیریت	برنامه ریزی	
مدیریت	برگه درخواست، گزارش کارشناسی، گزارش شرکت راهبر	دفتر فنی	بررسی و تصمیم گیری	
تدارکات / امور مالی	برگه درخواست راه اندازی، هزینه، مشخصات راندازی	مدیریت	بررسی و تصمیم گیری	
معاونت پشتیبانی (اگرهزینه خیلی بالا باشد)	گزارش دفتر فنی، براورد هزینه	مدیریت	بررسی و تصمیم گیری	
مدیریت	کل اطلاعات و گزارش و صدور مجوز	معاونت پشتیبانی	بررسی و تصمیم گیری	
دفتر فنی	گزارش دفتر فنی یا گزارش شرکت راه انداز، مجوز	مدیریت	بررسی و تصمیم گیری / درخواست مدیریت	
مدیریت	نحوه کار، شرایط مناقصه	دفتر فنی	مناقصه	
امور مالی	دستور اجرا نحوه کار، شرایط مناقصه	مدیریت	مناقصه	
معاونت پشتیبانی	انتخاب پیمانکارو عقد قرارداد	مدیریت و کمیسیون مناقصه	مناقصه	

				ایران
پیمانکار	ابلاغ یک نسخه از قرارداد	مدیریت	مناقصه	
دفاتر اعضای کمیسیون	یک نسخه از قرارداد	مدیریت	مناقصه	
مدیریت، دفتر فنی	گزارش انجام عملیات راه اندازی	شرکت راه انداز	راه اندازی	
مدیریت، دفتر فنی	گزارشها و تست های عملیات	شرکت راه انداز	اتمام عملیات راه اندازی	
مدیریت، شرکت راه انداز	بررسی تست و راه اندازی عملیات	دفتر فنی	بررسی گزارشات	
مدیریت	تنظیم صورت جلسه عملیات موضوع قرارداد	دفتر فنی،، شرکت راه انداز	تحویل دستگاه راه اندازی شده	
مشتری، استفاده کننده، واحد	دستورالعمل نگهداری،اطلاعات آموزشی	شرکت راه انداز	آموزش	
امور مالی	دستور پرداخت	مدیریت	پیش پرداخت	
شرکت راه انداز، پیمانکار	انجام تعهدات مالی موضوع قرارداد	امور مالی	پیش پرداخت	
مدیریت	کل اطلاعات و سوابق	شرکت راه انداز، دفتر فنی، مدیر واحد	تحویل نهایی	
امور مالی	دستور پرداخت نهایی	مدیریت	تسویه حساب	
شرکت راه انداز، پیمانکار	انجام تعهدات نهایی مالی موضوع قرارداد	امور مالی	تسویه حساب	
مدیر واحد،مدیریت، حراست	کل سوابق	امور مالی، دفتر فنی	مستندسازی	
				نگهداشت
مدیریت	برنامه ریزی کارنگهداشت و درخواست	دفتر فنی	برنامه ریزی	
دفتر فنی، معاونت پشتیبانی	بررسی درخواست، برنامه ریزی کارنگهداشت	مدیریت	برنامه ریزی	
امور مالی	برگه درخواست نگهداشت، هزینه، مشخصات نگهداشت	مدیریت	بررسی و تصمیم گیری	
مدیریت دفتر فنی، معاونت	کل اطلاعات و گزارش و صدور مجوز	معاونت پشتیبانی	بررسی و تصمیم گیری	
مدیریت	نحوه کار، شرایط مناقصه	دفتر فنی	مناقصه	
امور مالی	دستور اجرا نحوه کار، شرایط مناقصه	مدیریت	مناقصه	
معاونت پشتیبانی	انتخاب پیمانکار و عقد قرارداد	مدیریت و کمیسیون مناقصه	مناقصه	
پیمانکار	ابلاغ یک نسخه از قرارداد	مدیریت	مناقصه	
دفاتر اعضای کمیسیون	یک نسخه از قرارداد	مدیریت	مناقصه	
دفتر فنی	گزارش عملیات	شرکت راهبر	نگهداشت	
مدیریت، شرکت راهبر	دستورالعمل نگهداشت	دفتر فنی	نگهداشت	
مدیریت	گزارش عملیات (ماهانه)	شرکت راهبر	نگهداشت	
دفتر فنی	گزارش عملیات (ماهانه)	مدیریت	نگهداشت	
مدیریت، شرکت راهبر	گزارش کارشناسی (تائید پرداخت)	دفتر فنی	نظارت بر عملکرد	
امور مالی	دستور پرداخت	مدیریت	پرداخت	
شرکت راه بر، پیمانکار	انجام تعهدات مالی موضوع قرارداد	امور مالی	پرداخت	
مدیریت	کل اطلاعات	دفتر فنی، امور مالی، شرکت راه بر،	مستندسازی	
				تعمیرات
مدیریت	اعلام خرابی	پیمانکار، نگهدارنده ناظردفتر فنی	درخواست	
پیمانکار(تعمیر جزئی)، دفتر فنی	درخواست تعمیر	مدیریت	بررسی و تصمیم گیری	
مدیریت	گزارش کارشناسی	دفتر فنی	بررسی	
تدارکات، امور مالی	دستور تعمیر	مدیریت	تصمیم گیری	
پیمانکار، شرکت تعمیر کننده	عقد قرارداد، فاکتور	مدیریت، دفتر فنی	قرارداد/ فاکتور	

ایران

تعمیرات	پیمانکار، شرکت تعمیر کننده	گزارش انجام تعمیرات	مدیریت، دفتر فنی
بررسی تعمیرات	دفتر فنی	گزارش کارشناسی و تست دستگاه	مدیریت
بررسی تعمیرات	دفتر فنی، پیمانکار	تنظیم صورت جلسه انجام تعمیرات	مدیریت
تحویل دستگاه تعمیر شده	شرکت تعمیرکننده، پیمانکار ، دفتر فنی	کل گزارش تحویل دستگاه ، صورت جلسه	پیمانکار، بهره بردار
پرداخت	مدیریت	دستور پرداخت	امور مالی
پرداخت	امور مالی	انجام تعهدات مالی موضوع قرارداد/ فاکتور	شرکت تعمیرکننده، پیمانکار
مستندسازی	دفتر فنی، امور مالی، شرکت تعمیرکننده،	کل اطلاعات	مدیریت
متوقف کردن			
درخواست متوقف کردن	دفتر فنی، پیمانکار، مدیر واحد	اعلام خرابی بیش از حد ، استهلاک، تعویض جهت ، حادثه، بهینه سازی	مدیریت
بررسی و تصمیم گیری	مدیریت	درخواست متوقف کردن	دفتر فنی، معاونت پشتیبانی
بررسی و تصمیم گیری	دفتر فنی	گزارش کارشناسی	مدیریت، معاونت پشتیبان
بررسی و تصمیم گیری	مدیریت	دستور متوقف نمودن دستگاه به زیرمجموعه ، دستور نهایی با گزارش	امور مالی، دفتر فنی، پیمانکارو بهره بردار
متوقف کردن	انبار اسقاط	صورت جلسه	امور مالی، دفتر فنی، بهره بردار
بررسی و تصمیم گیری	امور مالی	صورت جلسه و گزارش	مدیریت
متوقف کردن	معاونت پشتیبانی	بخشنامه	واحدهای تابعه، تدارکات
تصمیم گیری نهایی	مدیریت	گزارش متوقف و انبار نمودن دستگاه	دفتر فنی، معاونت پشتیبانی، حراست
مستندسازی	دفتر فنی، امور مالی، بهره بردار	کل اطلاعات	مدیریت

APPENDIX - A16

Questionnaire Laundry

Universität Karlsruhe (TH)
Prof. Dr.-Ing. Kunibert Lennerts
Facility Management

Datenerhebung zur Wäscheversorgung im Krankenhaus
(bitte Ausfüllen bzw. Zutreffendes ankreuzen)
Diskussionsgrundlage

1.   Planbettenanzahl   _____________

2.   Anzahl OP-Säle   _____________
     Anzahl Stationen   _____________
     Anzahl Kostenstellen   _____________

3.   Summe der abgerechneten Pflegetage für das Jahr 2004   ___________
3.1 Durchschnittliche Verweildauer in 2004   ___________
4.   Gebäudetypologie
4.1 Anzahl Gebäude gesamt   ___________
4.2 Anzahl Gebäude mit max. 4 Geschossen (<12,5m^1)   ___________
4.3 Anzahl Gebäude mit mehr als 7 Geschossen (>22m^2) ___________

5.   Wäscheverbrauch:
5.1 Gesamt-Wäschemenge pro Jahr   ___________kg
5.2 OP-Wäsche pro Jahr   ___________kg
5.3 Stationswäsche pro Jahr   ___________kg
5.4 Berufsbekleidung pro Jahr   ___________kg
5.5 Mangelwäsche pro Jahr   ___________kg

6.   Anzahl mit Berufskleidung versorgter Mitarbeiter (in Vollzeit)___________
     davon tätig als/im Bereich:
6.1 Pflegedienst   _____________
6.2 Ärzteschaft   _____________
6.3 Hauswirtschaft   _____________
6.4 Funktionsbereich   _____________
6.5 Reinigung   _____________
6.6 Technik   _____________

7.   Wäscheart:
7.1 OP-Wäsche: Einweg   ☐
                       Mehrweg   ☐
7.2 Stationswäsche: Eigene Wäsche   ☐
                       Mietwäsche   ☐
7.3 Berufskleidung: Eigene Wäsche   ☐
                       Mietwäsche   ☐
                       Anteil Poolbekleidung   ___________ %
                       Anteil persönliche Berufskleidung ___________ %

8.   Versorgungsmodell:
8.1 Vollversorgung durch Dienstleister   ☐
8.2 externe Logistik und Wäscherei durch DL, interne Logistik selbst   ☐
8.3 eigene Wäscherei, externe und interne Logistik selbst   ☐

---

[1] Landesbauordnung für Baden-Württemberg, Teil 5, §29
[2] Landesbauordnung für Baden-Württemberg, Teil 1, §2

Universität Karlsruhe (TH)
Prof. Dr.-Ing. Kunibert Lennerts
Facility Management

9.   Verteilungssystem
9.1   Lieferung dezentral auf Station (Einzelkostenstellenbezogen)   ☐
9.2   Lieferung in zentrales Wäschelager, danach Sammelverteilung   ☐

10.   Vorhaltesystem
10.1   Stationswäsche   fest eingebautes Schranksystem   ☐
                Wäschewagen in Schranknutzung   ☐
10.2   Berufsbekleidung   offener Zugang durch Schrankfächer   ☐
                Lieferung in Spinde   ☐
                Automaten   ☐

11.   Kommissionierung
11.1   intern   ☐
11.2   extern   ☐

12.   Flächenbedarf
12.1   interne Kommissionierung   _____________m²
12.2   interne Näherei   _____________m²
12.3   Wäschekammer   _____________m²
12.4   Zentrales Schmutzwäschelager   _____________m²
12.5   Dezentrale Schmutzwäschelager auf
    Kostenstellen im Ø   _____________m²

13.   Liefer-/Abholrhythmus
13.1   Frischwäsche   _____________Mal pro Woche
13.2   Schmutzwäsche   _____________Mal pro Woche

14.   Bedarfserfassung
14.1   Wer erfasst Wäschebedarf für anstehende Lieferung?
    Stationspersonal   ☐
    Inhouse-Logistik   ☐
    Dienstleister   ☐
14.2   Wie werden die Daten übermittelt ?
    Telefonisch   ☐
    Per email   ☐
    Schriftlich   ☐
14.3   Zufriedenheit hinsichtlich Übermittlungszeit und Vollständigkeit?
    Zufrieden   ☐
    Unzufrieden   ☐

15.   Inhouse-Logistik/Transport
15.1   Durchschnittliche Dauer der Frischwäschelieferung   _____________min.
    , dabei werden _________ (Anzahl) Kostenstellen beliefert.
15.2   Durchschnittliche Dauer der Schmutzwäschesammlung_________ min.
    , dabei werden _________ (Anzahl) Kostenstellen abgedeckt.
15.3   Sind Frischwäschelieferung und Schmutzwäschelieferung logistisch
    verknüpft?
                ☐ Ja
                ☐ Nein

Universität Karlsruhe (TH)
Prof. Dr.-Ing. Kunibert Lennerts
Facility Management

16. Anzahl der für die Wäsche verantwortlichen/ausführenden
Mitarbeiter (in Vollzeit) __________
16.1 Mitarbeiter Wäschekammer __________
16.2 Mitarbeiter Verwaltung __________
16.3 Mitarbeiter Hol-/Bringdienst (Inhouse Logistik) __________
16.4 Mitarbeiter Näherei __________
16.5 Mitarbeiter Einkauf __________
16.6 Mitarbeiter Wäscherei __________
16.7 Mitarbeiter Kommissionierung __________

17. Durchschnittlicher Wäschezukauf
für 2003 __________Euro
für 2004 __________Euro

18. Geschätzter Wert des Wäschebestandes __________Euro

19. Preis pro kg Schmutzwäsche __________Euro

20. Wäschebestand/Sortiment
(gemäß letzter Inventur) Anzahl
Stationswäsche
20.1 Bettbezug 140 x 200cm __________
20.2 Kopfkissenbezug 80 x 80cm __________
20.3 Hanselkissen 40 x 40cm __________
20.4 Bettuch __________
20.5 Halbspannlaken __________
20.6 Stecklaken __________
20.7 Frottierhandtuch 50 x 100cm __________
20.8 Badetuch 90 x 135cm __________
20.9 Waschhandschuh __________
20.10 Gläsertuch __________
20.11 Vorbinder __________
20.12 Schwesternüberwurfschürze __________
20.13 Nässeschutzunterlage __________
20.14 Wickelsack __________
Babywäsche
20.20 Kopfstreifen __________
20.21 Baby-Bettbezug __________
20.22 Strampler __________
20.23 Body __________
20.24 Hemdchen __________
20.25 Slip __________
20.26 Jäckchen __________
20.27 Baby-Infusionshemd __________
20.28 Zwirnkörperwindeln __________
20.29 Lätzchen __________
Kinderwäsche
20.30 Bettuch 110 x 180cm __________
20.31 Bettbezug 110 x 140cm __________
20.32 Kopfkissenbezug 40 x 50cm __________

Universität Karlsruhe (TH)
Prof. Dr.-Ing. Kunibert Lennerts
Facility Management

20.33	Seiflappen	__________
20.34	Unterhemd	__________
20.35	Unterhose	__________
20.36	T-shirt 1/1 Arm	__________
20.37	Kinder-Patientenhemd	__________
20.38	Schlafjacke	__________
20.39	Schlafhose	__________

Tischwäsche

20.40	Serviettee	50 x 50cm	__________
20.41	Deckserviette	80 x 80cm	__________
20.42	Tischdecke	130 x 130cm	__________
20.43	Tischdecke	130 x 160cm	__________

Inlays

20.50	Einziehdecke	135 x 200cm	__________
20.51	Kissenfüllung	80 x 80cm	__________
20.52	Kissenfüllung	40 x 40cm	__________
20.53	Kinder-Einziehdecke	100 x 135cm	__________
20.54	Baby-Einziehdecke	100 x 135cm	__________

21.   Berufskleidung
Ausstattung Berufskleidung je Person (oder Gesamtbestand nach letzter Inventur): Anzahl **Hose**, **Kasack**, **Mantel**, **Hem**d

21.1	Pflegedienst	m__________	w__________
21.2	Ärzteschaft	m__________	w__________
21.3	Hauswirtschaft	m__________	w__________
21.4	Funktionsbereich	m__________	w__________
21.5	Reinigung	m__________	w__________
21.6	Technik	m__________	w__________

تهیه ارقام برای فرایند رختشویی در بیمارستان

1. تعداد تخت مصوب شده: __________

2. تعداد اطاق عمل: __________

  2.1 تعداد بخشها: __________

  2.2 تعداد واحدهای هزینه ای: __________

3. مجموع روز های بستری بیمارستان در سال 1384: __________

متوسط روز های بستری / اقامت در سال 1384: __________

4. نما و شکل ساختمان:

  4.1 تعداد کل ساختمانها: __________

5. مصرف رخت خرید شده در سال:

  5.1 مجموع مقدار رخت در سال: __________ kg

  5.2 لباس جراحی در سال: __________ kg

  5.3 لباس بخشها در سال: __________ kg

  5.4 لباس کاری/ یونیفرم در سال: __________ kg

چنانچه مقدار لباسها در دسترس نیست سوال 23 را جواب دهید

6. مصرف رخت موجودی در سال:

  6.1 مجموع مقدار رخت در سال: __________ kg

  6.2 لباس جراحی در سال: __________ kg

  6.3 لباس بخشها در سال: __________ kg

  6.4 لباس کاری در سال: __________ kg

چنانچه مقدار لباسها در دسترس نیست سوال 23 را جواب دهید

بهینه سازی و تجزیه و تحلیل فرآیندهای بیمارستانی

OPK

7. تعداد همکارانی که لباسهای شغلی در اختیارشان قرار داده می شود (تمام وقت):

7.1. پرستاری: ________________

7.2. پزشکان: ________________

7.3. فنی: ________________

7.4. نظافت: ________________

8. انواع رخت ها؛

8.1. لباس های جراحی: ☐ یک بار مصرف   ☐ چند بار مصرف

☐ رخت خصوصی   ☐ رخت اجاره شد

8.2. رخت بخش ها؛

☐ رخت خصوصی   ☐ رخت اجاره شد

8.3. رخت شغلی:

☐ رخت خصوصی   ☐ رخت اجاره شد

8.4. نسبت رختهای عمومی به کل رختهای موجود: ________________%

8.5. نصبت رختهای شغلی خصوصی کل رختهای موجود: ________________%

9. مدل تامین

9.1. ارایه کامل بوسیله: ☐ خدمات بخش خصوصی   ☐ خود بیمارستان

9.2. حمل و نقل بیرون بوسیله: ☐ خدمات بخش خصوصی   ☐ خود بیمارستان

9.3. حمل و نقل درون بوسیله: ☐ خدمات بخش خصوصی   ☐ خود بیمارستان

10. سیستم تقسیم رخت ها؛

☐ مرکزی گرفته می شود   ☐ مستقیم به بخشها می رود

11. نگهداری رخت ها

11.1	رخت بخش:	☐ در کمد دیواری	☐ در کمد کنار تخت
11.2	رخت شغلی:	☐ توزیع از انبار	☐ از قفسه
		☐ در قفسه مخصوص	☐ در کمد مخصوص

12. بخش بازرسی و کنترل کالا:

☐ داخل بیمارستان   ☐ خارج از بیمارستان

13. نیاز فضا:

13.1   بخش بازرسی و کنترل کالا داخلی :   __________ m²

13.2   بخش خیاطی داخلی: __________ m²

13.3   انبار رخت بخش: __________ m²

13.4   انبار مرکزی رخت: __________ m²

13.5   انبار رخت کثیف: __________ m²

13.6   انبار مرکزی رخت کثیف: __________ m²

14. دوره زمانی ارسال و دریافت

رخت تمیز: __________ بار در هفته

رخت کثیف: __________ بار در هفته

15. تشخیص نیاز:

15.1   چه کسی نیاز مصرف رخت را تنظیم میکند؟

☐ بخش پرسنل

☐ بخش حمل و نقل

☐ شرکت خدمت

326

بهینه سازی و تجزیه و تحلیل فرآیند های بیمارستانی

15.2   اعلام درخواست چگونه است؟

☐ تلفنی

☐ از طریق e-mail

☐ مکتبــــــی

15.3   آیا از زمان اطلاع رسانی تا انجام آن راضی هستید؟

☐ بلی   خیر ☐   درصد رضایت ___________ %

16. حمل و نقل داخل بیمارستان

16.1   حدودا چند دقیقه برای رخت تازه صرف می شود؟ _______min

16.2   چه مقداری/به چند بخش رخت داده می شود؟ ___________

16.3   حدودا چند دقیقه برای رخت کثیف صرف می شود؟ _______min

16.4   چه مقداری/از چند بخش رخت گرفته می شود؟ ___________

16.5   آیا حمل و نقل رخت از نظر پشتیبانی با هم ارتباط دارند؟

☐ بلی   ☐ خیر

17. رخت خراب یا با اشکال

17.1   تعریف خراب یا با اشکال در بیمارستان شما چیست؟

17.2   این رخت چگونه از سیکل خارج میشود؟

☐ در یک کیسه مخصوص

☐ جدا نمی شود (با رخت کثیف جمع آوری میشود)

17.3   در صد رخت خارج شده از سیکل در بیمارستان ___________ %

18. تعداد پرسنل شاغل در بخش رختشویخانه (تمام وقت)

18.1   پرسنل در انبار   ___________

18.2   پرسنل در بخش اداری   ___________

18.3   پرسنل حمل و نقل داخلی   ___________

18.4   پرسنل خیاطی   ___________

18.5   پرسنل خرید   ___________

18.6   پرسنل رختشویی   ___________

18.7   پرسنل بازرسی و کنترل   ___________

19. حدود متوسط خرید رخت در سال: ___________ ریال

20. تخمین ارزش رخت موجود در حال حاضر: ___________ ریال

21. هزینه شستشوی هر کیلو رخت کثیف   ___________ ریال

برای بیمارستان هایی که بصورت کیلوگرم محاسبه نمی کنند لطفا سوال شماره 23 را پاسخ دهید

22. تفاوت بین پیشبینی شده و تعداد موجود در سال   ___________ %

23. رخت موجود

مصرف	قیمت تکه	تعداد	رخت در بخش	
________	________	________	ملافه (140 cm * 200)	23.1
________	________	________	روکش بالش (80*80 cm)	23.2
________	________	________	روکش بالش کوچک (40 cm *40)	23.3
________	________	________	لحاف	23.4
________	________	________	روکش نخت	23.5
________	________	________	حوله ست (100 cm * 50)	23.6
________	________	________	حوله حمام (135 cm * 90)	23.7

328

رخت میز

_______	_______	_______	23.27 دستمال سفره (50* 50 cm)
_______	_______	_______	23.28 دستمال سفره میزی (80* 80 cm)
_______	_______	_______	23.29 رومیزی (130* 130 cm)
_______	_______	_______	23.30 رومیزی (160* 160 cm)

رخت مربوط به خواب - این لی

_______	_______	_______	23.31 لحاف (135* 200 cm)
_______	_______	_______	23.32 بالش (80* 80 cm)
_______	_______	_______	23.33 بالش (40* 40 cm)
_______	_______	_______	23.34 پتو بچه (100* 135 cm)
_______	_______	_______	23.35 پتو نوزاد (100* 135 cm)

لباس اتاق عمل

_______	_______	_______	23.36 شلوار اتاق عمل
_______	_______	_______	23.37 پیراهن اتاق عمل

لباسهای شغلی

تعداد تعویض در هفته		تعداد لباسهای شغلی هر فرد	
_______	_______ آقایان	_______ خانمها	23.38 پرستاری:
_______	_______ آقایان	_______ خانمها	23.39 پزشکان:
_______	_______ آقایان	_______ خانمها	23.40 فنی:
_______	_______ آقایان	_______ خانمها	23.41 نظافت:

APPENDIX - A17

Process Comparison- Laundry Management

Deutschland

Rahmenprozess Wäscheversorgung

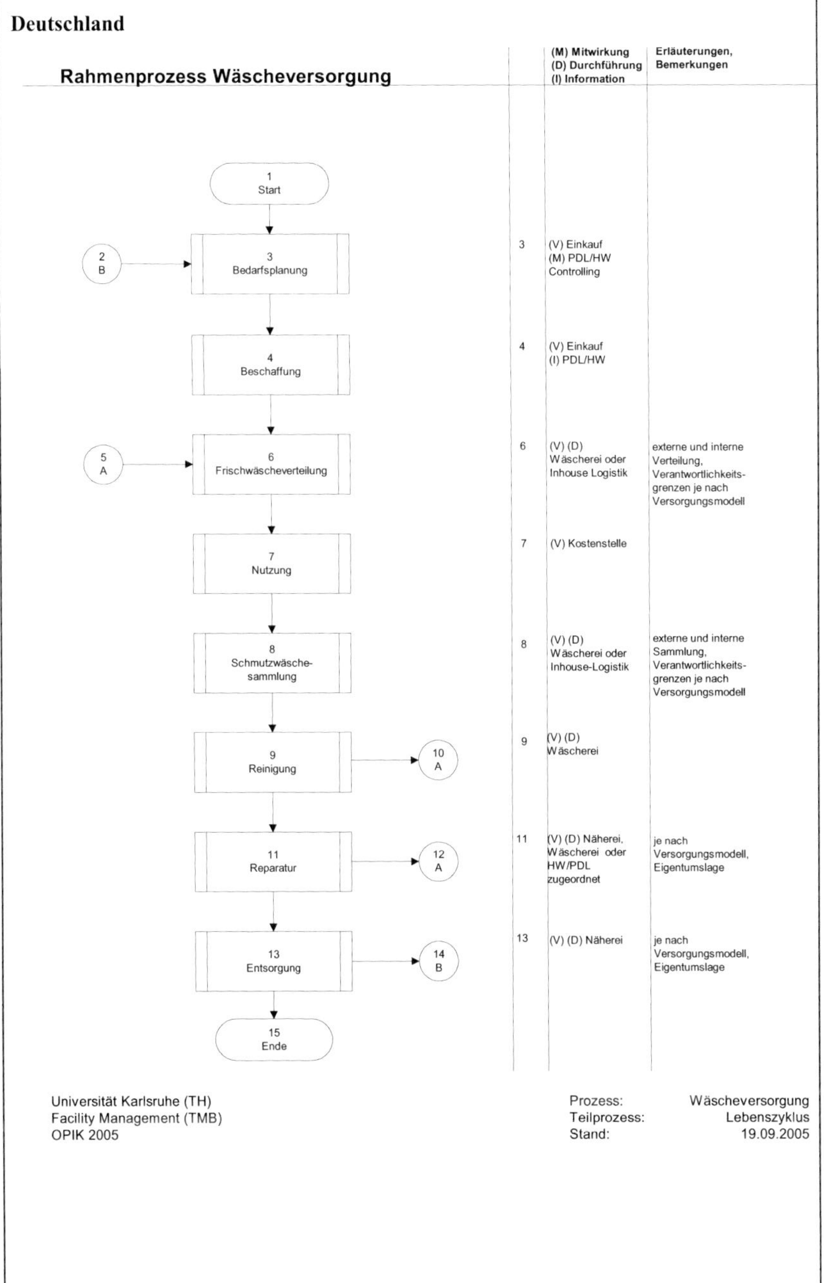

Iran

Rahmenprozess
Wäscheversorgung

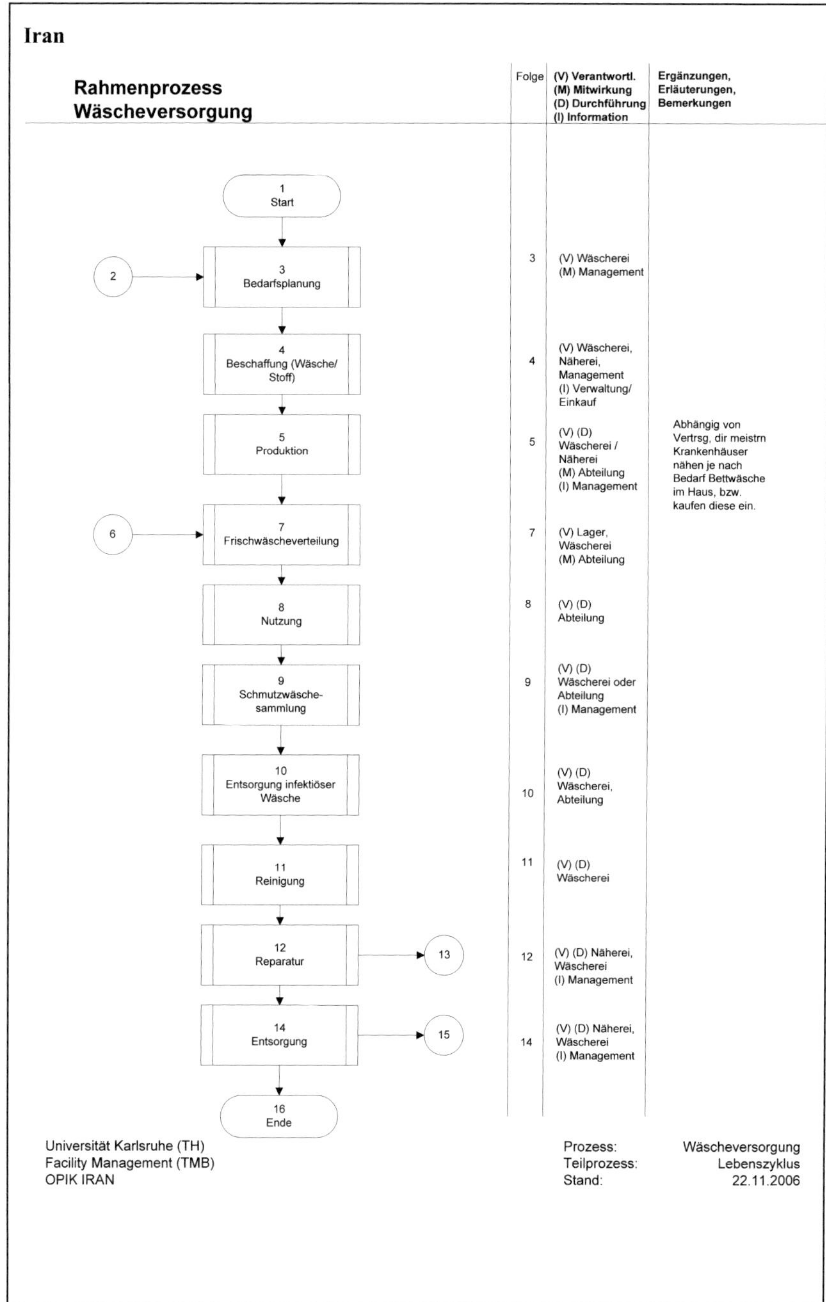

Deutschland

Beschaffung Wäsche

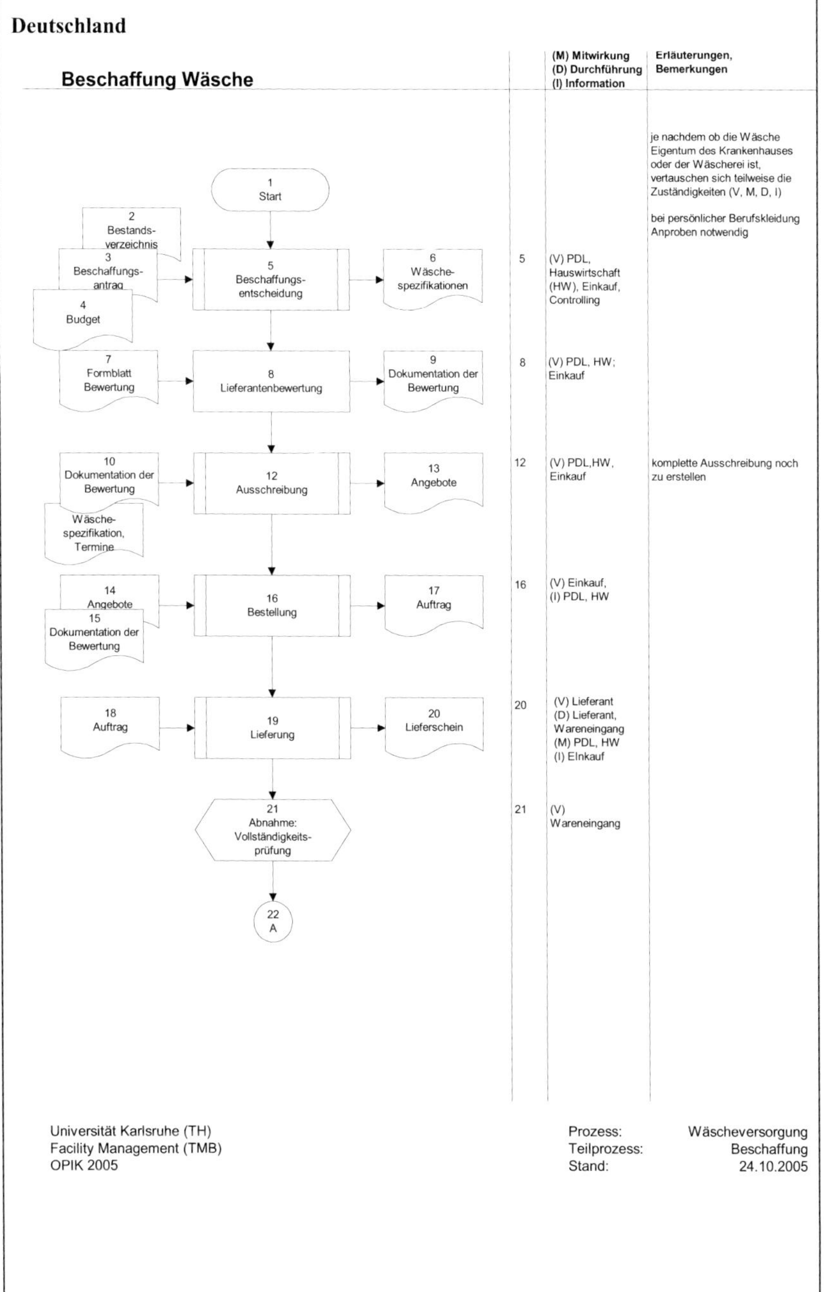

Universität Karlsruhe (TH)
Facility Management (TMB)
OPIK 2005

Prozess: Wäscheversorgung
Teilprozess: Beschaffung
Stand: 24.10.2005

Iran

Planung und Beschaffung Wäsche

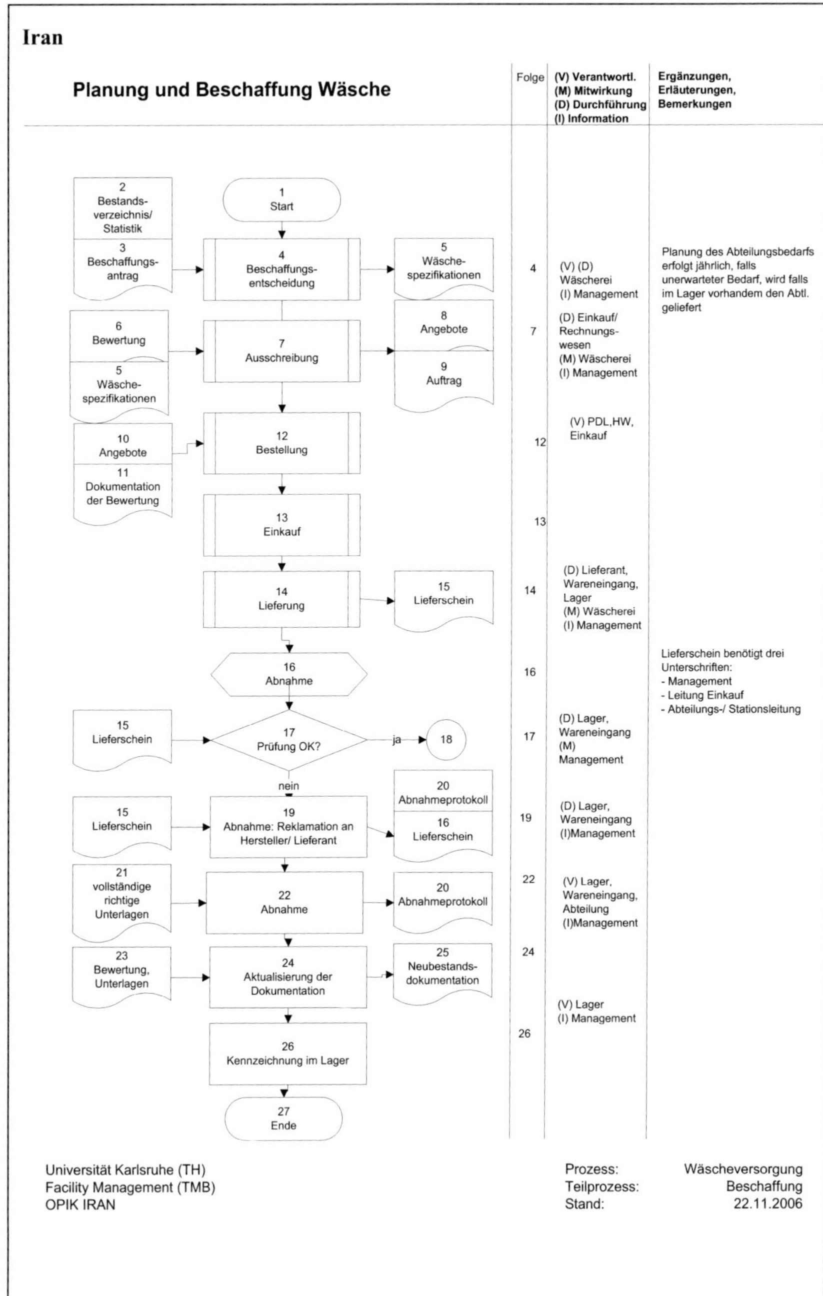

Deutschland

Beschaffung 2

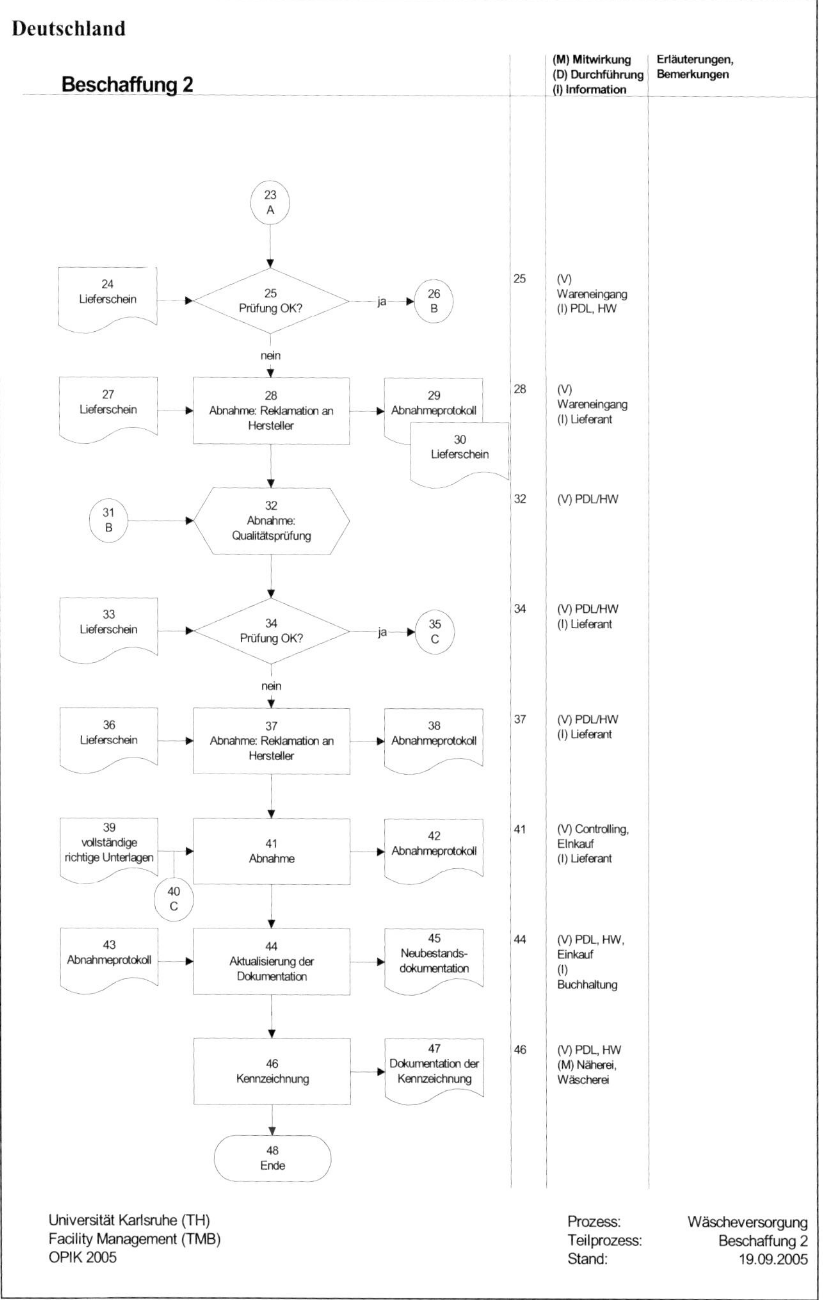

Universität Karlsruhe (TH)	Prozess: Wäscheversorgung
Facility Management (TMB)	Teilprozess: Beschaffung 2
OPIK 2005	Stand: 19.09.2005

Iran

Produktion

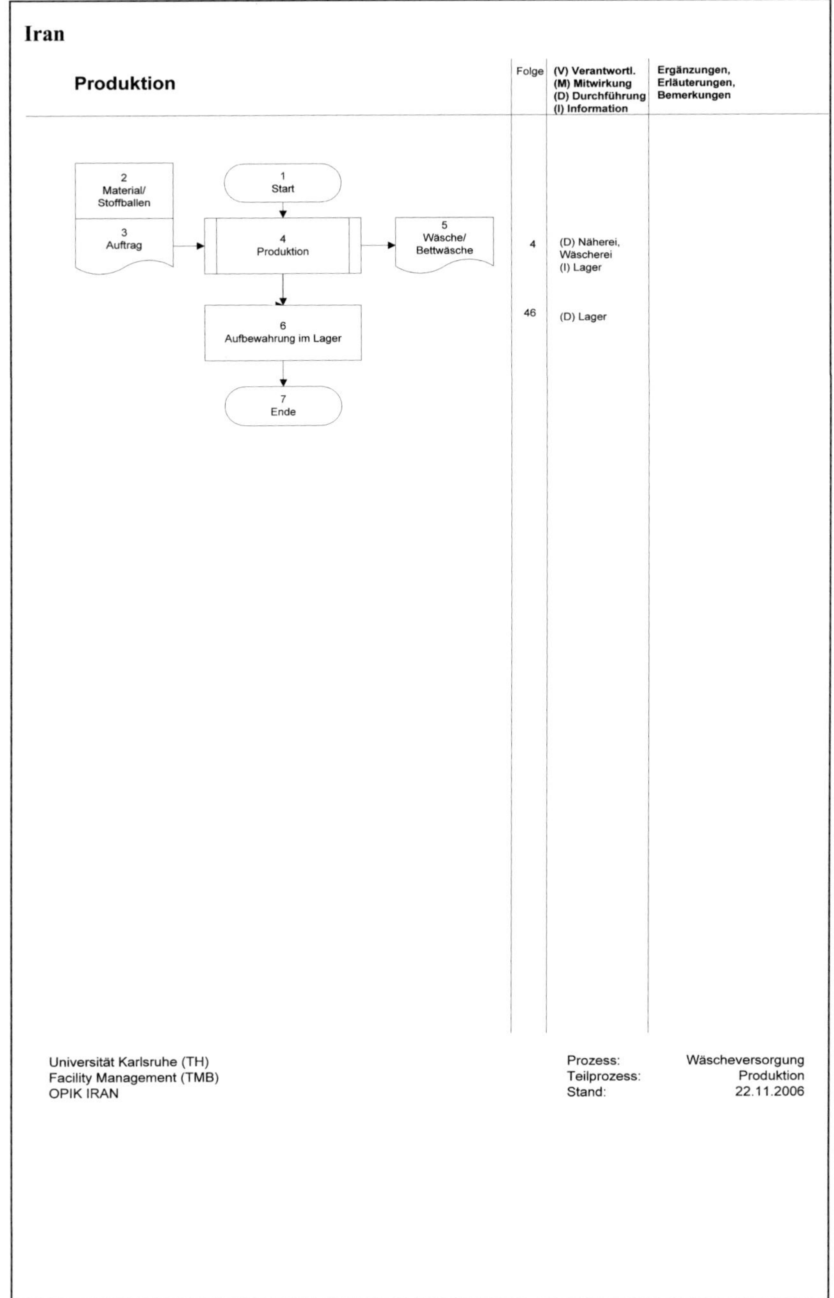

Folge	(V) Verantwortl. (M) Mitwirkung (D) Durchführung (I) Information	Ergänzungen, Erläuterungen, Bemerkungen
4	(D) Näherei, Wäscherei (I) Lager	
46	(D) Lager	

Universität Karlsruhe (TH)
Facility Management (TMB)
OPIK IRAN

Prozess: Wäscheversorgung
Teilprozess: Produktion
Stand: 22.11.2006

Deutschland

Frischwäscheverteilung - zyklisch

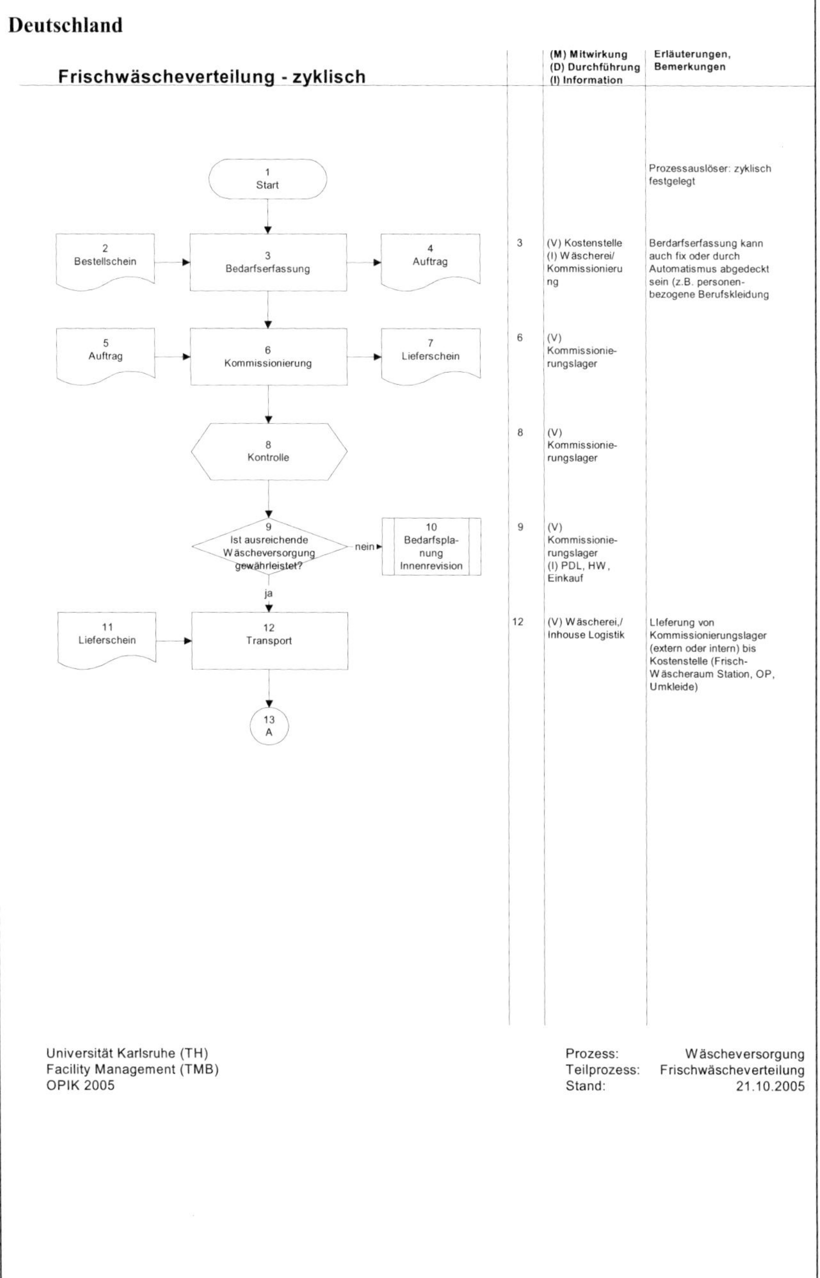

	(M) Mitwirkung (D) Durchführung (I) Information	Erläuterungen, Bemerkungen
		Prozessauslöser: zyklisch festgelegt
3	(V) Kostenstelle (I) Wäscherei/ Kommissionierung	Berdarfserfassung kann auch fix oder durch Automatismus abgedeckt sein (z.B. personen-bezogene Berufskleidung
6	(V) Kommissionie-rungslager	
8	(V) Kommissionie-rungslager	
9	(V) Kommissionie-rungslager (I) PDL, HW, Einkauf	
12	(V) Wäscherei,/ Inhouse Logistik	Lieferung von Kommissionierungslager (extern oder intern) bis Kostenstelle (Frisch-Wäscheraum Station, OP, Umkleide)

Universität Karlsruhe (TH)
Facility Management (TMB)
OPIK 2005

Prozess: Wäscheversorgung
Teilprozess: Frischwäscheverteilung
Stand: 21.10.2005

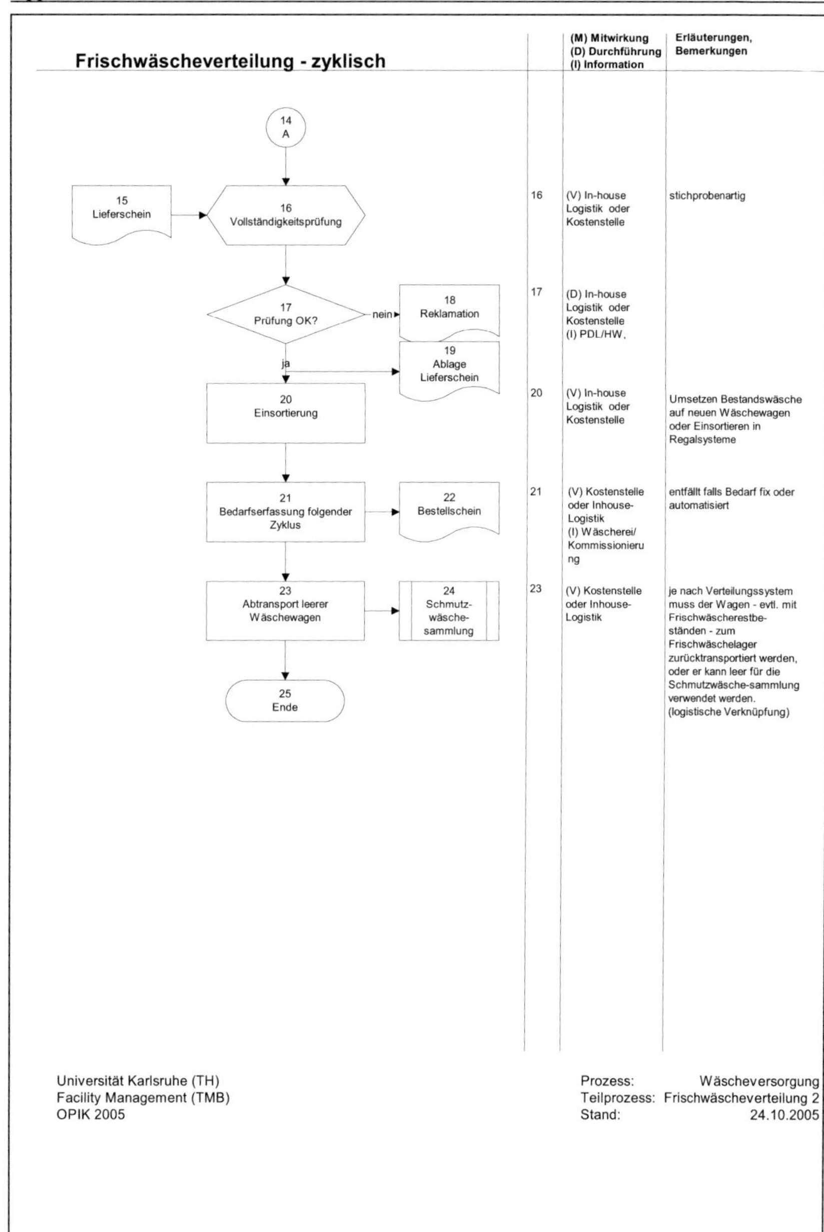
Frischwäscheverteilung - zyklisch

(M) Mitwirkung
(D) Durchführung
(I) Information

Erläuterungen,
Bemerkungen

14
A

15
Lieferschein

16
Vollständigkeitsprüfung

16
(V) In-house
Logistik oder
Kostenstelle

stichprobenartig

17
Prüfung OK?

nein

18
Reklamation

17
(D) In-house
Logistik oder
Kostenstelle
(I) PDL/HW,

ja

19
Ablage
Lieferschein

20
Einsortierung

20
(V) In-house
Logistik oder
Kostenstelle

Umsetzen Bestandswäsche
auf neuen Wäschewagen
oder Einsortieren in
Regalsysteme

21
Bedarfserfassung folgender
Zyklus

22
Bestellschein

21
(V) Kostenstelle
oder Inhouse-
Logistik
(I) Wäscherei/
Kommissionieru
ng

entfällt falls Bedarf fix oder
automatisiert

23
Abtransport leerer
Wäschewagen

24
Schmutz-
wäsche-
sammlung

23
(V) Kostenstelle
oder Inhouse-
Logistik

je nach Verteilungssystem
muss der Wagen - evtl. mit
Frischwäscherestbe-
ständen - zum
Frischwäschelager
zurücktransportiert werden,
oder er kann leer für die
Schmutzwäsche-sammlung
verwendet werden.
(logistische Verknüpfung)

25
Ende

Universität Karlsruhe (TH)
Facility Management (TMB)
OPIK 2005

Prozess: Wäscheversorgung
Teilprozess: Frischwäscheverteilung 2
Stand: 24.10.2005

Iran

Frischwäscheverteilung - zyklisch

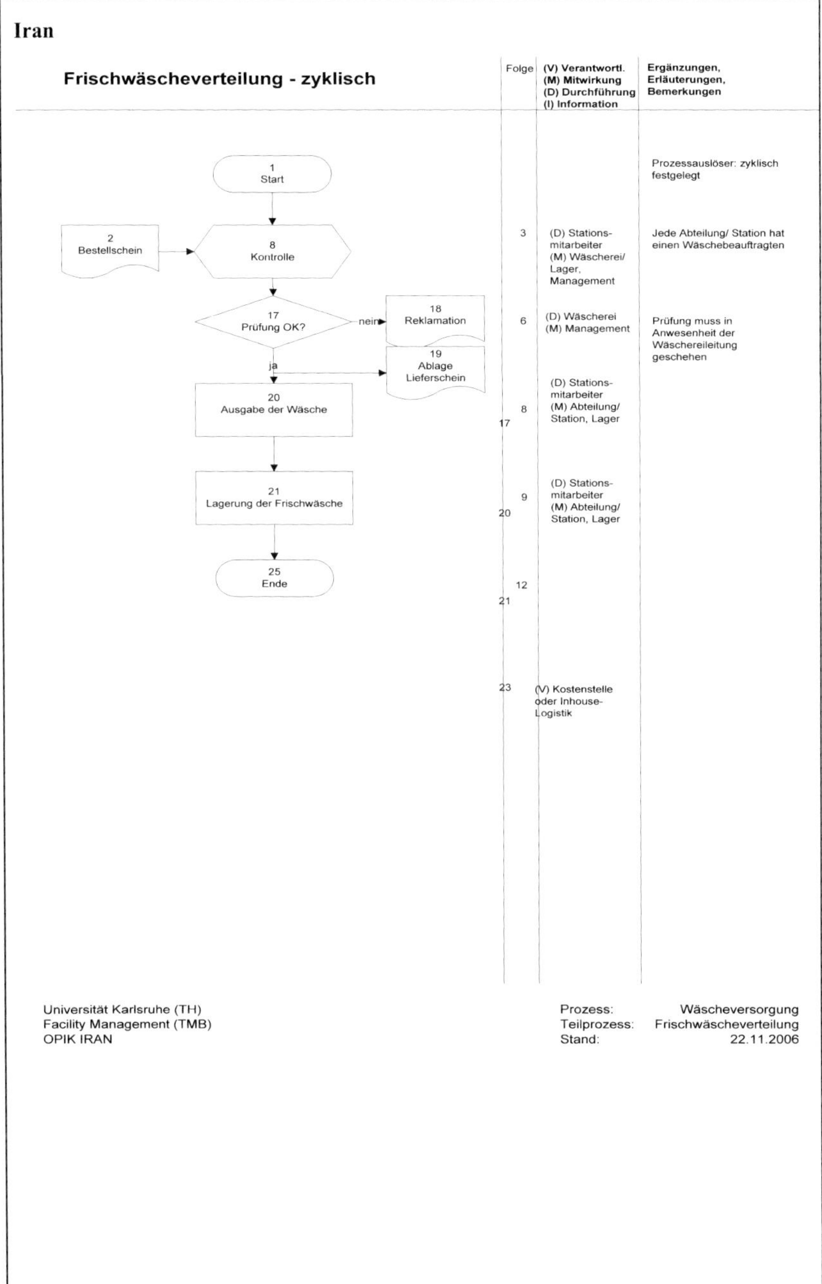

Folge	(V) Verantwortl. (M) Mitwirkung (D) Durchführung (I) Information	Ergänzungen, Erläuterungen, Bemerkungen
		Prozessauslöser: zyklisch festgelegt
3	(D) Stations- mitarbeiter (M) Wäscherei/ Lager, Management	Jede Abteilung/ Station hat einen Wäschebeauftragten
6	(D) Wäscherei (M) Management	Prüfung muss in Anwesenheit der Wäschereileitung geschehen
8 17	(D) Stations- mitarbeiter (M) Abteilung/ Station, Lager	
9 20	(D) Stations- mitarbeiter (M) Abteilung/ Station, Lager	
12 21		
23	(V) Kostenstelle oder Inhouse- Logistik	

Universität Karlsruhe (TH)
Facility Management (TMB)
OPIK IRAN

Prozess: Wäscheversorgung
Teilprozess: Frischwäscheverteilung
Stand: 22.11.2006

Deutschland

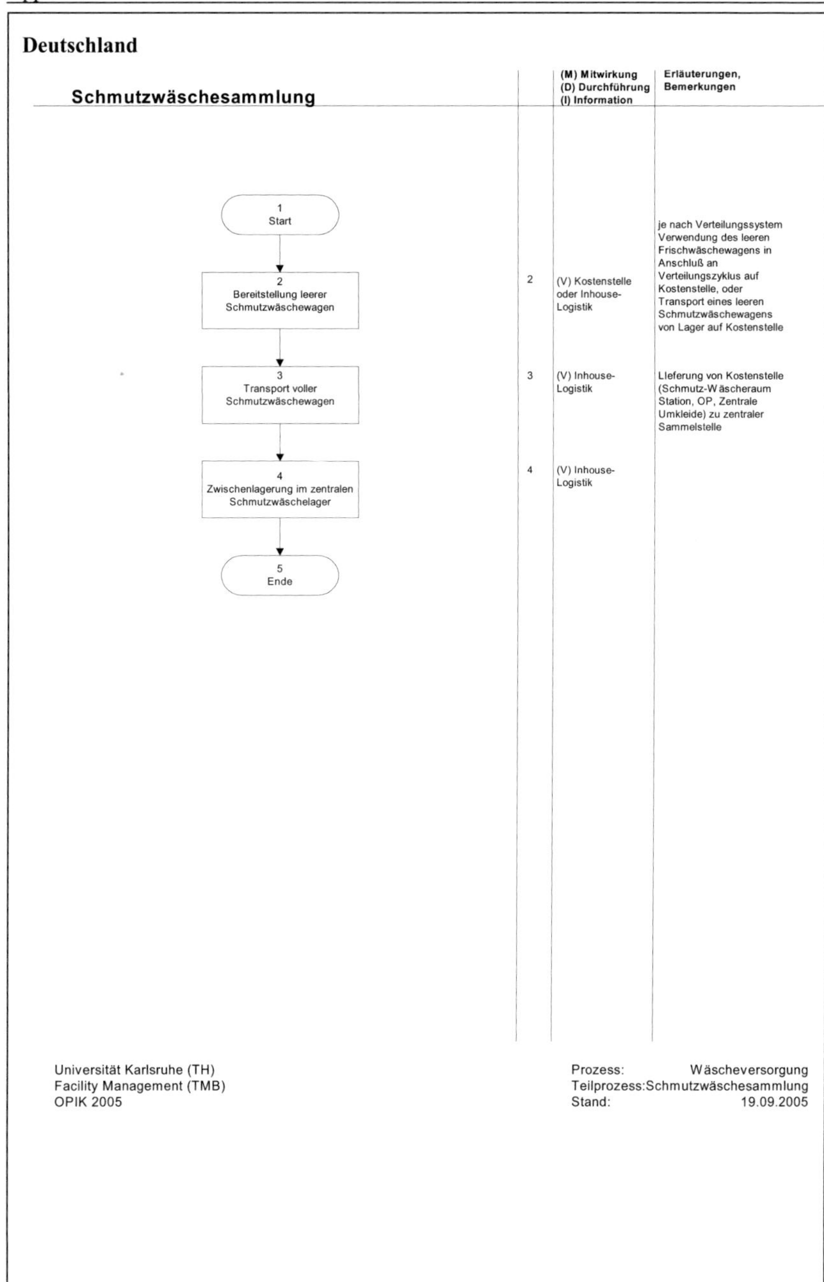

Universität Karlsruhe (TH)
Facility Management (TMB)
OPIK 2005

Prozess: Wäscheversorgung
Teilprozess:Schmutzwäschesammlung
Stand: 19.09.2005

Iran

Schmutzwäschesammlung

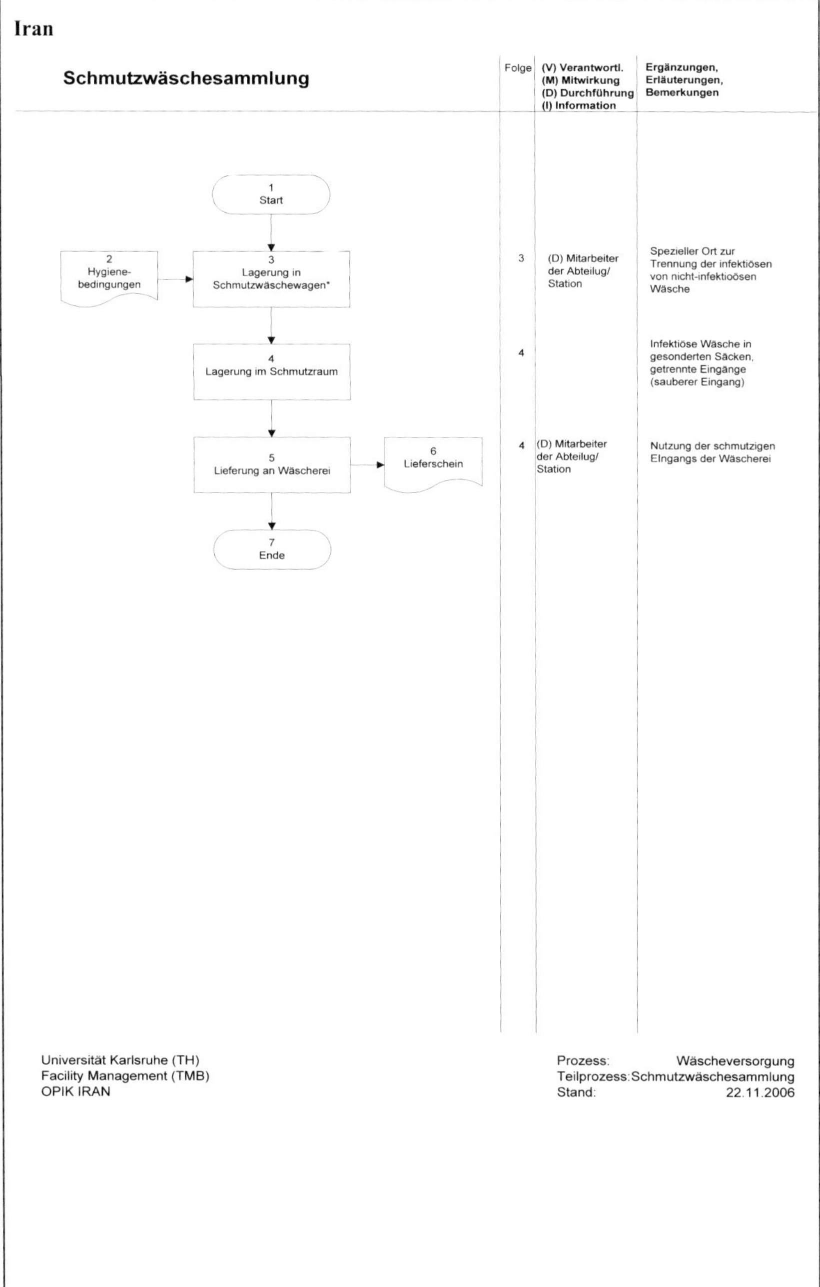

Universität Karlsruhe (TH)
Facility Management (TMB)
OPIK IRAN

Prozess: Wäscheversorgung
Teilprozess:Schmutzwäschesammlung
Stand: 22.11.2006

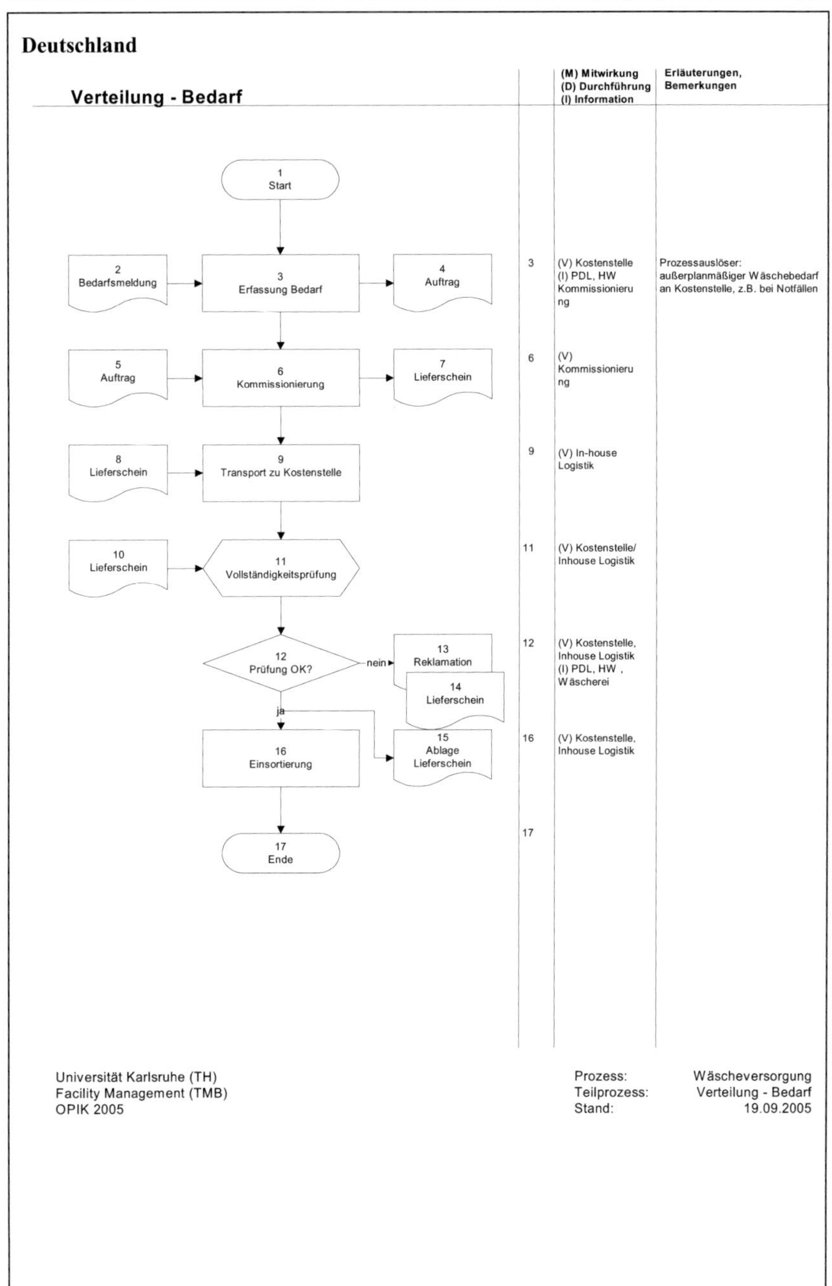
Deutschland
Verteilung - Bedarf
(M) Mitwirkung
(D) Durchführung
(I) Information
Erläuterungen, Bemerkungen
1
Start
2
Bedarfsmeldung
3
Erfassung Bedarf
4
Auftrag
3
(V) Kostenstelle
(I) PDL, HW
Kommissionierung
Prozessauslöser:
außerplanmäßiger Wäschebedarf
an Kostenstelle, z.B. bei Notfällen
5
Auftrag
6
Kommissionierung
7
Lieferschein
6
(V) Kommissionierung
8
Lieferschein
9
Transport zu Kostenstelle
9
(V) In-house Logistik
10
Lieferschein
11
Vollständigkeitsprüfung
11
(V) Kostenstelle/ Inhouse Logistik
12
Prüfung OK?
nein
13
Reklamation
14
Lieferschein
12
(V) Kostenstelle, Inhouse Logistik
(I) PDL, HW , Wäscherei
ja
16
Einsortierung
15
Ablage Lieferschein
16
(V) Kostenstelle, Inhouse Logistik
17
Ende
17
Universität Karlsruhe (TH)
Facility Management (TMB)
OPIK 2005
Prozess:
Teilprozess:
Stand:
Wäscheversorgung
Verteilung - Bedarf
19.09.2005

Deutschland

Nutzung

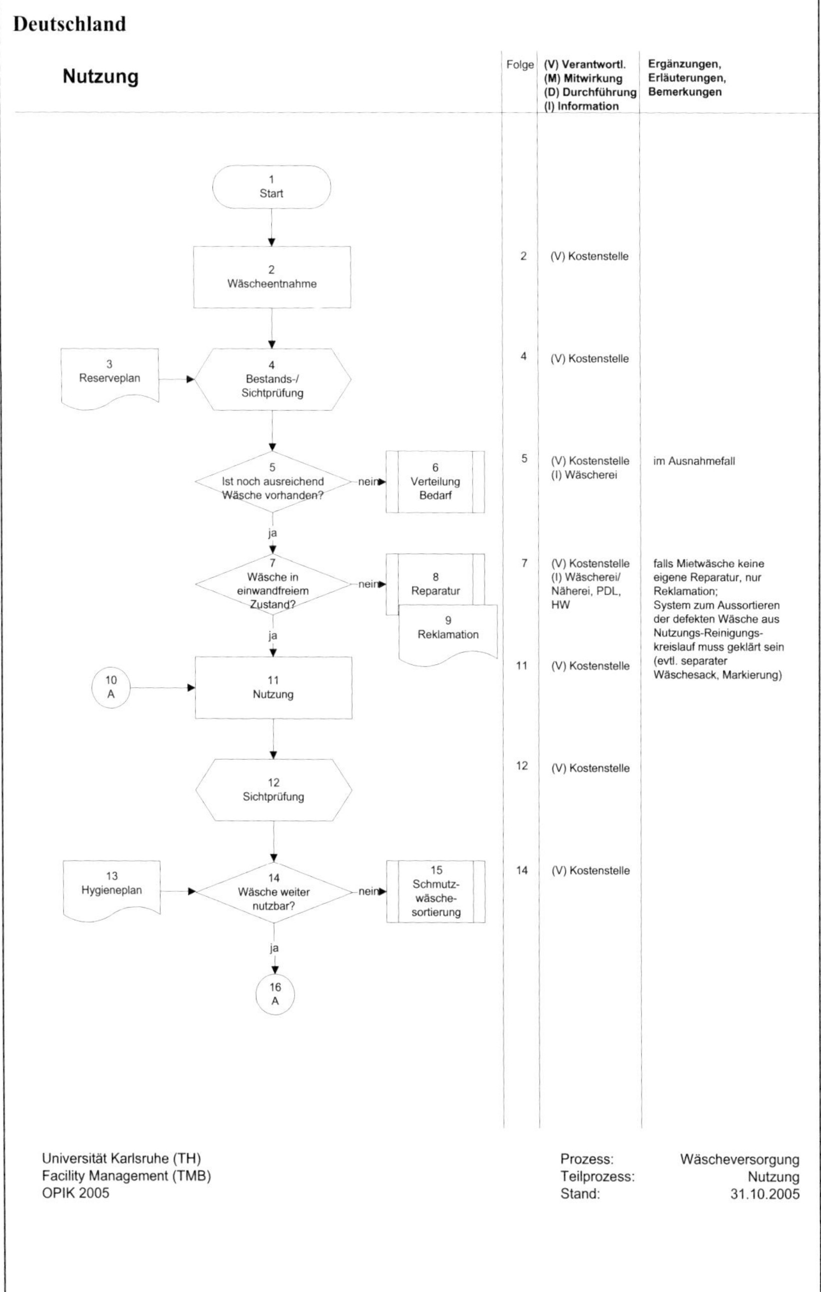

Universität Karlsruhe (TH)
Facility Management (TMB)
OPIK 2005

Prozess: Wäscheversorgung
Teilprozess: Nutzung
Stand: 31.10.2005

Iran

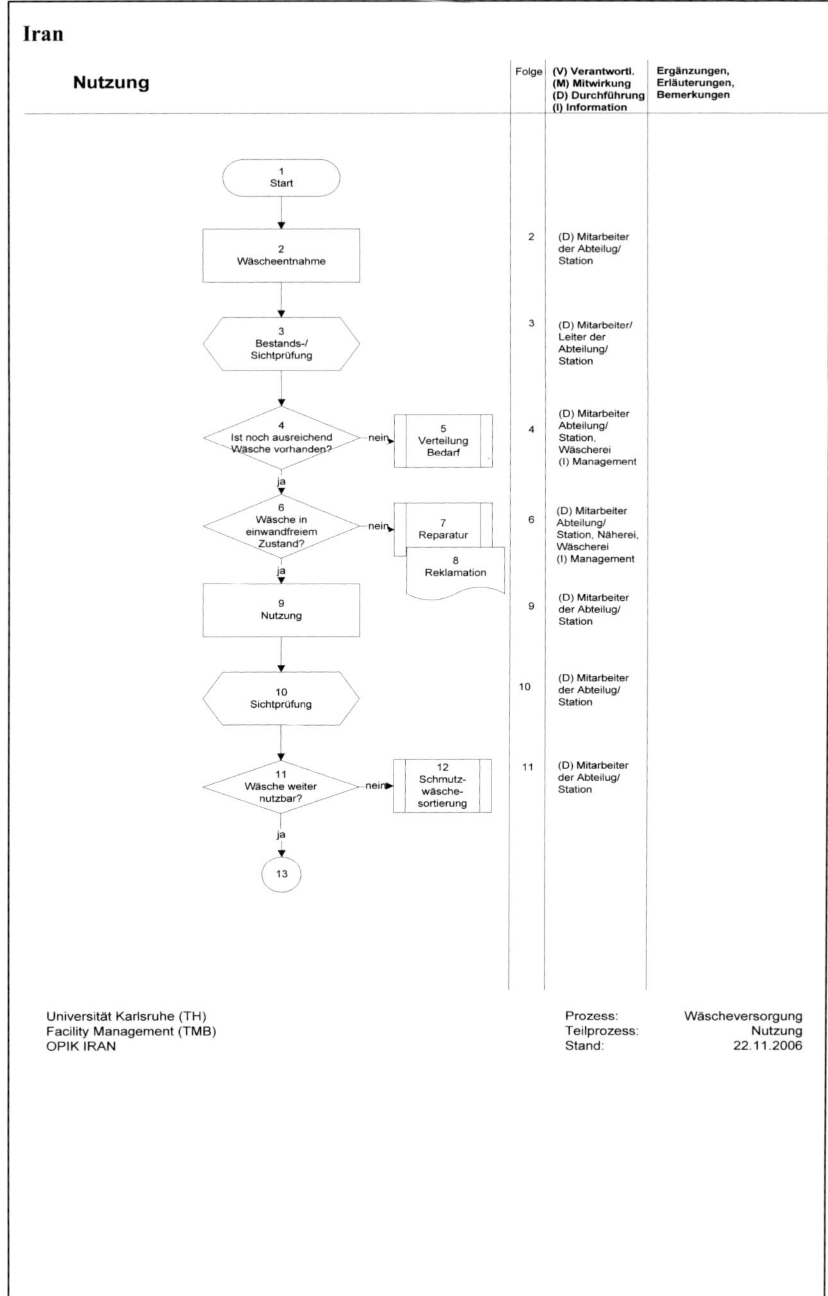

Universität Karlsruhe (TH)
Facility Management (TMB)
OPIK IRAN

Prozess: Wäscheversorgung
Teilprozess: Nutzung
Stand: 22.11.2006

Deutschland

Frischwäscheverteilung - zyklisch

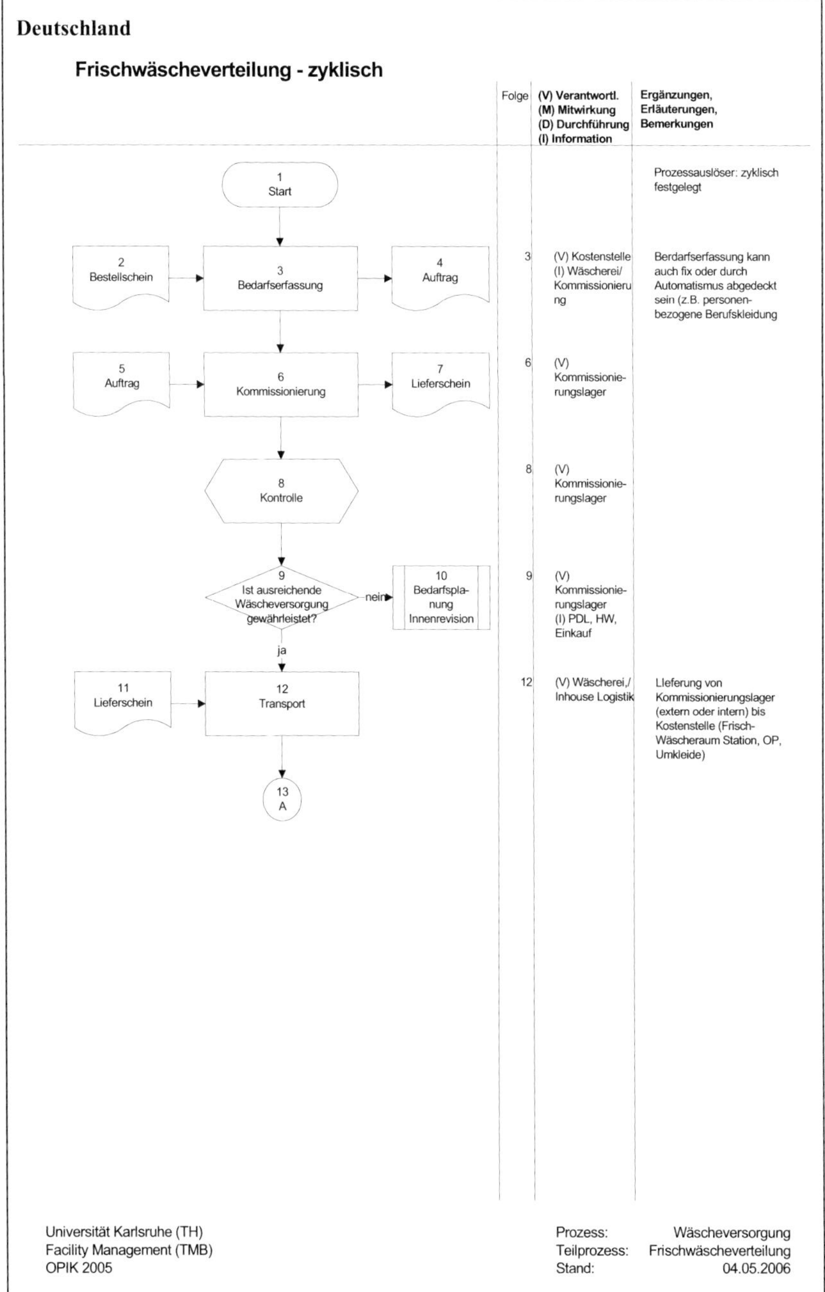

Deutschland

Reinigung

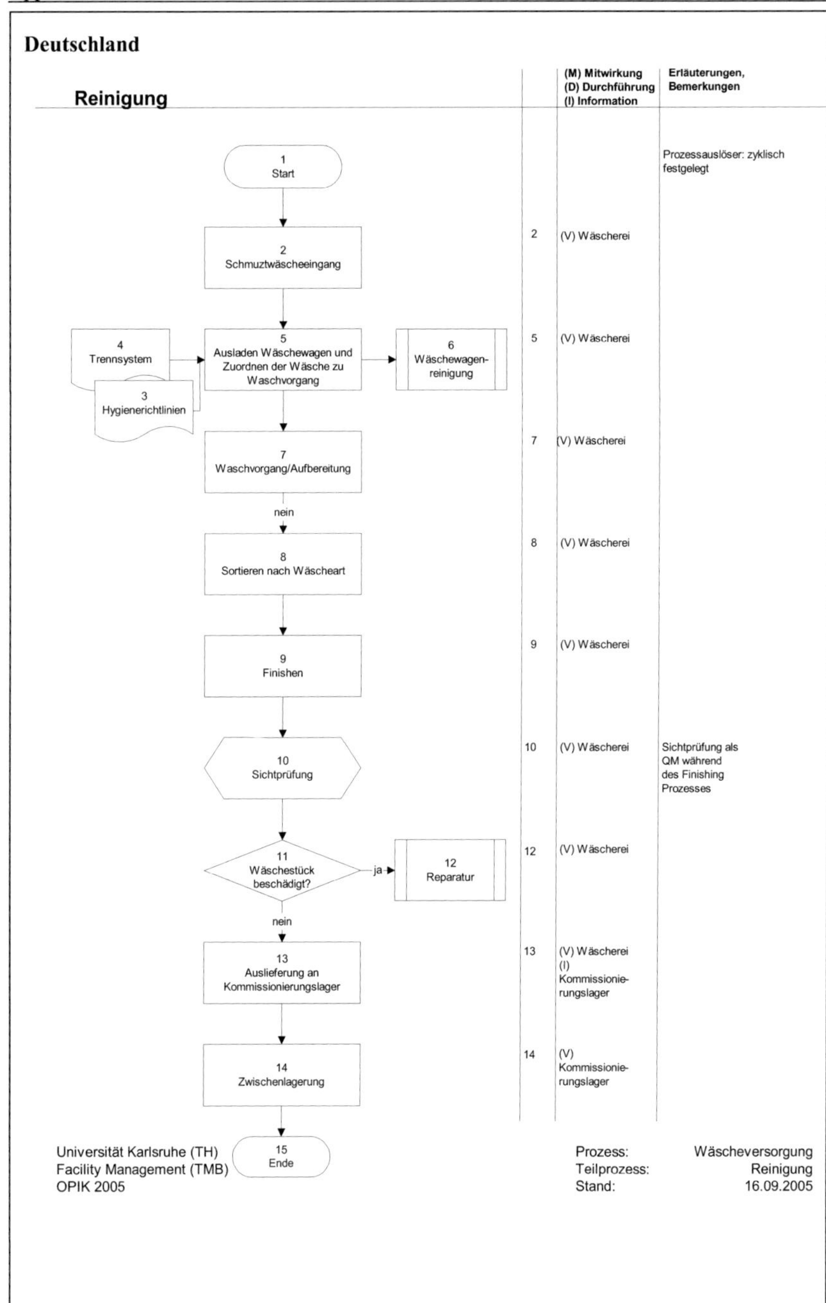

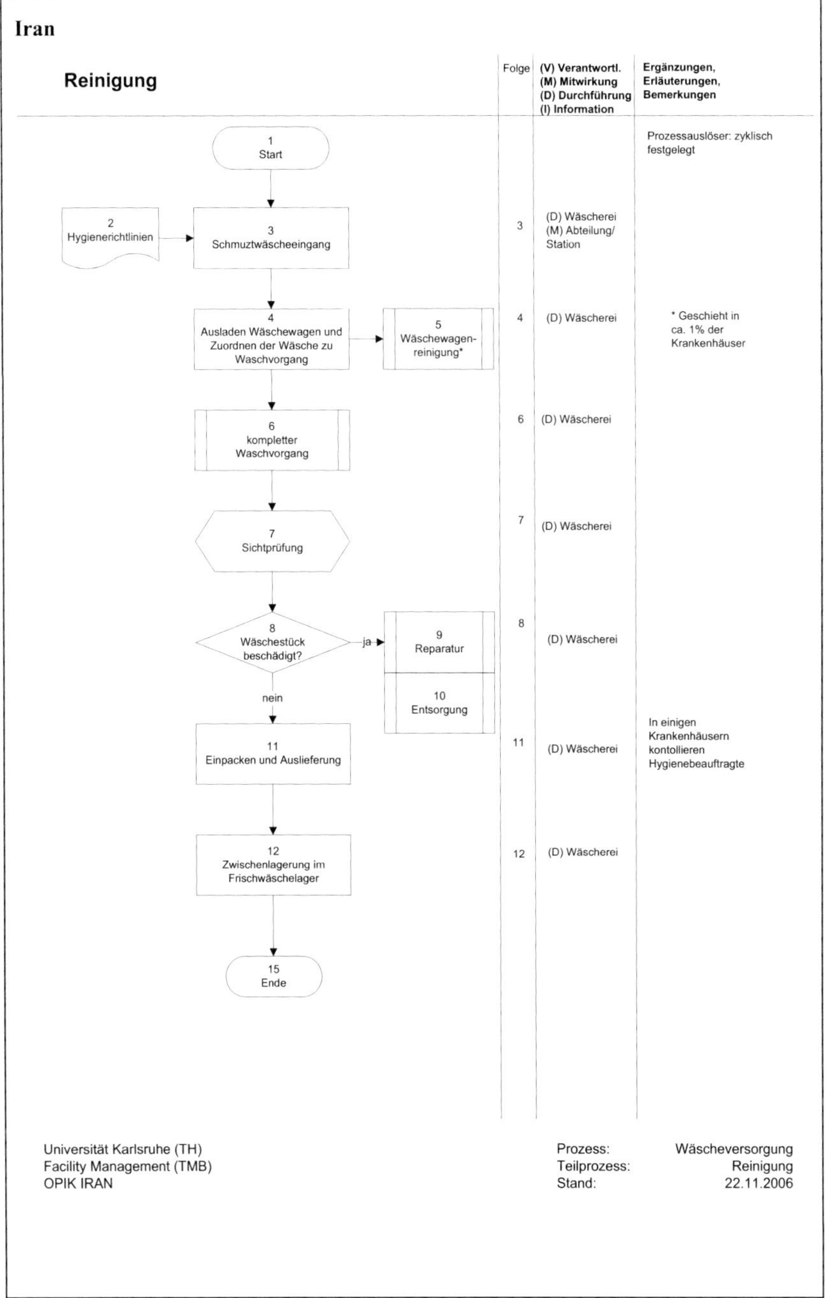
Iran
Reinigung
Folge
(V) Verantwortl.
(M) Mitwirkung
(D) Durchführung
(I) Information
Ergänzungen,
Erläuterungen,
Bemerkungen
1
Start
Prozessauslöser: zyklisch
festgelegt
2
Hygienerichtlinien
3
Schmuztwäscheeingang
3
(D) Wäscherei
(M) Abteilung/
Station
4
Ausladen Wäschewagen und
Zuordnen der Wäsche zu
Waschvorgang
5
Wäschewagen-
reinigung*
4
(D) Wäscherei
* Geschieht in
ca. 1% der
Krankenhäuser
6
kompletter
Waschvorgang
6
(D) Wäscherei
7
Sichtprüfung
7
(D) Wäscherei
8
Wäschestück
beschädigt?
ja
9
Reparatur
8
(D) Wäscherei
nein
10
Entsorgung
11
Einpacken und Auslieferung
11
(D) Wäscherei
In einigen
Krankenhäusern
kontollieren
Hygienebeauftragte
12
Zwischenlagerung im
Frischwäschelager
12
(D) Wäscherei
15
Ende
Universität Karlsruhe (TH)
Facility Management (TMB)
OPIK IRAN
Prozess:
Teilprozess:
Stand:
Wäscheversorgung
Reinigung
22.11.2006

Deutschland

Durchführung Reparatur

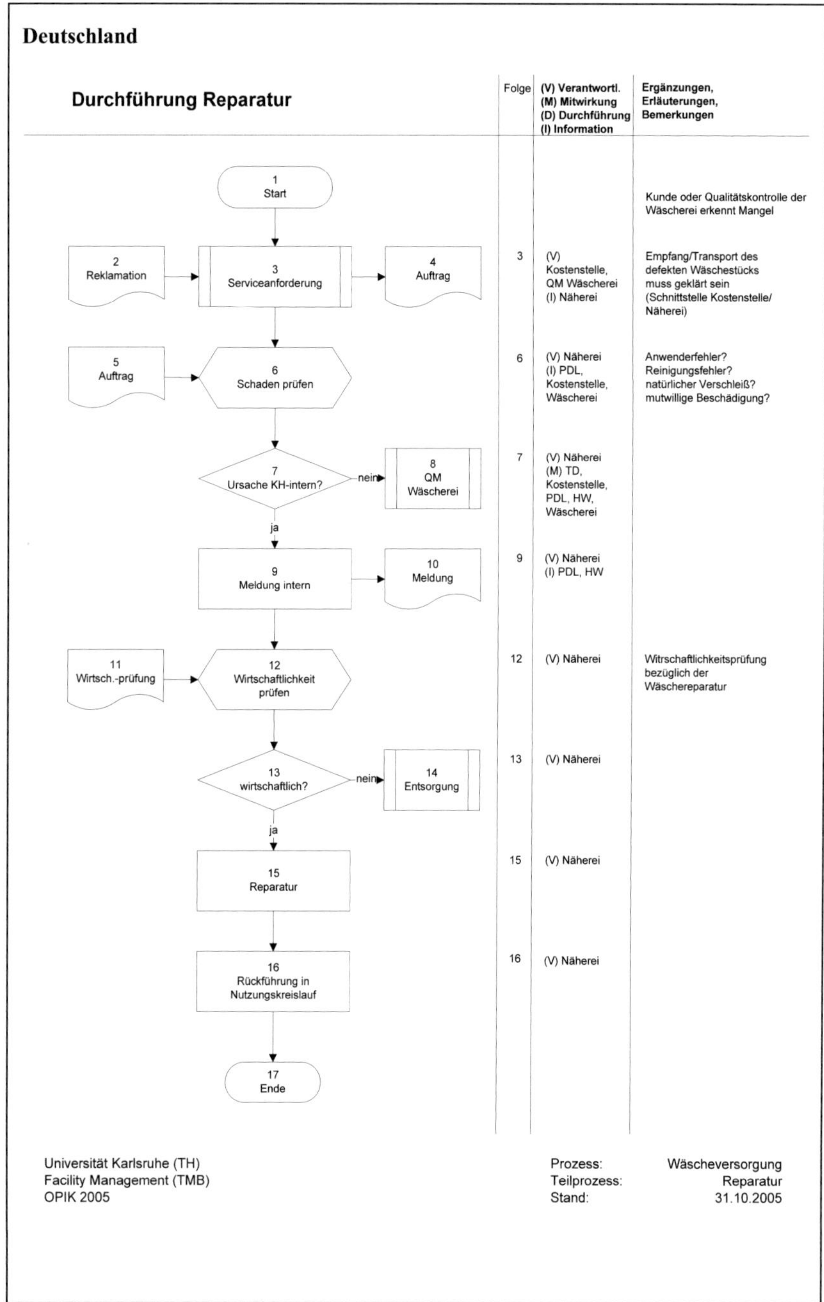

Universität Karlsruhe (TH)
Facility Management (TMB)
OPIK 2005

Prozess: Wäscheversorgung
Teilprozess: Reparatur
Stand: 31.10.2005

Iran

Durchführung Reparatur

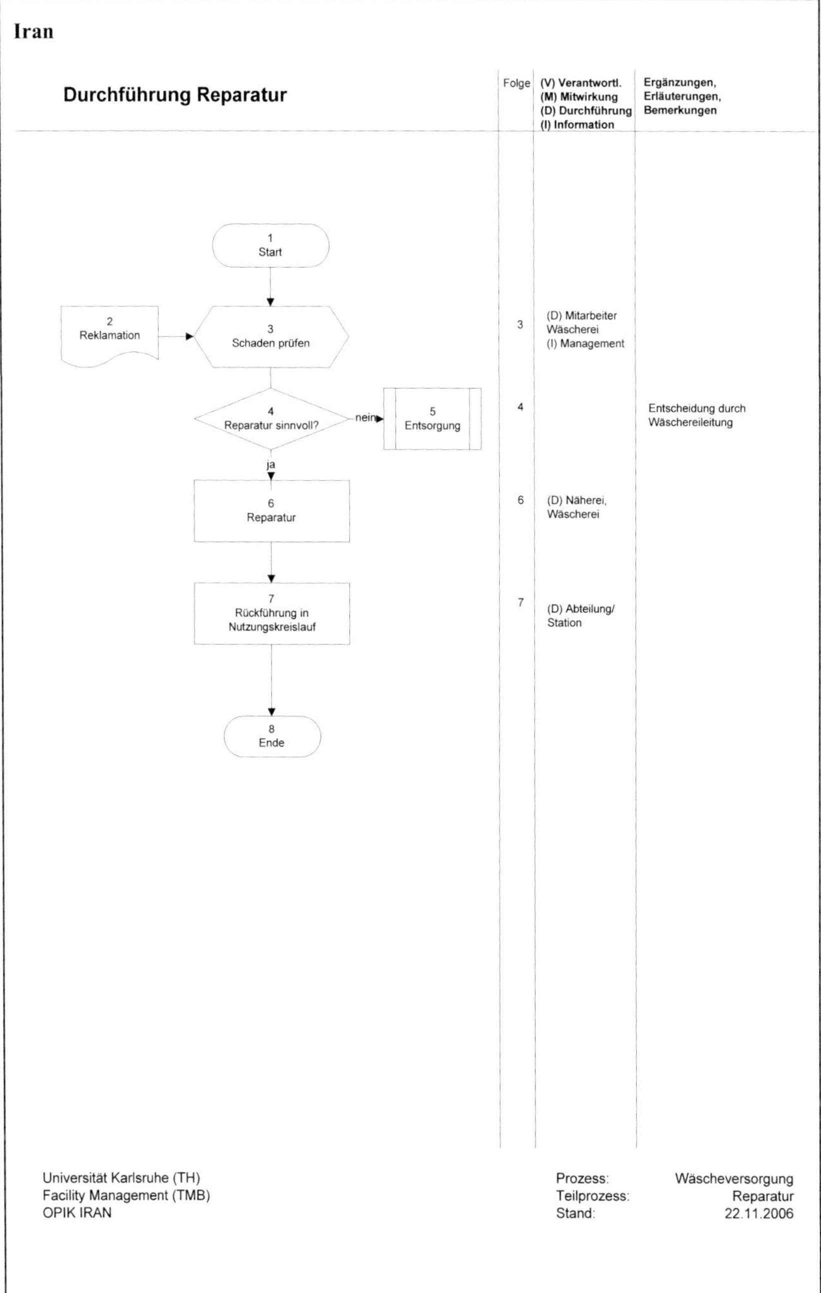

	Folge	(V) Verantwortl. (M) Mitwirkung (D) Durchführung (I) Information	Ergänzungen, Erläuterungen, Bemerkungen
	3	(D) Mitarbeiter Wäscherei (I) Management	
	4		Entscheidung durch Wäschereileitung
	6	(D) Näherei, Wäscherei	
	7	(D) Abteilung/ Station	

Universität Karlsruhe (TH) Facility Management (TMB) OPIK IRAN	Prozess: Wäscheversorgung Teilprozess: Reparatur Stand: 22.11.2006

Deutschland

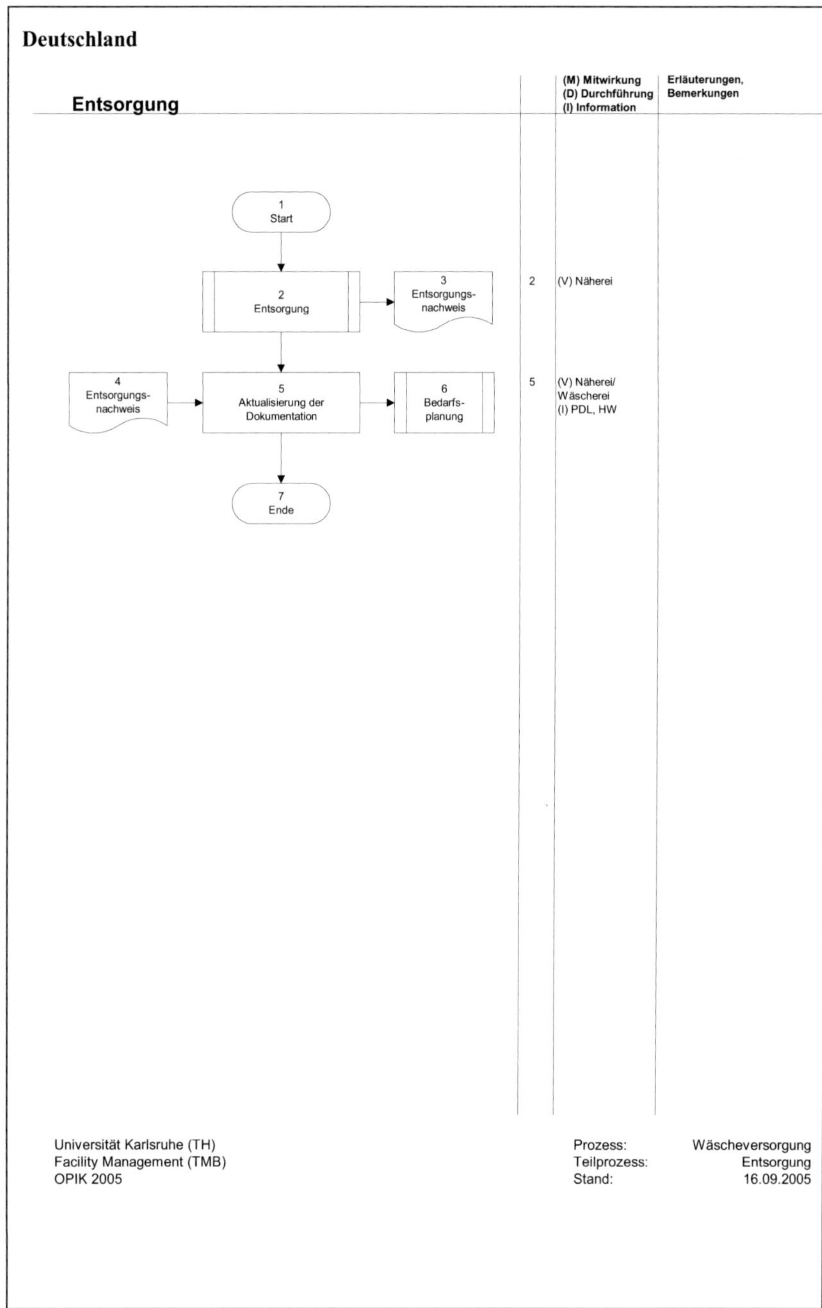

Iran

Entsorgung

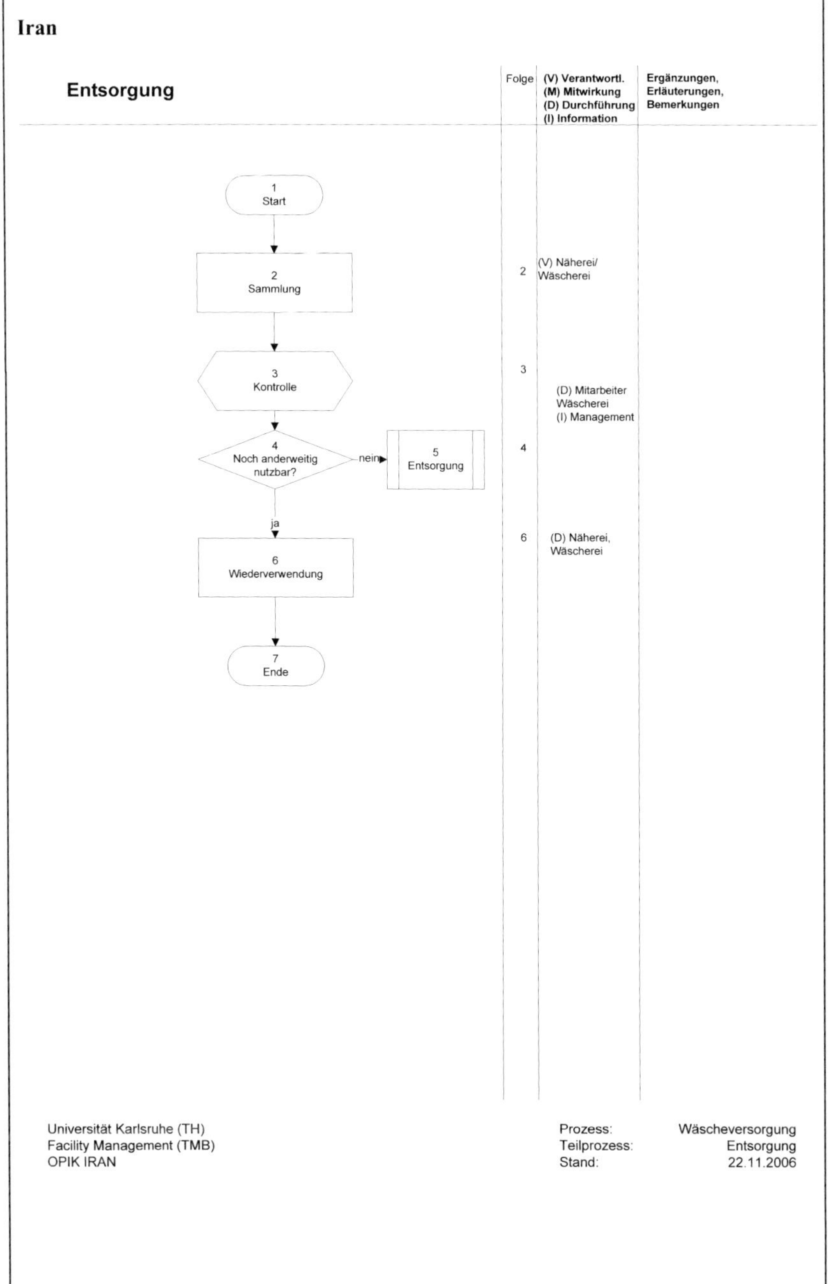

Universität Karlsruhe (TH)
Facility Management (TMB)
OPIK IRAN

Prozess: Wäscheversorgung
Teilprozess: Entsorgung
Stand: 22.11.2006

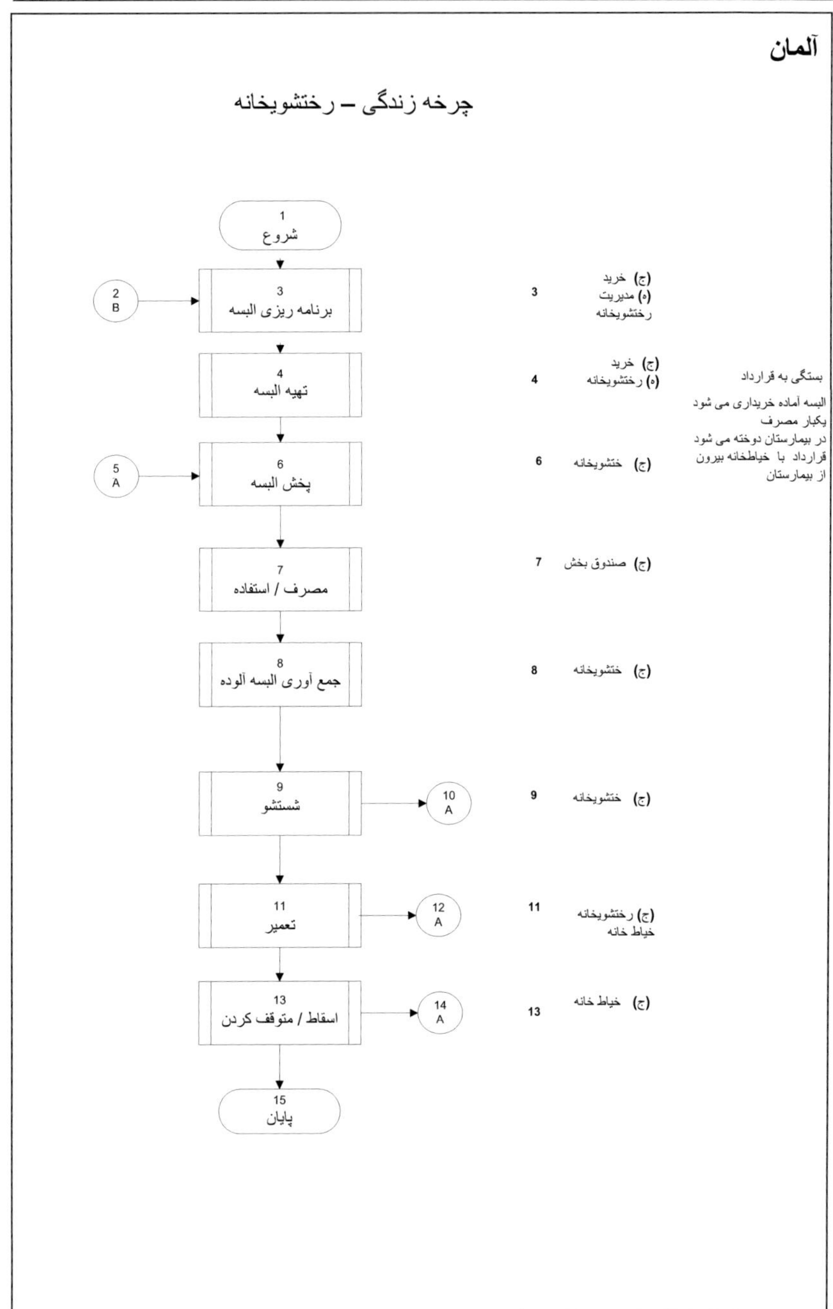
آلمان
چرخه زندگی – رختشویخانه
1
شروع
2
B
3
برنامه ریزی البسه
4
تهیه البسه
5
A
6
پخش البسه
7
مصرف / استفاده
8
جمع آوری البسه آلوده
9
شستشو
10
A
11
تعمیر
12
A
13
اسقاط / متوقف کردن
14
A
15
پایان
3   (ج) خرید
(ه) مدیریت رختشویخانه
4   (ج) خرید
(ه) رختشویخانه
6   (ج) ختشویخانه
7   (ج) صندوق بخش
8   (ج) ختشویخانه
9   (ج) ختشویخانه
11   (ج) رختشویخانه
خیاط خانه
13   (ج) خیاط خانه
بستگی به قرارداد
البسه آماده خریداری می شود
یکبار مصرف
در بیمارستان دوخته می شود
قرارداد با خیاطخانه بیرون
از بیمارستان

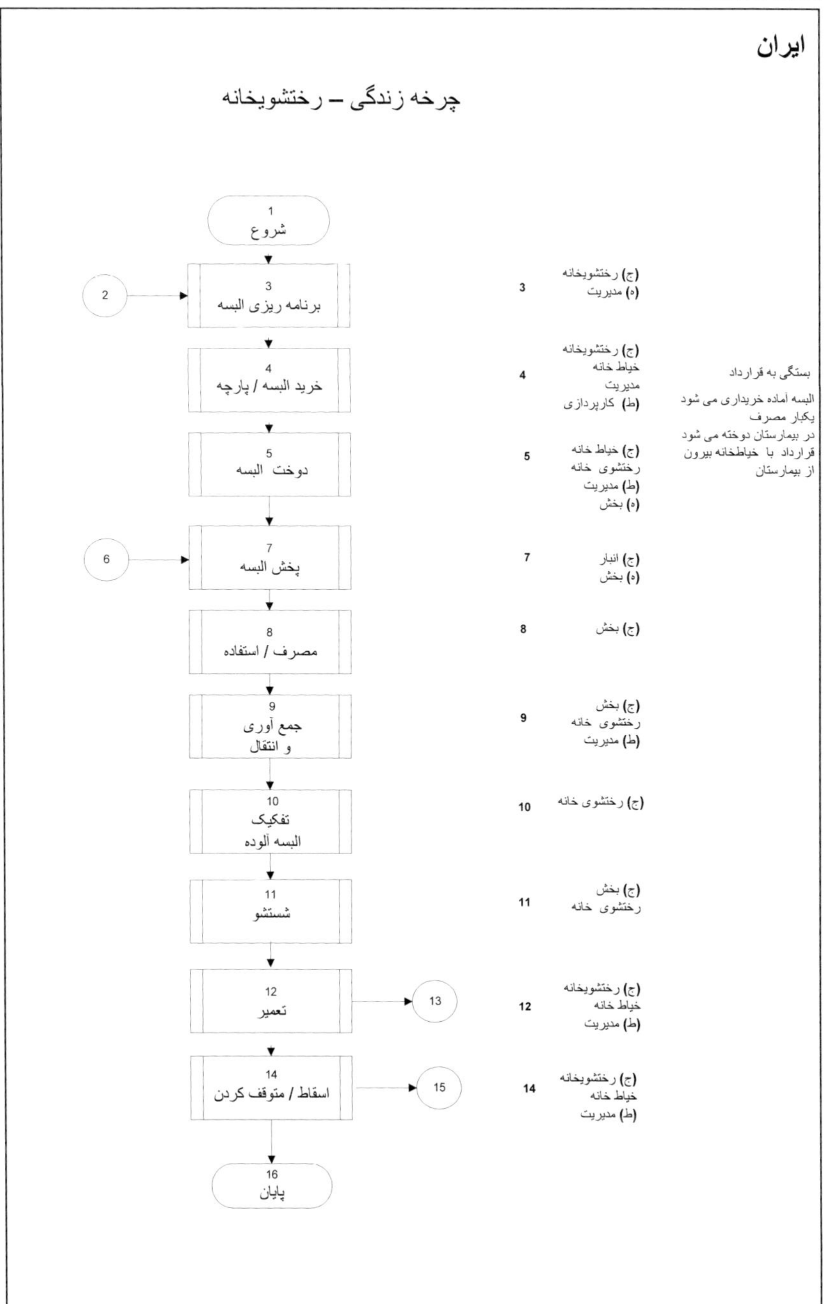
ایران

چرخه زندگی – رختشویخانه

شروع
1

2

3
برنامه ریزی البسه

4
خرید البسه / پارچه

5
دوخت البسه

6

7
پخش البسه

8
مصرف / استفاده

9
جمع آوری
و انتقال

10
تفکیک
البسه آلوده

11
شستشو

12
تعمیر

13

14
اسقاط / متوقف کردن

15

16
پایان

3
(ج) رختشویخانه
(ه) مدیریت

4
(ج) رختشویخانه
خیاط خانه
مدیریت
(ط) کارپردازی

5
(ج) خیاط خانه
رختشوی خانه
(ط) مدیریت
(ه) بخش

7
(ج) انبار
(ه) بخش

8
(ج) بخش

9
(ج) بخش
رختشوی خانه
(ط) مدیریت

10
(ج) رختشوی خانه

11
(ج) بخش
رختشوی خانه

12
(ج) رختشویخانه
خیاط خانه
(ط) مدیریت

14
(ج) رختشویخانه
خیاط خانه
(ط) مدیریت

بستگی به قرارداد
البسه آماده خریداری می شود
یکبار مصرف
در بیمارستان دوخته می شود
قرارداد با خیاطخانه بیرون
از بیمارستان

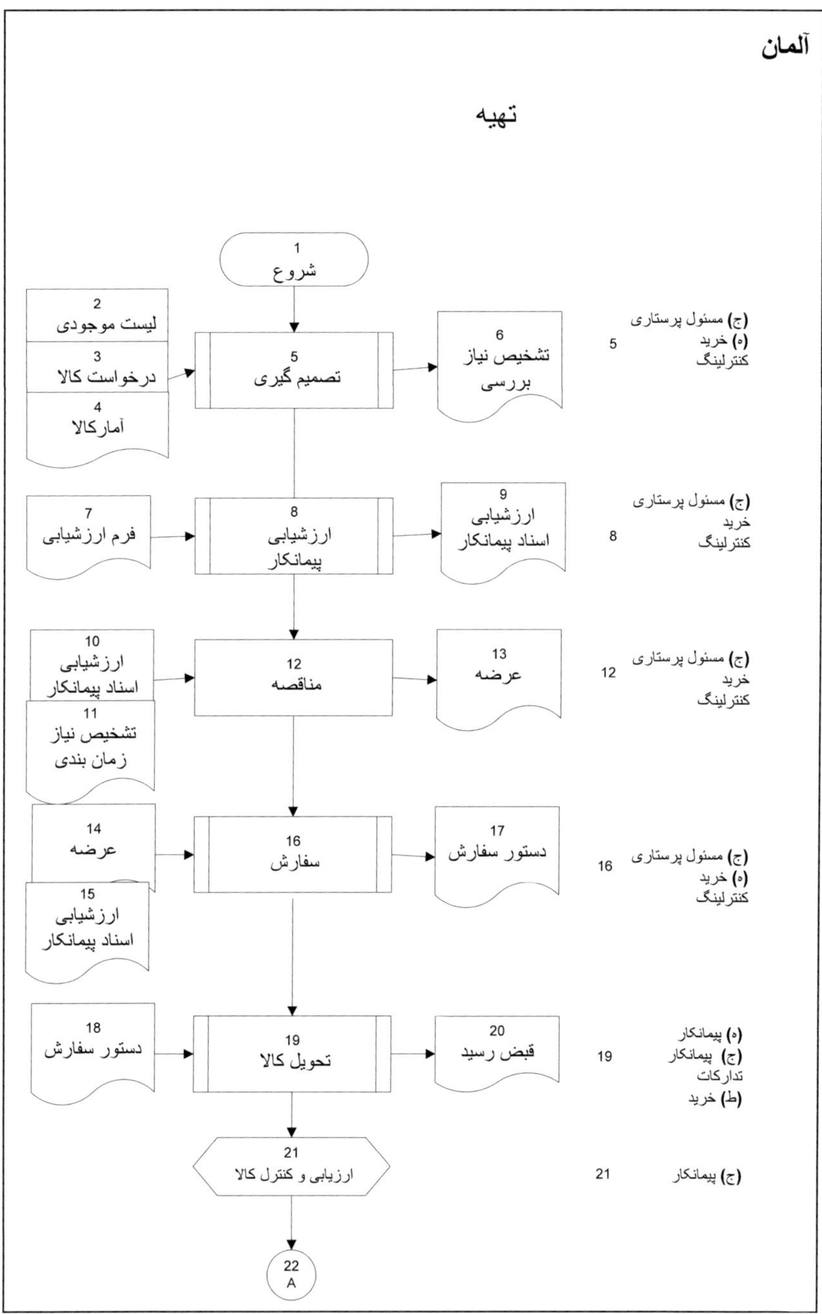
آلمان

تهیه

۱
شروع

۲
لیست موجودی

۳
درخواست کالا

۴
آمارکالا

۵
تصمیم گیری

۶
تشخیص نیاز
بررسی

(ج) مسئول پرستاری
(ه) خرید
کنترلینگ
۵

۷
فرم ارزشیابی

۸
ارزشیابی
پیمانکار

۹
ارزشیابی
اسناد پیمانکار

(ج) مسئول پرستاری
خرید
کنترلینگ
۸

۱۰
ارزشیابی
اسناد پیمانکار

۱۱
تشخیص نیاز
زمان بندی

۱۲
مناقصه

۱۳
عرضه

(ج) مسئول پرستاری
خرید
کنترلینگ
۱۲

۱۴
عرضه

۱۵
ارزشیابی
اسناد پیمانکار

۱۶
سفارش

۱۷
دستور سفارش

(ج) مسئول پرستاری
(ه) خرید
کنترلینگ
۱۶

۱۸
دستور سفارش

۱۹
تحویل کالا

۲۰
قبض رسید

(ه) پیمانکار
(ج) پیمانکار
تدارکات
(ط) خرید
۱۹

۲۱
ارزیابی و کنترل کالا

(ج) پیمانکار
۲۱

۲۲
A

آلمان

تهیه ۲

23
A
24
قبض رسید
25
ارزیابی OK?
بلی
26
B
خیر
(ج) تدارکات
(ط) خرید
مسئول پرستاری
25
27
قبض رسید
28
تلخیز
شکایت به شرکت تولید کننده
29
گزارش تلخیز
30
قبض رسید
(ج) تدارکات
(ط) پیمانکار
28
31
B
32
تلخیز
ارزیابی کیفیت
(ج) تدارکات
مسئول پرستاری
32
33
قبض رسید
34
ارزیابی OK?
بلی
35
C
خیر
(ج) خرید
مسئول پرستاری
(ط) پیمانکار
34
36
قبض رسید
37
تلخیز
شکایت به شرکت تولید کننده
38
گزارش تلخیز
(ج) خرید
مسئول پرستاری
(ط) پیمانکار
37
39
مدارک کامل
41
تلخیز
42
گزارش تلخیز
40
C
(ج) کنترلینگ
مسئول پرستاری
(ط) پیمانکار
41
43
گزارش تلخیز
44
بروز کردن مستندات
45
مستندات بروز
(ج) خرید
مسئول پرستاری
(ط) حسابداری
44
46
علامتگزاری
47
مستندات
علامتگزاری
(ج) خرید
مسئول پرستاری
(ه) خیاطی
رختشویخانه
46
48
پایان

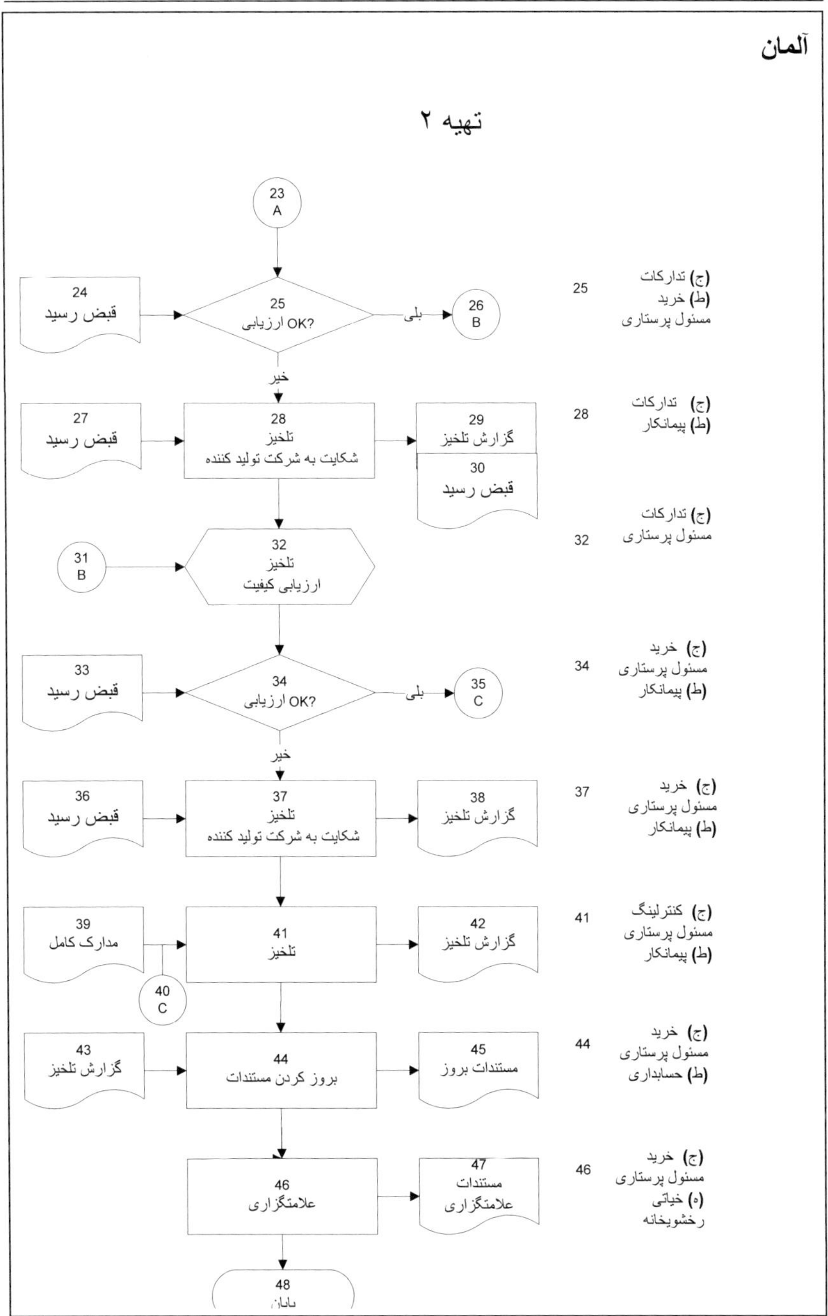

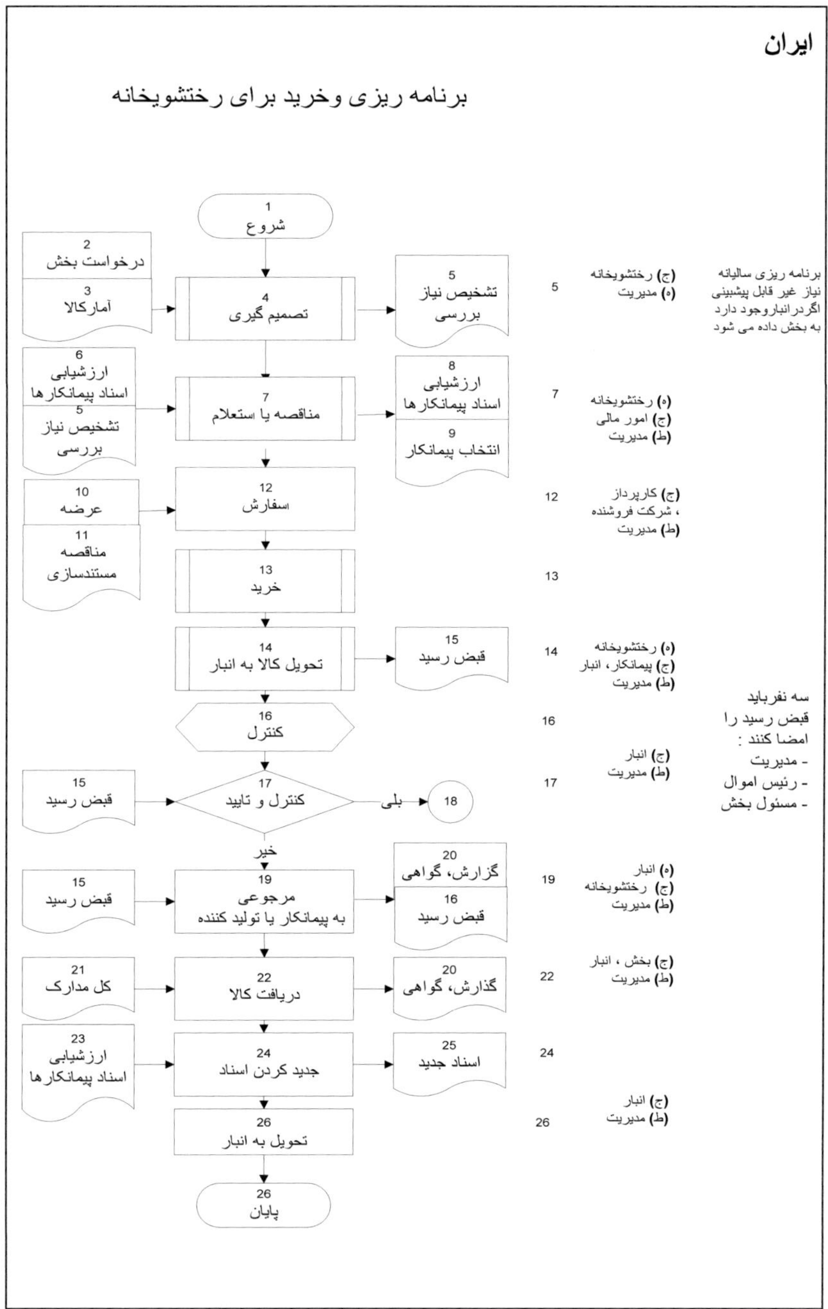
ایران
برنامه ریزی وخرید برای رختشویخانه
1 شروع
2 درخواست بخش
3 آمارکالا
4 تصمیم گیری
5 تشخیص نیاز بررسی
5 (ج) رختشویخانه (ه) مدیریت
برنامه ریزی سالیانه نیاز غیر قابل پیشبینی اگر درانباروجود دارد به بخش داده می شود
6 ارزشیابی اسناد پیمانکارها
5 تشخیص نیاز بررسی
7 مناقصه یا استعلام
8 ارزشیابی اسناد پیمانکارها
9 انتخاب پیمانکار
7 (ه) رختشویخانه (ج) امور مالی (ط) مدیریت
10 عرضه
11 مناقصه مستندسازی
12 اسفارش
12 (ج) کارپرداز ، شرکت فروشنده (ط) مدیریت
13 خرید
13
14 تحویل کالا به انبار
15 قبض رسید
14 (ه) رختشویخانه (ج) پیمانکار ، انبار (ط) مدیریت
16 کنترل
16 (ج) انبار (ط) مدیریت
سه نفرباید قبض رسید را امضا کنند : - مدیریت - رئیس اموال - مسئول بخش
15 قبض رسید
17 کنترل و تایید
بلی
18
17 (ط) مدیریت
خیر
15 قبض رسید
19 مرجوعی به پیمانکار یا تولید کننده
20 گزارش، گواهی
16 قبض رسید
19 (ه) انبار (ج) رختشویخانه (ط) مدیریت
21 کل مدارک
22 دریافت کالا
20 گزارش، گواهی
22 (ج) بخش ، انبار (ط) مدیریت
23 ارزشیابی اسناد پیمانکارها
24 جدید کردن اسناد
25 اسناد جدید
24
26 تحویل به انبار
26 (ج) انبار (ط) مدیریت
26 پایان

ایران

دوخت البسه

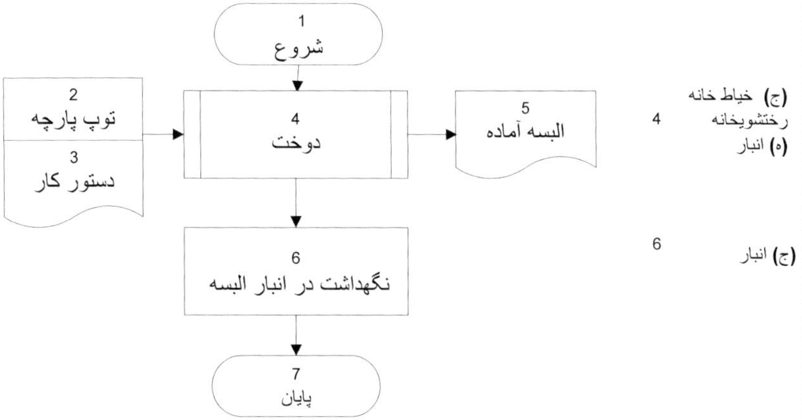

```
            ┌──────────────┐
            │      1        │
            │    شروع       │
            └──────┬───────┘
                   │
  ┌──────────┐  ┌──▼──────────┐   ┌──────────────┐
  │    2     │  │     4        │   │     5         │
  │ توپ پارچه│─▶│    دوخت      │──▶│  البسه آماده  │
  │    3     │  │              │   │               │
  │ دستور کار│  └──────┬───────┘   └──────────────┘
  └──────────┘         │
                       │
              ┌────────▼────────────┐
              │        6            │
              │ نگهداشت در انبار البسه│
              └────────┬────────────┘
                       │
              ┌────────▼───────┐
              │      7          │
              │    پایان        │
              └────────────────┘
```

(ج) خیاط خانه   4
رختشویخانه
(ه) انبار

(ج) انبار   6

آلمان

پخش البسه

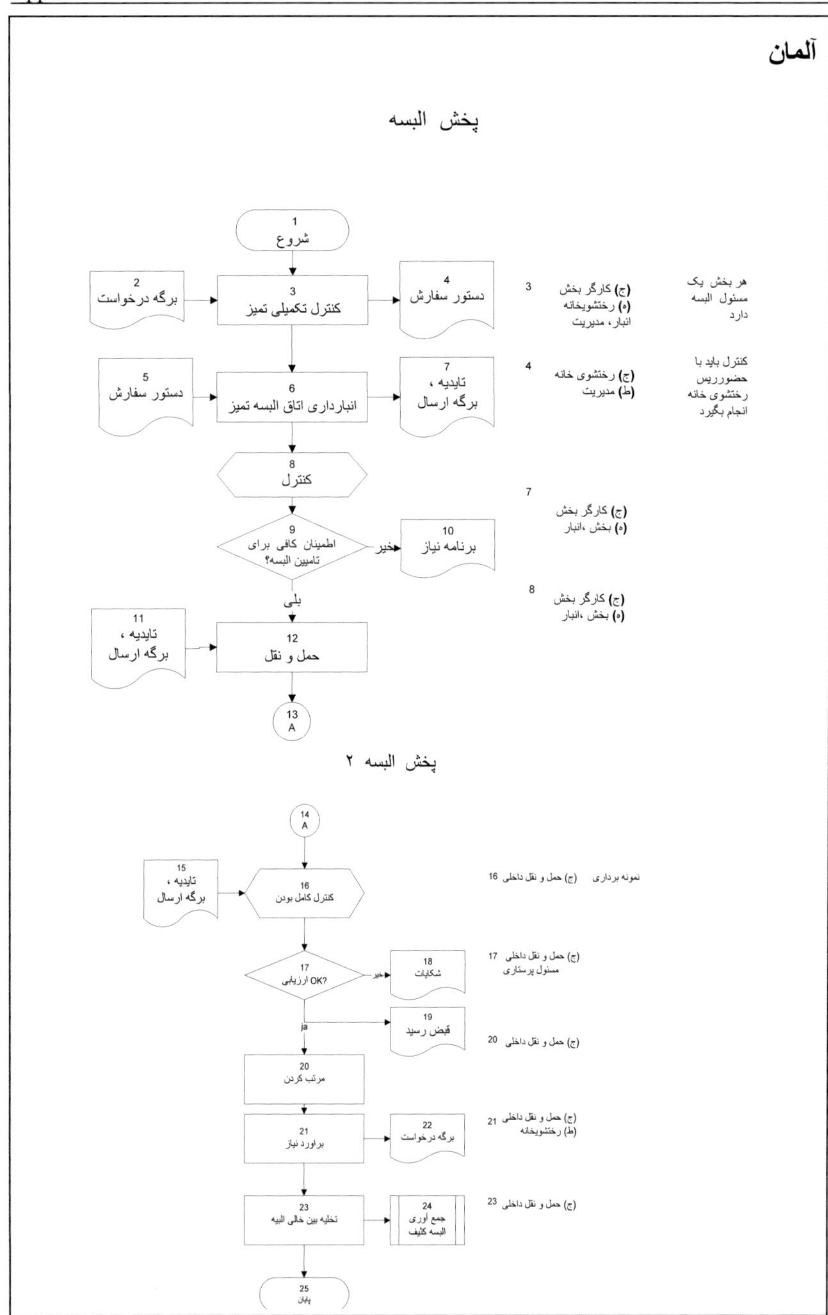

پخش البسه ۲

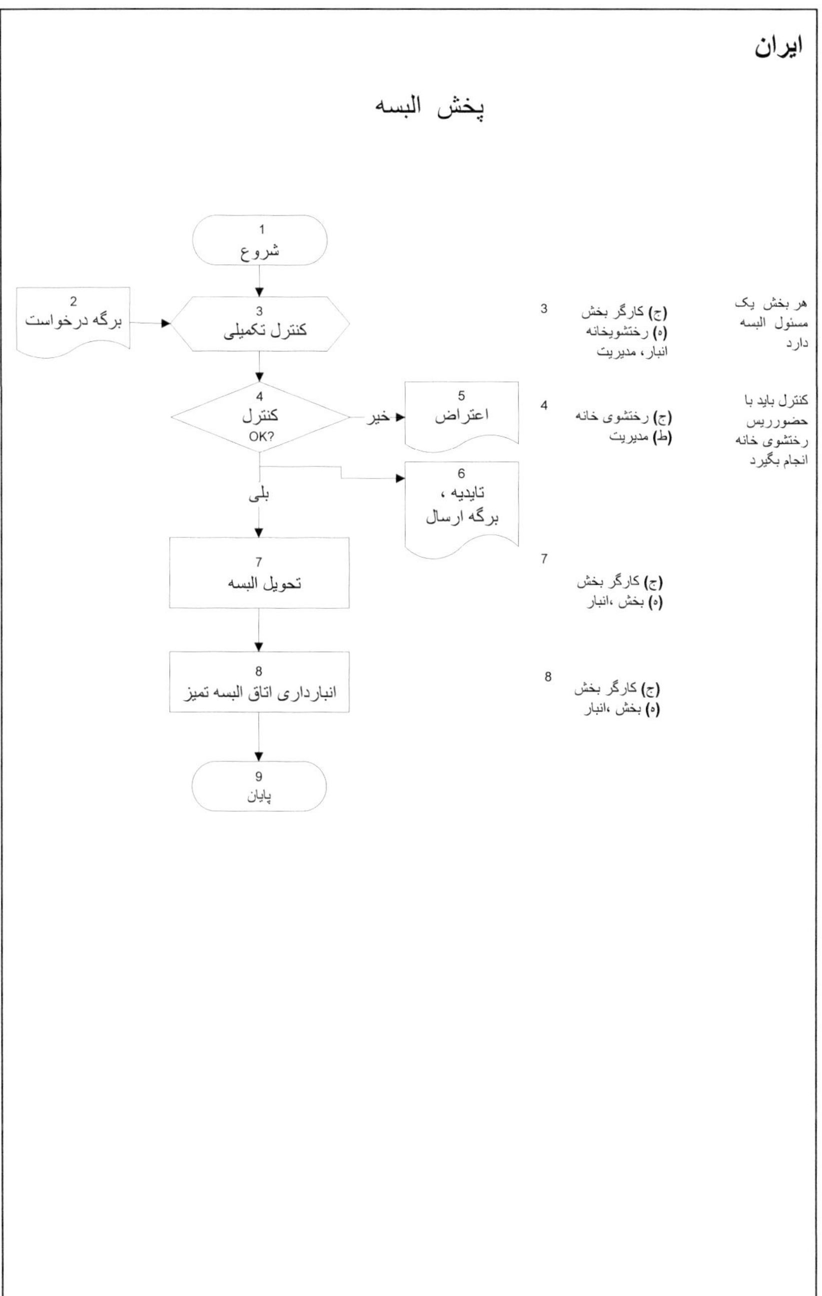
ایران
پخش البسه
1
شروع
2
برگه درخواست
3
کنترل تکمیلی
4
کنترل
OK?
خیر
5
اعتراض
بلی
6
تاییدیه ،
برگه ارسال
7
تحویل البسه
8
انبارداری اتاق البسه تمیز
9
پایان
3
(ج) کارگر بخش
(ه) رختشویخانه
انبار ، مدیریت
هر بخش یک
مسئول البسه
دارد
4
(ج) رختشوی خانه
(ط) مدیریت
کنترل باید با
حضور رئیس
رختشوی خانه
انجام بگیرد
7
(ج) کارگر بخش
(ه) بخش ،انبار
8
(ج) کارگر بخش
(ه) بخش ،انبار

آلمان

جمع آوری البسه استفاده شده

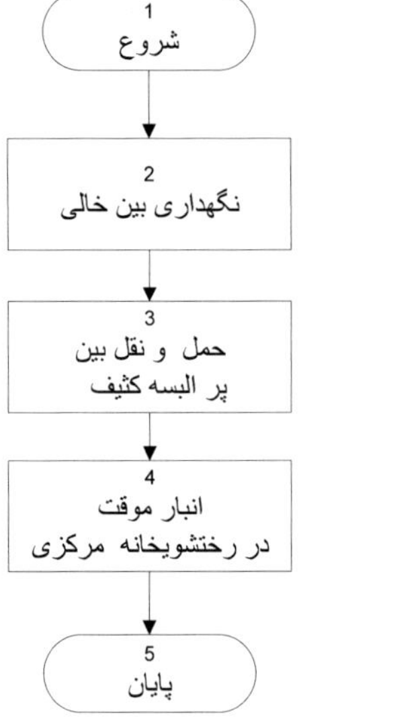

(ج) حمل و نقل داخلی   2

(ج) حمل و نقل داخلی   3

(ج) حمل و نقل داخلی   4

ایران

جمع آوری البسه استفاده شده

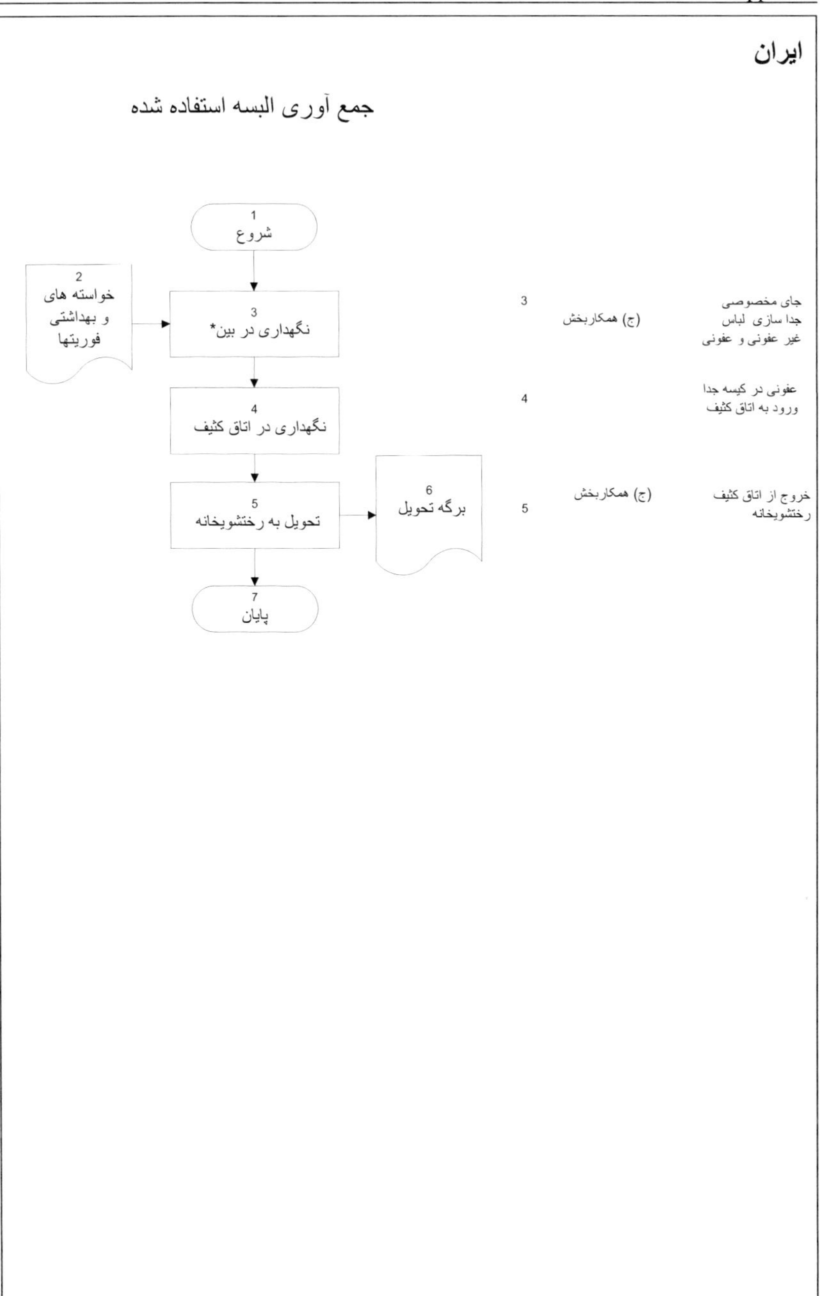

آلمان

پخش البسه

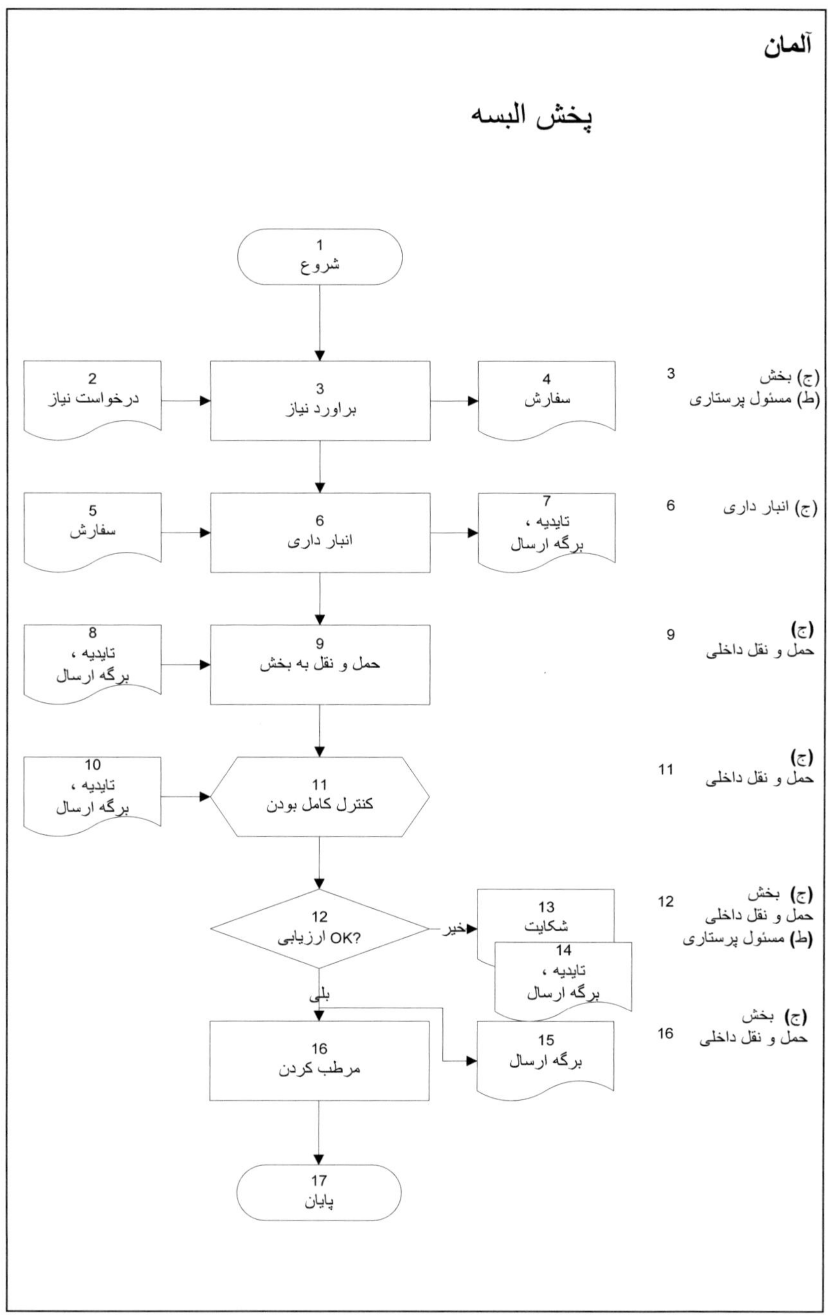

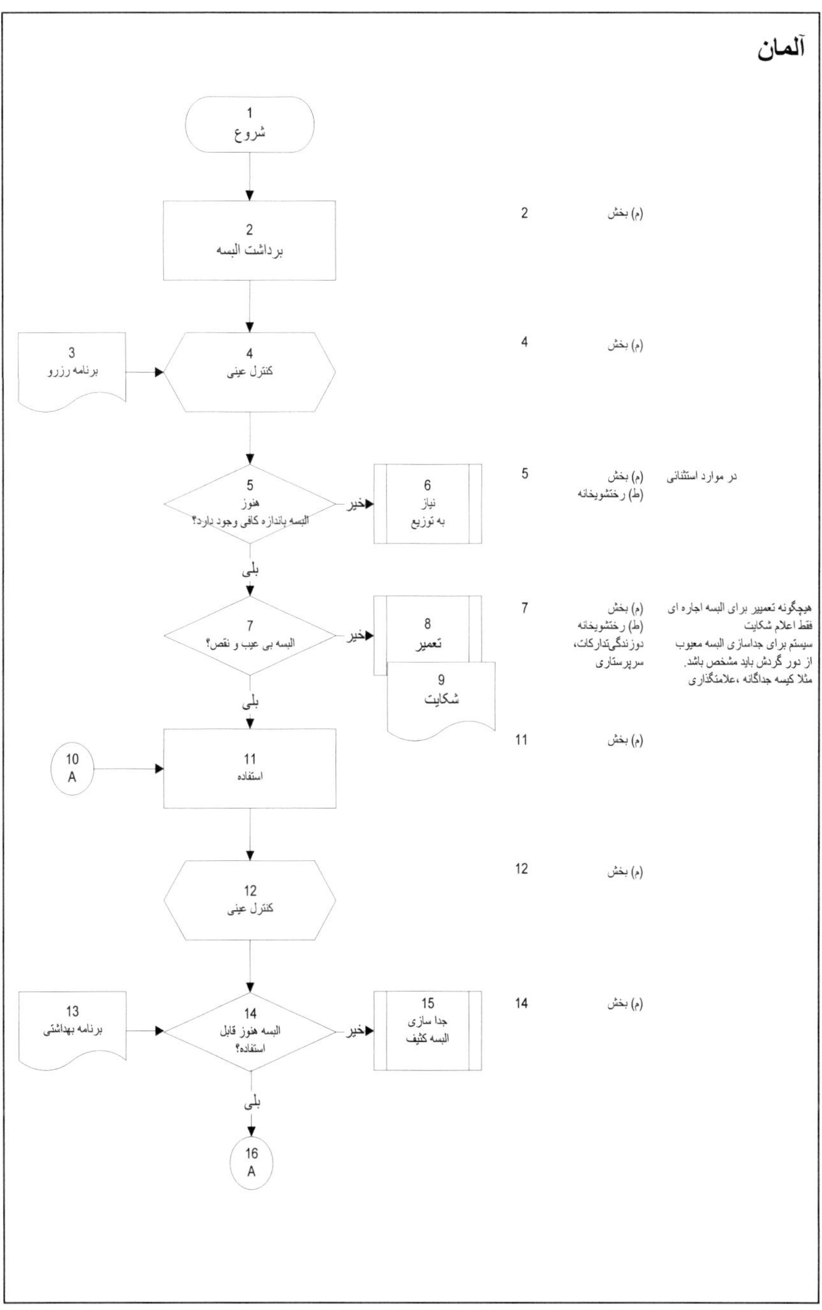
آلمان

۱
شروع

۲
برداشت البسه

۳
برنامه رزرو

۴
کنترل عینی

۵
هنوز
البسه باندازه کافی وجود دارد؟
خیر

۶
نیاز
به توزیع

بلی

۷
البسه بی عیب و نقص؟
خیر

۸
تعمیر

۹
شکایت

بلی

۱۰
A

۱۱
استفاده

۱۲
کنترل عینی

۱۳
برنامه بهداشتی

۱۴
البسه هنوز قابل
استفاده؟
خیر

۱۵
جدا سازی
البسه کثیف

بلی

۱۶
A

۲   (م) بخش

۴   (م) بخش

۵   (م) بخش   در موارد استثنائی
           (ط) رختشویخانه

۷   (م) بخش   هیچگونه تعمیر برای البسه اجاره ای
           (ط) رختشویخانه   فقط اعلام شکایت
           دوزندگی،تدارکات،   سیستم برای جداسازی البسه معیوب
           سرپرستاری   از دور گردش باید مشخص باشد.
                            مثلا کیسه جداگانه ،علامتگذاری

۱۱   (م) بخش

۱۲   (م) بخش

۱۴   (م) بخش

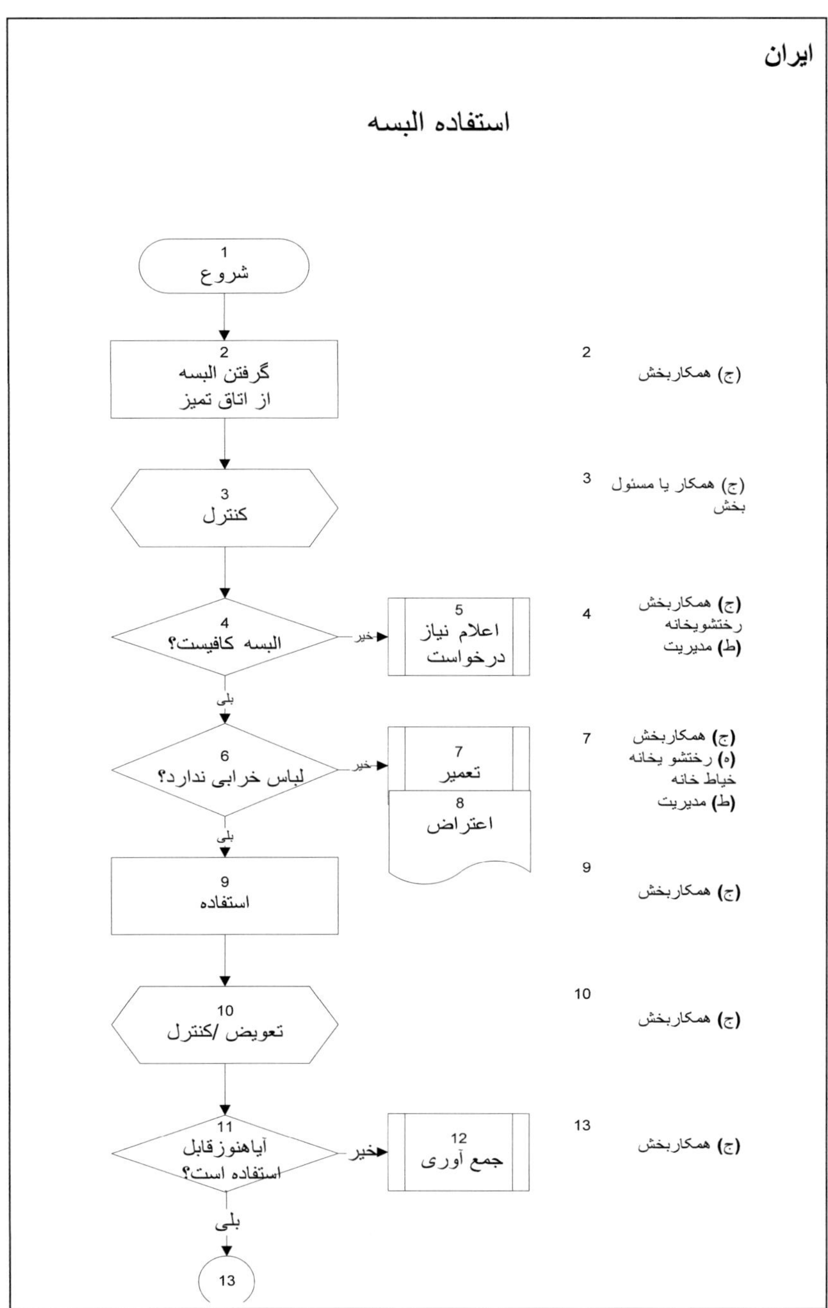
ایران

استفاده البسه

۱
شروع

۲
گرفتن البسه
از اتاق تمیز

۳
کنترل

۴
البسه کافیست؟

۵
اعلام نیاز
درخواست

خیر

بلی

۶
لباس خرابی ندارد؟

خیر

۷
تعمیر

۸
اعتراض

بلی

۹
استفاده

۱۰
تعویض /کنترل

۱۱
آیاهنوزقابل
استفاده است؟

خیر

۱۲
جمع آوری

بلی

۱۳

۲
(ج) همکاربخش

۳
(ج) همکار یا مسئول
بخش

۴
(ج) همکاربخش
رختشویخانه
(ط) مدیریت

۷
(ج) همکاربخش
(ه) رختشو یخانه
خیاط خانه
(ط) مدیریت

۹
(ج) همکاربخش

۱۰
(ج) همکاربخش

۱۳
(ج) همکاربخش

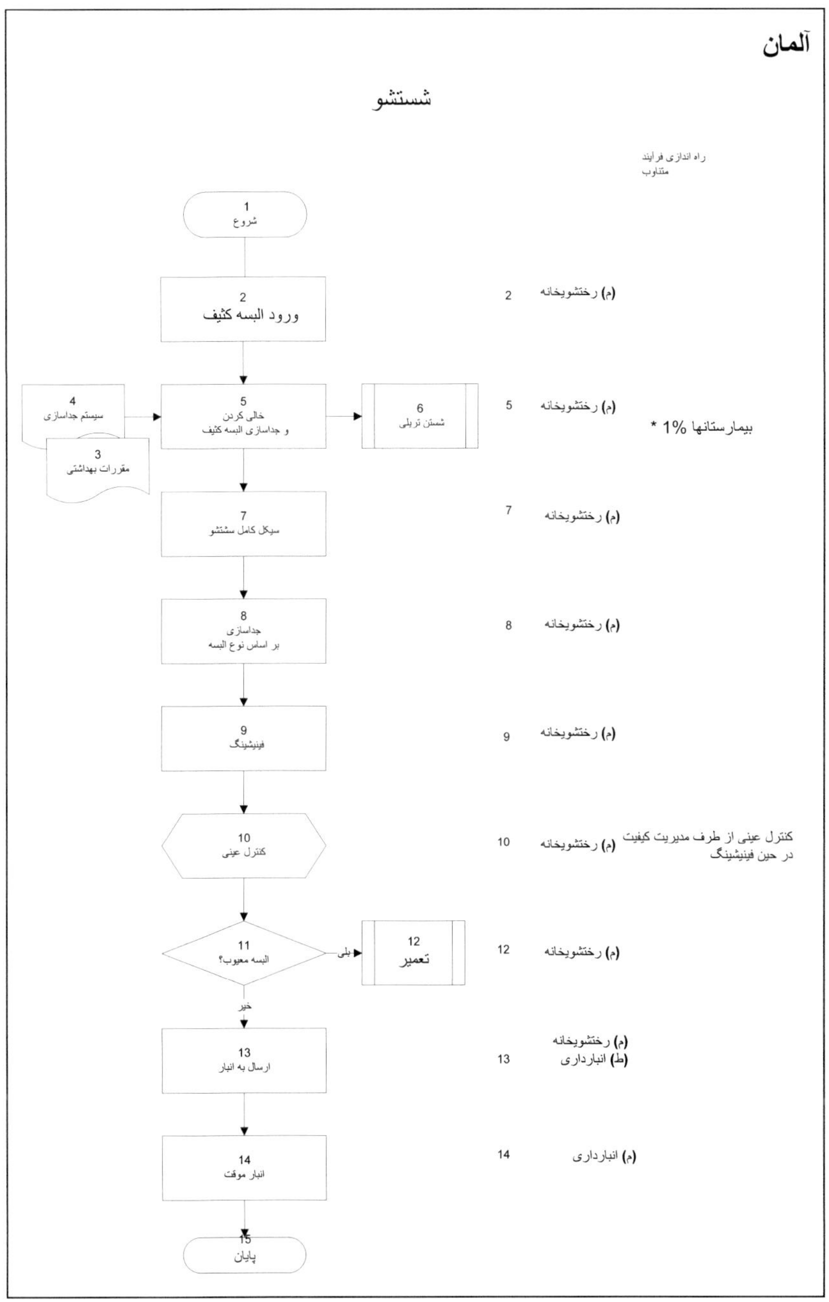
آلمان

شستشو

راه اندازی فرآیند
متناوب

1
شروع

2
ورود البسه کثیف

(م) رختشویخانه   2

4
سیستم جداسازی

3
مقررات بهداشتی

5
خالی کردن
و جداسازی البسه کثیف

6
شستن ترپلی

(م) رختشویخانه   5

بیمارستانها %1 *

7
سیکل کامل شستشو

(م) رختشویخانه   7

8
جداسازی
بر اساس نوع البسه

(م) رختشویخانه   8

9
فینیشینگ

(م) رختشویخانه   9

10
کنترل عینی

(م) رختشویخانه   10

کنترل عینی از طرف مدیریت کیفیت
در حین فینیشینگ

11
البسه معیوب؟

بلی

12
تعمیر

(م) رختشویخانه   12

خیر

13
ارسال به انبار

(م) رختشویخانه
(ط) انبارداری   13

14
انبار موقت

(م) انبارداری   14

15
پایان

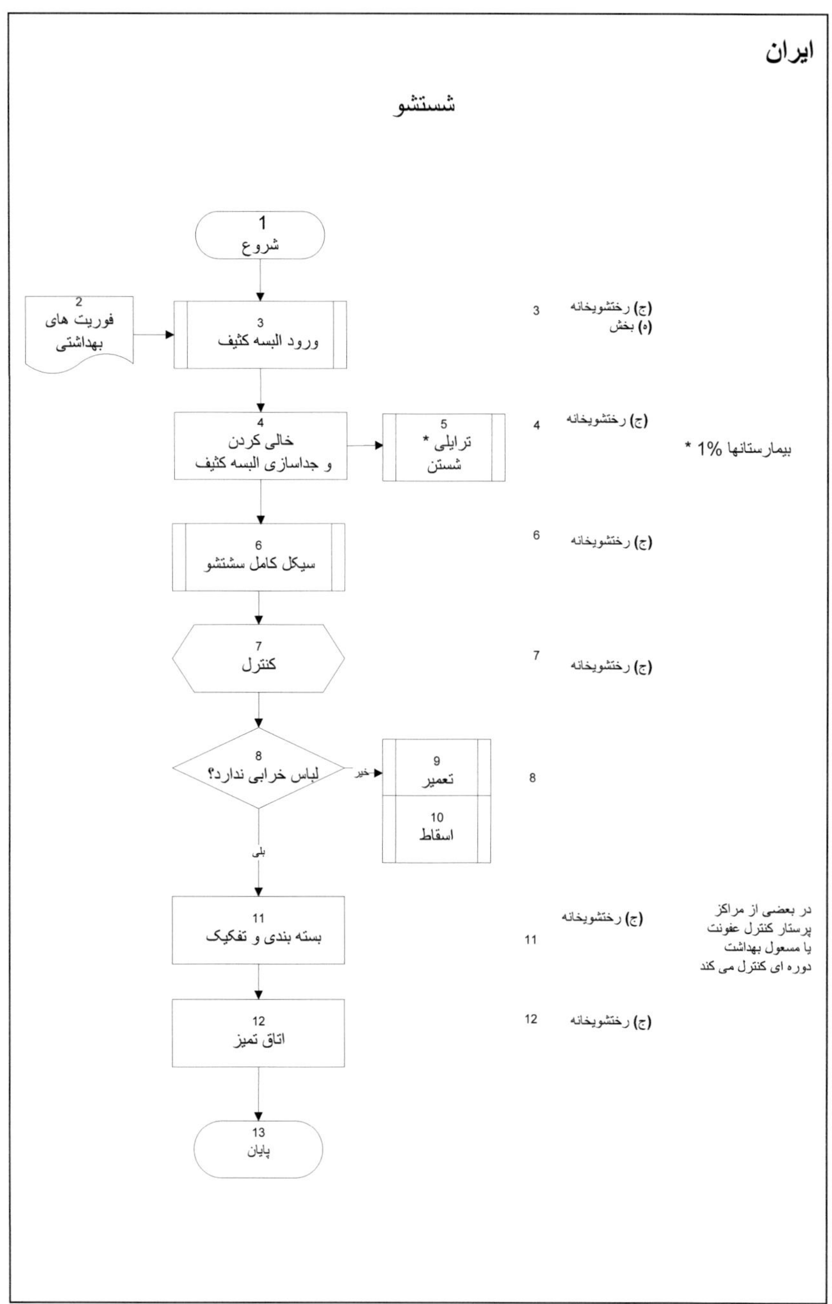
ایران
شستشو
1
شروع
2
فوریت های بهداشتی
3
ورود البسه کثیف
(ج) رختشویخانه
(ه) بخش
3
4
خالی کردن و جداسازی البسه کثیف
(ج) رختشویخانه
4
5
ترایلی *
شستن
بیمارستانها 1% *
6
سیکل کامل شستشو
(ج) رختشویخانه
6
7
کنترل
(ج) رختشویخانه
7
8
لباس خرابی ندارد؟
خیر
بلی
9
تعمیر
10
اسقاط
8
11
بسته بندی و تفکیک
(ج) رختشویخانه
11
در بعضی از مراکز
پرستار کنترل عفونت
یا مسعول بهداشت
دوره ای کنترل می کند
12
اتاق تمیز
(ج) رختشویخانه
12
13
پایان

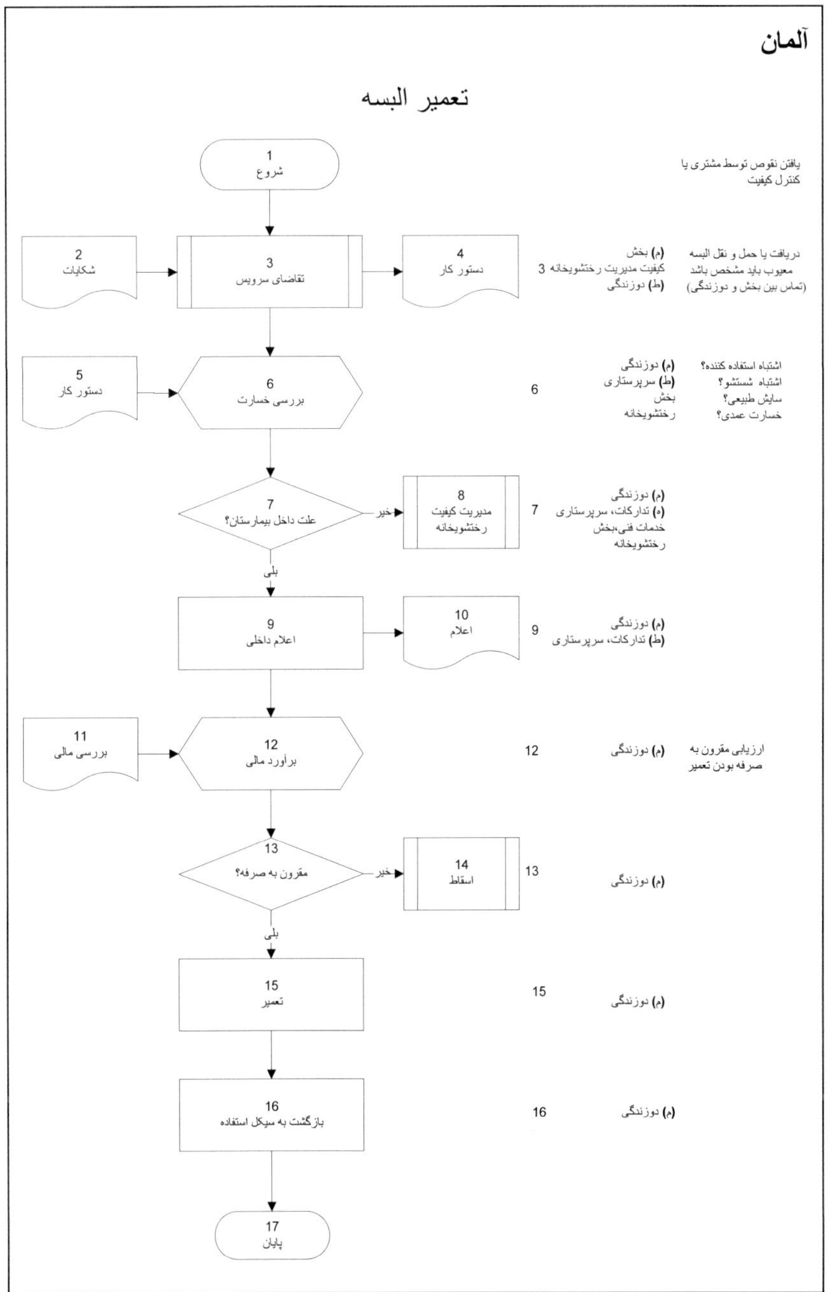
آلمان

تعمیر البسه

۱
شروع

یافتن نقوص توسط مشتری یا
کنترل کیفیت

۲
شکایات

۳
تقاضای سرویس

۴
دستور کار

(م) بخش
کیفیت مدیریت رختشویخانه ۳
(ط) دوزندگی

دریافت یا حمل و نقل البسه
معیوب باید مشخص باشد
(تماس بین بخش و دوزندگی)

۵
دستور کار

۶
بررسی خسارت

۶

(م) دوزندگی
(ط) سرپرستاری
بخش
رختشویخانه

اشتباه استفاده کننده؟
اشتباه شستشو؟
سایش طبیعی؟
خسارت عمدی؟

۷
علت داخل بیمارستان؟

خیر

۸
مدیریت کیفیت
رختشویخانه

۷

(م) دوزندگی
(ه) تدارکات، سرپرستاری
خدمات فنی،بخش
رختشویخانه

بلی

۹
اعلام داخلی

۱۰
اعلام

۹

(م) دوزندگی
(ط) تدارکات، سرپرستاری

۱۱
بررسی مالی

۱۲
برآورد مالی

۱۲

(م) دوزندگی

ارزیابی مقرون به
صرفه بودن تعمیر

۱۳
مقرون به صرفه؟

خیر

۱۴
اسقاط

۱۳

(م) دوزندگی

بلی

۱۵
تعمیر

۱۵

(م) دوزندگی

۱۶
بازگشت به سیکل استفاده

۱۶

(م) دوزندگی

۱۷
پایان

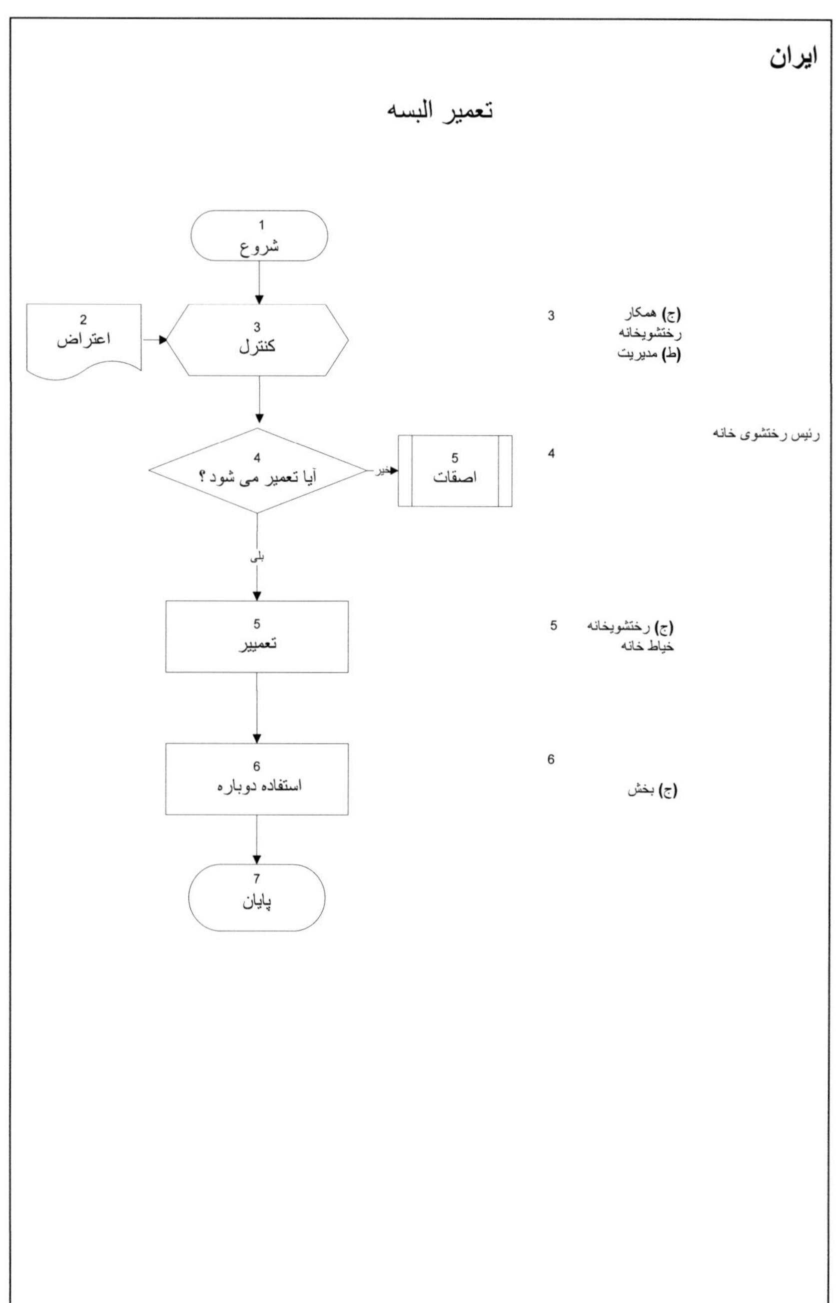
ایران
تعمیر البسه
۱
شروع
۲
اعتراض
۳
کنترل
۴
آیا تعمیر می شود؟
خیر
بلی
۵
اصقات
۵
تعمییر
۶
استفاده دوباره
۷
پایان
۳
(ج) همکار
رختشویخانه
(ط) مدیریت
رئیس رختشوی خانه
۴
۵
(ج) رختشویخانه
خیاط خانه
۶
(ج) بخش

آلمان

اسقاط البسه

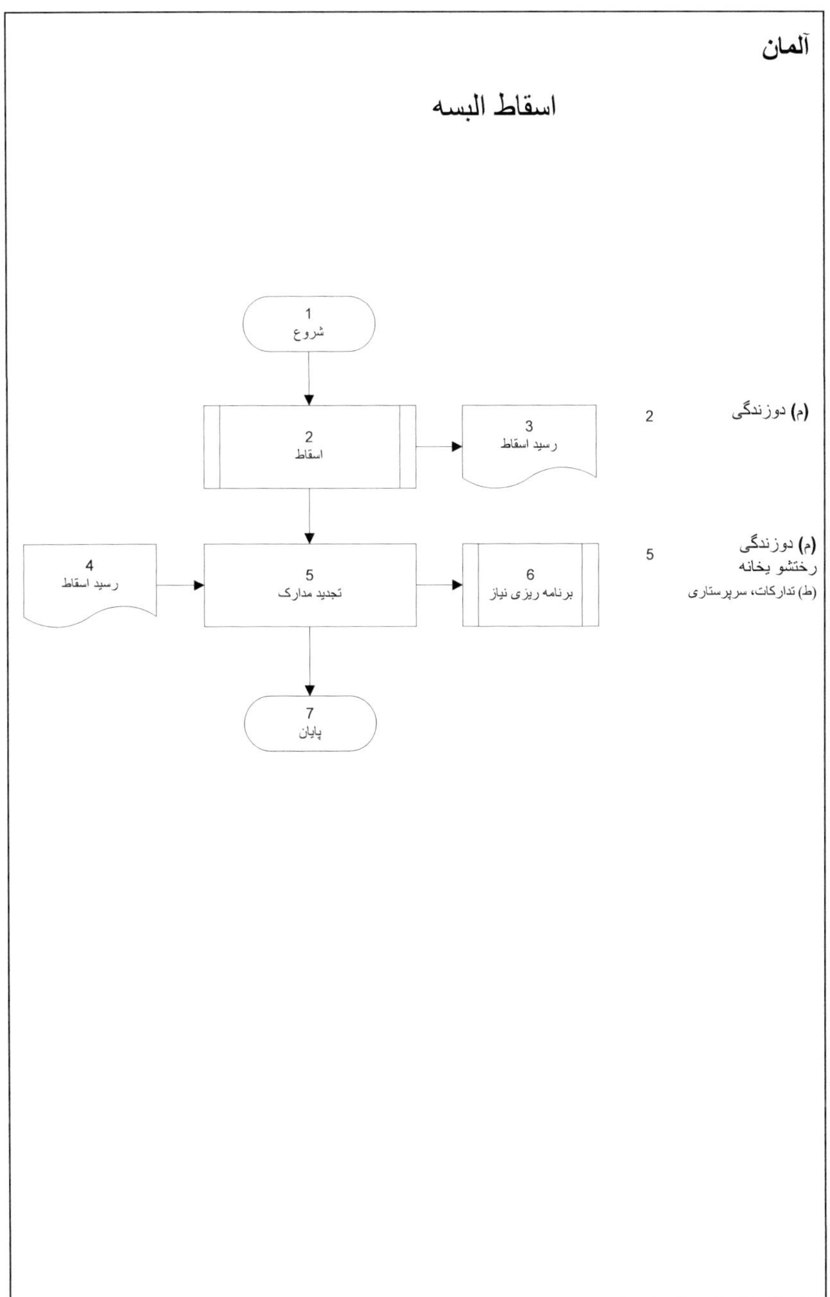

ایران

البسه اصقات

<table>
<tr><td>شروع
1</td><td></td></tr>
<tr><td>جمع آوری
2</td><td>(ج) رختشو یخانه   2</td></tr>
<tr><td>کنترل
3</td><td>3

(ج) رختشو یخانه</td></tr>
<tr><td>آیا هنوز قابل استفاده است؟ *
4   خیر →   اصقاط
5</td><td>(ج) رختشو یخانه   4   * به جهت استفاده دیگر
(ط) مدیریت</td></tr>
<tr><td>بلی
استفاده دوباره
6</td><td>(ج) بخش   6</td></tr>
<tr><td>پایان
7</td><td></td></tr>
</table>

APPENDIX - A18

Interfaces of the Process - Laundry Management

Deutschland

Prozess	Prozessschritt	Kostenfaktor[1]	Wertung*	Qualitätsfaktor[2]	Wertung*	Zeitfaktor[3]	Wertung*
	Bedarfsplanung	Planungskosten bezogen auf Investitionskosten, Budget	B	Gewährleistung der Versorgung: Anzahl Versorgungsengpässe, Beschwerden durch Kostenstellen	A	Dauer der Planung	B
	Beschaffung	Anschaffungskosten bezogen auf Wäschewert	A	Anzahl der Beanstandungen	A	Dauer der Beschaffung	B
	Frischwäscheverteilung	Dauer der Verteilung bezogen auf Wäschevolumen in Abhängigkeit von Gebäudestruktur, Verantwortlichkeit (Schnittstelle: hauseigenes Personal, Wäscherei-Personal)	A	Nutzung/Auslastung der Transportmittel (Aufzüge) und -wege, Anzahl Mengenreklamationen, Versorgungsmodell	B	Dauer	B
	Nutzung OP-Wäsche	Dauer der Nutzung bezogen auf Wäschevolumen	B	Hygienemaßstab, hauseigene Standards	B	Dauer	B
	Nutzung Stationswäsche	Dauer der Nutzung bezogen auf Wäschevolumen	A	Hygienemaßstab, hauseigene Standards	A	Dauer	B
	Nutzung Berufskleidung	Dauer der Nutzung bezogen auf Wäschevolumen	A	Hygienemaßstab, hauseigene Standards	B	Dauer	B
	Schmutzwäscheabholung	Dauer der Sammlung bezogen auf Wäschevolumen in Abhängigkeit von Gebäudestruktur, Verantwortlichkeit (Schnittstelle: hauseigenes Personal, Wäscherei-Personal)	B	Nutzung/Auslastung der Transportmittel (Aufzüge) und -wege, Versorgungsmodell	B	Dauer	B
	Reinigung	Dauer des Reinigungsprozesses, Auslastung der Anlagen	B	Anzahl der Beanstandungen durch Kunden, Einhaltung der Richtlinien	B	Dauer	B
	Reparatur	Reparaturkosten bezogen auf Wiederbeschaffungswert	B	Dauer der Reparatur	B	Dauer der Instandsetzung	B
	Entsorgung	Entsorgungskosten bezogen auf Anschaffungswert	C	Anzahl der nicht dokumentierten Entsorgungen; Dauer des Prozesses	C	Dauer	C
	Reinigung	Dauer des Reinigungsprozesses, Auslastung der Anlagen	B	Anzahl der Beanstandungen durch Kunden, Einhaltung der Richtlinien	B	Dauer	B
	Reparatur	Reparaturkosten bezogen auf Wiederbeschaffungswert	B	Dauer der Reparatur	B	Dauer der Instandsetzung	B
	Entsorgung	Entsorgungskosten bezogen auf Anschaffungswert	C	Anzahl der nicht dokumentierten Entsorgungen; Dauer des Prozesses	C	Dauer	C

[1]Kostenfaktor:	Zähler: Welche Einheit bestimmt die Größe der Kosten Nenner: Welche Bezugsgröße kann zum Vergleich herangezogen werden
[2]Qualitätsfaktor:	Welches quantifizierbare Ereignis gibt die Abweichung von - oder die Erfüllung der geforderten Qualität wieder
[3]Zeitfaktor:	Dauer des jeweiligen Prozessschrittes
*Wertung	Wertungen reichen jeweils unterschieden für die Kosten-, Qualitäts- und Zeitfaktoren von A für besonders hohen Einfluss auf den Gesamtprozess bis C für besonders wenig den Gesamtprozess beeinflussend.

Iran

Prozess	Prozessschritt	Kostenfaktor[1]	Qualitätsfaktor[2]	Zeitfaktor[3]	
	Planung und Beschaffung				
	Bedarfsplanung und Entscheidung	Planungskosten bezogen auf Investitionskosten, Budget	Gewährleistung der Versorgung: Anzahl Versorgungsengpässe, Beschwerden	Dauer der Planung	
		Entscheidungszeit für Bestellung	Kostenvergleich im Zuge der Nutzung	Dauer der Entscheidung	
		Dauer der Ausschreibung	Kenntnis über Lieferer	Zeit für Informationsammlung über Lieferer	
	Bestellung	Dauer der Lieferung	Anzahl der fehlgeschlagenen Lieferungen, Anzahl der Reklamationen	Dauer	
	Bestellung/ Einkauf	Kosten der Lieferung/ Einkaufs bezogen auf den Wert der Ware	Anzahl der Reklamationen bzgl. des Vertrages	Dauer	
	Lieferung der Ware	Arbeitszeit bezogen auf den Wert der Ware	Anzahl bis zur richtigen Lieferung, Einhaltung der Termine	Dauer	
	Kontrolle	Arbeitszeit bezogen auf den Wert der Ware	Wirkliche Durchführung der Kontrolle	Dauer der Kontrolle	
		Arbeitszeit pro Reklamation	Anzahl der Reklamationen	Dauer	
	Annahme	Arbeitszeit im Vergleich zum Einkaufspreis	Anzahl der Beanstandungen	Dauer	
	Dokumentation	Arbeitszeit pro Dokument	Anzahl der fehlerhaften Dokumentation	Dauer	
	Produktion/ Näheren				
		Arbeitszeit für die Produktion	Anzahl der fehlerhaften Produkte	Dauer	
	Frischwäscheverteilung				
	Bedarfsbestellung	Dauer der Nutzung bezogen auf Wäschevolumen	Hygienemaßstab, hauseigene Standards	Dauer	
	Nutzung Stationswäsche	Dauer der Nutzung bezogen auf Wäschevolumen	Hygienemaßstab, hauseigene Standards	Dauer	
	Nutzung Berufskleidung	Dauer der Nutzung bezogen auf Wäschevolumen	Hygienemaßstab, hauseigene Standards	Dauer	
	Schmutzwäschesammlung				
	Vorhandensein von Bin (Sammelwagen)	Arbeitszeit pro Bin (Sammelwagen)	Anzahl der unbeschädigten Bin (Sammelwagen)	Dauer	
	Transport	Transportzeit	Hindernisse während des Transports	Dauer	
	Sammlung im Schmutzwäscheraum	Arbeitszeit für Sammlung, Raumkosten im Verhältnis der Schmutzwäsche	Dauer der Sammlung, Trennung von sauberen Wäsche	Dauer	
	Reinigung				
	Eingang der Schmutzwäsche	Arbeitszeit pro kg Wäsche	Arbeitszeit	Dauer	
	Leeren des Bin und Trennung der Wäsche	Arbeitszeit im Vergleich zum Wäschevolumen	Anzahl der internen Beschwerden	Dauer	
	Kontrolle	Arbeitszeit pro Kontrollgang	Überzeugung der Kontrolldurchführung	Kontrolldauer	
	Zusammenlegung und Sortierung	Arbeitszeit pro Zusammenlegung eines Wäschestücks	Dauer des Zusammenlegens und Sortierens	Dauer	
	Reparatur				
	Kontrolle	Arbeitszeit bezogen auf Wiederbeschaffungswert	Analyse der Schadensentstehung, Anzahl der Beschwerden	Dauer	
	Reparatur	Arbeitszeit bezogen auf Wäschevolumen	Weitere Nutzungsdauer	Dauer	
	Nutzung	Arbeitszeit bezogen auf Wiederbeschaffungswert	Dauer	Dauer	
	Entsorgung				
		Arbeitszeit pro Entsorgung	Entsorgungskosten	Dauer	

آلمان

فرایند	گام فرایند	فاکتور های هزینه ای[1]	*ارزیابی	فاکتور های کیفیت[2]	*ارزیابی	فاکتور های زمانی[3]	*ارزیابی
	برنامه ریزی نیاز ها	برآورد هزینه برنامه ریزی در مقایسه با اعتبار داده شده، بودجه	B	میزان کمبود، تعداد اعتراض	A	طول برنامه ریزی	B
	خرید	هزینه تهیه در مقایسه با ارزش البسه	A	تعداد اعتراض	A	مدت خرید	B
	توزیع البسه	مدت توزیع در مقایسه با مقدار البسه بسته به ساختار ساختمان، مسئولیتها	A	میزان استفاده از ظرفیت آسانسور ها و راهها، تعداد اعتراضات، چگونگی تدارکات	B	مدت	B
	استفاده البسه اتاق عمل	مدت زمان صرف شده در مقایسه با حجم البسه	B	میزان رعایت بهداشت، استانداردهای داخلی	B	مدت	B
	استفاده البسه بخش	مدت زمان صرف شده مقایسه در با حجم البسه	A	میزان رعایت بهداشت، استانداردهای داخلی	A	مدت	B
	استفاده البسه کارکنان	مدت زمان صرف شده درمقایسه با حجم البسه	A	میزان رعایت بهداشت، استانداردهای داخلی	B	مدت	B
	جمع آوری البسه استفاده شده	مدت جمع آوری در مقایسه با مقدار البسه بسته به ساختار ساختمان، مسئولیتها	B	میزان استفاده از ظرفیت آسانسور ها و راهها، تعداد اعتراضات، چگونگی تدارکات	B	مدت	B
	شستشو	مدت شستشو ،میزان استفاده از ظرفیت ماشین		تعداد اعتراض، رعایت مقررات		مدت	
	تعمیرات	هزینه تعمیرات در مقایسه با ارزش تهیه مجدد		مدت تعمیرات		مدت	
	اسقاط	هزینه اسقاط در مقایسه با ارزش تهیه		تعداد اسقاط ثبت نشده، مدت فرآیند		مدت	

فاکتور های هزینه ای[1]	صورت: چه واحدی تعیین کننده میزان هزینه است؟ مخرج: چه مرجعی میتواند برای مقایسه در نظر گرفته شود؟
فاکتور های کیفیت[2]	چه کمیت قابل اندازه گیری میتواند بازگوی رعایت یا انحراف از کیفیت باشد؟
فاکتور های زمانی[3]	مدت لازم برای گام فرآیند
*ارزیابی	نمرات نشانگر ارزیابی فاکتور های مختلف هستند. A نشانگر تاثیر زیاد بر روی فرآیند و C نشانگر تاثیر کم

ایران

فرایند	گام فرایند	فاکترهای هزینه ای	فاکترهای کیفیت	فاکترهای زمان
برنامه ریزی				
	بررسی و تصمیم گیری	براورد هزینه برنامه ریزی در مقایسه با اعتبار داده شده، بودجه	میزان کمبود، تعداد اعتراض	طول برنامه ریزی
		زمان تصمیم گیری برای سفارش	مقایسه هزینه های هین مصرف کالا	مدت تصمیم گیری
		مدت زمان برگزاری مناقصه	تعداد ادم اگاهی کارفرما از پیمانکار	مدت زمان صرف شده جهت جمع آوری اطلاعات از پیمانکار
	سفارش	مدت زمان صرف شده جهت سفارش کالا	تعداد لغو سفارش / تعداد اعتراض	مدت زمان سفارش
	سفارش/ خرید	هزینه سفارش/ خرید در مقایسه با ارزش کالا/ البسه	تعداد اعتراض به بند های قرارداد بین پیمانکار و کارفرما	زمان انجام سفارش
	دریافت کالا	مقایسه زمان اداری سرف شده دریافت با ارزش کالا	تعداد تا دریافت کل کالا / دریافت به موقع کالا	زمان طی شده جهت دریافت
	کنترل	زمان اداری در مقایسه با ارزش کالا	اطمینان از انجام کنترل کالا	زمان طی شده جهت کنترل
		زمان برای هر اعتراض	تعداد اعتراض های غیر قابل غبول	زمان اعتراض
	تحویل	زمان اداری در مقایسه با کل هزینه خرید	تعداد موارد ناقص دیده نشده	طول زمان
	مستندسازی	زمان اداری برای هر مستند	تعداد خطا انبارگردانی	طول زمان
دوخت				
		زمان اداری برای دوخت	تعداد دوخت اشتباه	طول زمان
پخش				
	درخواست	زمان صرف شده برای هر درخواست	درخواست به موقع، تعداد نیاز های استراری	طول زمان درخواست
	کنترل تکمیلی	زمان اداری برای هر کنترل	تعداد شکایات هر واحد	طول زمان هر کنترل
	تحویل به واحد	زمان حمل و نقل	تعداد شکایات، زمان، موانع در تی حمل ونقل	طول زمان
	کنترل	زمان اداری برای هر کنترل	اطمینان از انجام کنترل	طول زمان هر کنترل
	چیدمان	زمان اداری برای هر تعداد البسه	طول زمان در مقایسه با سیمتم انبارداری	طول زمان
جمع آوری البسه				
	موجودی بین	زمان اداری برای هر بین	تعداد بین سالم	طول زمان
	حمل و نقل	زمان حمل و نقل	موانع در تی حمل ونقل	طول زمان
	نگهداری در اناق کثیف	زمان اداری برای نگهداری، هزینه فضا در مقایسه با هجم البسه	طول زمان نگهداشت، جداسازی با البسه تمیز	طول زمان
شستشو				
	ورود البسه کثیف	زمان اداری برای هر کیلو	زمان صرف شده	طول زمان
	خالی کردن بین و جدا سازی	زمان اداری در مقایسه با هجم البسه	تعداد شکایت داخلی،	طول زمان
	کنترل	زمان اداری برای هر کنترل	اطمینان از انجام کنترل	طول زمان هر کنترل
	بسته بندی و تفکیک	زمان صرف شده برای بسته بندی	طول زمان بسته بندی	طول زمان
تعمیر				
	کنترل	زمان اداری در مقایسه با هزینه خرید البسه	آنالیز شناسایی علت خرابی، تعداد اعتراض بخش	طول زمان
	تعمیر	زمان اداری در مقایسه با هجم البسه	طول زمان استفاده مجدد	طول زمان
	استفاده دوباره	زمان اداری در مقایسه با هزینه خرید	طول زمان	طول زمان
اسقاط/ متوقف کردن				
	اسقاط/ متوقف کردن	زمان اداری هر اسقاط	هزینه اسقاط کردن	طول زمان

APPENDIX - A19

Identity Values – Laundry Management

Deutschland

Prozess	Prozessschritt	Sender	Information/Inhalt	Empfänger
Wäscheversorgung				
Wäscheversorgung Beschaffung				
	Beschaffungsentscheidung	HW/PDL	Wäschebedarf	technische Einkauf / Wäscherei
	Beschaffungsentscheidung	Einkauf	Budgetabstimmung	Finanzabteilung / Controlling
	Lieferantenbewertung	PDL/HW	technische Bewertung	Einkauf
	Lieferantenbewertung	Einkauf	wirtschaftliche Bewertung	PDL/HW
	Ausschreibung	PDL/HW	Wäschespezifikationen, Termine	Einkauf
	Ausschreibung	Einkauf	Wäschespezifikationen, Termine	Lieferanten
	Ausschreibung	Lieferanten	Abgabe Angebote Muster / Urmuster	Einkauf
	Ausschreibung	Einkauf	Muster / Urmuster	PDL/HW
	Ausschreibung	PDL/HW	Auftrag Musterbewertung, Urmuster	Wäscherei
	Ausschreibung	Wäscherei	Musterbewertung	PDL/HW
	Ausschreibung	PDL/HW	Musterbewertung	Einkauf
	Bestellung	Einkauf	Kosten, Wäschespezifikationen, Termine	Lieferant
	Bestellung	Lieferant	Terminbestätigung	Einkauf
	Lieferung	Lieferant	Lieferschein	Wareneingang
	Abnahme	Wareneingang	Bescheinigung des Warenempfangs	Lieferant
	Abnahme/ Reklamation	Wareneingang	Reklamation Menge	Lieferant
	Abnahme	Wareneingang	Lieferschein	PDL/HW (Lager)
	Abnahme / Reklamation	PDL/HW (Lager)	Reklamation Qualität	Lieferant
	Abnahme	PDL/HW (Lager)	Lieferschein	Finanzabteilung / Controlling
	Abnahme	Finanzabteilung / Controlling	Freigabe der Rechnung	Einkauf
	Kennzeichnung (nach Bedarf)	Näherei	Bestandsdaten	PDL,HW
Wäscheversorgung Verteilung zyklisch				
	Erfassung Bedarf	Kostenstelle (Besteller)	Auftrag: Art und Anzahl Wäschestücke	Kommissionierung
	Transport: externe Logistik	Kommissionierung	Lieferschein	Inhouse Logistik
	Transport zu Kostenstelle, Inhouse Logistik	Inhouse Logistik	Lieferschein	Kostenstelle
	Vollständigkeitsprüfung	Kostenstelle	Lieferschein, evtl. Reklamation	PDL,HW
	Bedarfserfassung folgender Zyklus	Kostenstelle oder Inhouse Logistik	Bestellschein: Art und Anzahl Wäschestücke	Kommissionierung
Wäscheversorgung Verteilung Bedarf				
	Erfassung Bedarf	Kostenstelle	Auftrag: Art und Anzahl Wäschestücke	Kommissionierung / Wäscherei
	Transport zu Kostenstelle, externe und Inhouse-Logistik	Kommissionierung	Lieferschein	Kostenstelle
	Vollständigkeitsprüfung	Kostenstelle	Lieferschein, evtl. Reklamation	PDL,HW
Reparatur				
	Serviceanforderung	Kostenstelle/Qualitätsmanagement Wäscherei	Wäschestück mit Angabe der Reklamation	Näherei
	Schaden prüfen	Näherei	Meldung externe Schadensursache	QM Wäscherei
	Schaden prüfen	Näherei	Meldung interne Schadensursache	PDL,HW
Entsorgung				
	Entsorgung	Näherei	Entsorgungsnachweis	PDL,HW; Wäscherei
	Aktualisierung der Dokumentation	PDL,HW; Wäscherei	theoretische Bestandsveränderung	Einkauf

Iran

Prozess	Prozessschritt	Sender	Information/Inhalt	Empfänger
Wäscheversorgung Planung und Beschaffung				
	Bedarfserfassung	Abteilung /Station	Wäschebedarf, Bestellschein	Leitung Wäscherei / Management
	Bedarfserfassung	Leitung Wäscherei	Bestellschein, Kosten, Wäschespezifikationen	Management
	Beschaffungsentscheidung	Leitung Wäscherei	fachmännische Bewertung / Bericht	Management
	Beschaffungsentscheidung	Management	fachmännische Bewertung / Bericht	Finanzabteilung
	Beschaffungsentscheidung	Finanzabteilung	fachmännische Bewertung / Bericht, finanzielle Bewertung	Einkauf
	Ausschreibung	Management, Leitung Wäscherei	Wäschespezifikationen	Management, Lieferanten
	Ausschreibung	Lieferanten	Abgabe Angebote Muster / Urmuster	Management, Leitung Wäscherei
	Ausschreibung	Management	Auswahl und Auftrag	Lieferanten
	Bestellung	Management, Leitung Wäscherei	Bestellung	Lieferanten
	Lieferung	Lieferant	Lieferschein	Lager, Leitung Wäscherei
	Lieferung	Lager, Leitung Wäscherei	Lieferschein	Einkauf, Finanzabteilung
	Abnahme/ Reklamation	Lager, Leitung Wäscherei	Reklamation, Bericht	Lieferant
	Abnahme	Lieferant	Lieferschein	Finanzabteilung
	Abnahme	Einkauf	Lieferschein (Warenankunft)	Management
	Aktualisierung der Dokumentation	Lager	Bestellscheine, Quittungen	Lager, Management
Produktion/ Nähen				
	Nähen	Management, Leitung Wäscherei	Wäschespezifikation	Näherei
Wäscheversorgung Verteilung Bedarf				
	Kontrolle	Abteilung/ Station	Bedarfsbestellung	Leitung Wäscherei
	Kontrolle	Leitung Wäscherei	Reklamation	Abteilung/ Station
	Kontrolle	Leitung Wäscherei	Bestätigung	Lager
Nutzung				
	Kontrolle	Mitarbeiter Abteilung/ Station	Bedarfsbestellung	Management, Leitung Wäscherei
		Mitarbeiter Abteilung/ Station	Reklamation, Reparatur	Management, Leitung Wäscherei
Schmutzwäscheabholung				
	Lieferung an Wäscherei	Mitarbeiter Abteilung/ Station	Wäscheanzahl	Management, Leitung Wäscherei
Reparatur				
		Leitung Wäscherei	Reparaturbedarf	Näherei
Entsorgung				
			Entsorgungsmenge	Management

آلمان

فرایند	گام فرایند	فرستنده	اطلاعات	گیرنده
تهیه البسه				
خرید				
	تصمیم خرید	تدارکات، سرپرستاری	میزان نیاز البسه	مسئول رختشویخانه/خرید فنی
	تصمیم خرید	خرید	بررسی بودجه	امور مالی
	ارزیابی شرکتهای فروشنده	تدارکات، سرپرستاری	برآورد فنی	خرید
	ارزیابی شرکتهای فروشنده	خرید	بررسی مالی	تدارکات، سرپرستاری
	مناقصه	تدارکات، سرپرستاری	دفترچه مشخصات، موعدها	خرید
	مناقصه	خرید	دفترچه مشخصات، موعدها	شرکت های فروشنده
	مناقصه	شرکت های فروشنده	اعلام قیمت، نمونه	خرید
	مناقصه	خرید	نمونه	تدارکات، سرپرستاری
	مناقصه	تدارکات، سرپرستاری	دستور ارزیابی نمونه ها	رختشویخانه
	مناقصه	رختشویخانه	ارزیابی نمونه ها	تدارکات، سرپرستاری
	مناقصه	تدارکات، سرپرستاری	ارزیابی نمونه ها	خرید
	سفارش	خرید	هزینه، دفترچه مشخصات موعدها	شرکت فروشنده
	سفارش	شرکت فروشنده	تایید موعد	خرید
	ارسال	شرکت فروشنده	فیش ارسال کالا	انبار ورودی کالا
	تحویل	انبار ورودی کالا	فیش ورود کالا	شرکت فروشنده
	تحویل / اعتراضات	انبار ورودی کالا	اعتراضات کمّی	شرکت فروشنده
	تحویل	انبار ورودی کالا	فیش ارسال کالا	تدارکات، سرپرستاری (انبار)
	تحویل / اعتراضات	تدارکات، سرپرستاری (انبار)	اعتراضات کیفی	شرکت فروشنده
	تحویل	تدارکات، سرپرستاری (انبار)	فیش ارسال کالا	امور مالی
	تحویل	امور مالی	تایید صورتحساب	خرید

فرایند	گام فرایند	فرستنده	اطلاعات	گیرنده
	تحویل	تدارکات، سرپرستاری (انبار)	فیش ارسال کالا	امور مالی
	تحویل	امور مالی	تایید صورتحساب	خرید
	علامت گذاری بر اساس نیاز	دوزندگی	موجودی	تدارکات، سرپرستاری
تدارک وپخش البسه متناوب				
	برآورد نیاز	بخش سفارش دهنده	تقاضا: نوع و تعداد البسه	انبار داری
	حمل و نقل، شرکت پیمانکار	انبار داری	فیش ارسال کالا	حمل و نقل داخلی
	حمل و نقل داخلی به بخش مربوطه	حمل و نقل داخلی	فیش ارسال کالا	بخش
	کنترل تکمیل بودن	بخش	فیش ارسال کالا، ایرادات	تدارکات، سرپرستاری
	برآورد نیاز برای دوره بعدی	بخش با حمل و نقل داخلی	سفارش: نوع و تعداد البسه	انبار داری
تدارک وپخش البسه مایحتاج				
	برآورد نیاز	بخش	تقاضا: نوع و تعداد البسه	انبار داری، رختشویخانه
	حمل و نقل خارجی و داخلی	انبار داری	فیش ارسال کالا	بخش
	کنترل تکمیل بودن	بخش	فیش ارسال کالا، ایرادات	تدارکات، سرپرستاری
تعمیرات				
	تقاضای سرویس	بخش/ کنترل کیفیت رختشویخانه	البسه با توصیف ایراد	دوزندگی
	بررسی خسارات	دوزندگی	اعلام عامل خارجی خسارت	کنترل کیفیت رختشویخانه
	بررسی خسارات	دوزندگی	اعلام عامل داخلی خسارت	تدارکات، سرپرستاری
اسقاط				
	اسقاط	دوزندگی	فیش اسقاط	تدارکات، سرپرستاری، رختشویخانه
	تجدید تدارک	تدارکات، سرپرستاری، رختشویخانه	تغییر موجودی انبار	خرید

ایران

گیرنده	اطلاعات	فرستنده	گام فرایند	فرایند
برنامه ریزی و خرید				
مسئول رختشویخانه/مدیریت	برگه درخواست	بخش	اعلام نیاز البسه	
مدیریت	برگه درخواست،هزینه،مشخصات البسه/پارچه	مسئول رختشویخانه	اعلام نیاز	
مدیریت	گزارش کارشناسی	مسئول رختشویخانه	بررسی و تصمیم گیری	
امور مالی	گزارش کارشناسی	مدیریت	بررسی و تصمیم گیری	
تدارکات	گزارش کارشناسی،برآورد مالی	امورمالی	بررسی و تصمیم گیری	
مسئول رختشویخانه/مدیریت	شرکت در مناقصه و معرفی کالا	مدیریت	مناقصه	
	اعلام نیاز	شرکت های فروشنده	مناقصه	
شرکت فروشنده*	انتخاب شرکت و هماهنگی	مدیریت	مناقصه	
شرکت فروشنده	فرم سفارش	مسئول رختشویخانه مدیریت	سفارش(انجام موضوع قرارداد)	
مسئول رختشویخانه مدیریت	تهیه	شرکت فروشنده	سفارش(انجام موضوع قرارداد)	
امور مالی/تدارکات	گواهی/تایید	انبار/مسئول رختشویخانه	دریافت کالا/البسه	
شرکت فروشنده	گزارش	انبار/مسئول رختشویخانه	اعتراض	
امور مالی	قبض انبار	شرکت فروشنده	تحویل کالا/البسه	
مدیریت	رسیدن کالا/البسه	تدارکات	تحویل کالا/البسه	
انبار(مدیریت)	قبض خرید کالا	انبار	جدید کردن اسناد	
دوخت				
حیاط خانه	دستورکار	مسئول رختشویخانه/مدیریت	دوخت	
پخش				
مسئول رختشویخانه	درخواست	بخش	کنترل	
بخش	اعتراض	مسئول رختشویخانه	کنترل	
انبار	تایید	مسئول رختشویخانه	کنترل	
استفاده				
رختشویخانه/مدیریت	اعلام نیاز	همکار بخش	کنترل	
رختشویخانه/مدیریت	اعتراض/تعمیر	همکار بخش		
جمع آوری				
رختشویخانه	تعداد البسه	بخش	تحویل به رختشویخانه	
شستشو				
تعمیر				
خیاط خانه	درخواست تعمیر	مسئول رختشویخانه		
اسقاط				
مدیریت	حجم/تعداد اسقاطی			

APPENDIX – A20

Expert Analyse

Bewertung der Einflussparameter

Einflussparameter		A	B	C
Wirtschaft/ Finanzstärke/ Monetäre Möglichkeiten	Makroebene			
	Mikroebene			
Politik	Makroebene			
	Mikroebene			
Management, Organisation, Verwaltung	Makroebene			
	Mikroebene			
Bildungs- und Forschungsstand	Makroebene			
	Mikroebene			
Judikative, Gesetzliche Bestimmungen	Makroebene			
	Mikroebene			
Soziale Strukturen	Makroebene			
	Mikroebene			
Öffentliche und Private Institutionen	Makroebene			
	Mikroebene			
Infrastruktur	Makroebene			
	Mikroebene			
Geographie, Klima, Umweltbedingungen	Makroebene			
	Mikroebene			

<table>
<tr><td align="right">بهینه سازی و تجزیه و تحلیل فرآیندهای بیمارستانی</td><td>OPIK</td></tr>
</table>

<h1 align="center">سنجش عوال موثر با روش تحلیل ABC</h1>

C	B	A		عوامل موثر
			صطح گسترش	اقتصاد، اقتصادی
			صطح محدود	بودن، امکانات مالی
			صطح گسترش	سیاست، برنامه ریزی
			صطح محدود	استراتژیک
			صطح گسترش	مدیریت، سازماندهی،
			صطح محدود	اداره جات
			صطح گسترش	سطح آموزش و
			صطح محدود	تحقیقات
			صطح گسترش	چارچوب قانونی
			صطح محدود	
			صطح گسترش	ساختار اجتماعی
			صطح محدود	
			صطح گسترش	موسسات عمومی و
			صطح محدود	خصوصی
			صطح گسترش	زیربنا
			صطح محدود	
			صطح گسترش	جغرافی، آب و هوا،
			صطح محدود	شرایط زست
				محطیتی

بیمارستان:		تاریخ:	دانشگاه علوم پزشکی تهران

APPENDIX – A21

Expert List

Experts List

NAME	ORGANIZATION	POSITION
Dr. Mokhber	Markazi Tebi Kudakan Hospital	Head of the Management
Dr. Jeddian	Shariati Hospital	Head of the Management
Mr. Salehzadeh	Vali Asr Hospital	Head of the Management
Dr. Rezai	Emam Khomeini Hospital	Head of the Management
Mr. Moghaddam,	Shariati Hospital	Head Medical Equipment
Mr. Farrokhdust,	Public Health Institute	Head Medical Equipment
Mr. Yaghmai	Markazi Tebi Kudakan Hospital	Head Medical Equipment
Mrs. Mohammadi	Vali Asr Hospital	Head Medical Equipment
Mr. Zarrinkolla	Public Health Institute	Member of the Medical Equipment Staff
Dr. Zare	Public Health Institute	Member of the Management Staff
Mr. Khaksari	Shariati Hospital	Head Laundry
Mr. Khoi	Emam Khomeini Hospital	Head Laundry
Mr. Haghighifar	Public Health Institute	Head of the Accounting
Mr. Kuseler	Vali Asr Hospital	Head Technical Facilities
Mr. Pedram	Markazi Tebi Kudakan Hospital	Head Technical Facilities
Mr. Bahramian	Shariati Hospital	Head Technical Facilities
Mr. Baba	Technical Center of Medical University of Tehran	Member of the Technical Staff
Prof. Dr. Banki	Allameh Tabatabai University Tehran	Head of the Institute of German Language and Literature Studies

Veröffentlichungen

des Instituts für Technologie und Management im Baubetrieb

Karlsruher Reihe Bauwirtschaft, Immobilien und Facility Management

KIT Scientific Publishing Karlsruhe, ISSN 1867-5867

Band 1	Jochen ABEL	2009

Ein produktorientiertes Verrechnungssystem für Leistungen des Facility Management im Krankenhaus

Band 2	Carolin BAHR	2008

Realdatenanalyse zum Instandhaltungsaufwand öffentlicher Hochbauten

Band 3	Karin DIEZ	2009

Ein prozessorientiertes Modell zur Verrechnung von Facility Management Kosten am Beispiel der Funktionsstelle Operationsbereich im Krankenhaus

Band 4	Mandana BANEDJ-SCHAFII	2010

System transferability of public hospital facility management between Germany and Iran

Die Bände sind unter www.uvka.de als PDF frei verfügbar oder als Druckausgabe bestellbar.

Reihe F – Forschung

Institutsintern verlegt bis einschließlich Heft 62, 2007

Heft 1   Hans PINNOW   1972
Vergleichende Untersuchungen von Tiefbauprojekten in offener Bauweise

Heft 2   Heinrich MÜLLER   1972
Rationalisierung des Stahlbetonbaus durch neue Schalverfahren und deren Optimierung beim Entwurf

Heft 3   Dieter KARLE   1972
Einsatzdimensionierung langsam schlagender Rammbäre aufgrund von Rammsondierungen

Heft 4   Wilhelm REISMANN   1973
Kostenerfassung im maschinellen Erdbau

Heft 5   Günther MALETON   1973
Wechselwirkungen von Maschine und Fels beim Reißvorgang

Heft 6   Joachim HORNUNG   1973
Verfahrenstechnische Analyse über den Ersatz schlagender Rammen durch die Anwendung lärmarmer Baumethoden

Heft 7   Thomas TRÜMPER / Jürgen WEID   1973
Untersuchungen zur optimalen Gestaltung von Schneidköpfen bei Unterwasserbaggerungen

Heft 8   Georg OELRICHS   1974
Die Vibrationsrammung mit einfacher Längsschwingwirkung – Untersuchungen über die Kraft- und Bewegungsgrößen des Systems Rammbär plus Rammstück im Boden

Heft 9   Peter BÖHMER   1974
Verdichtung bituminösen Mischgutes beim Einbau mit Fertigern

Heft 10   Fritz GEHBAUER   1974
Stochastische Einflußgrößen für Transportsimulationen im Erdbau

Heft 11   Emil MASSINGER   1976
Das rheologische Verhalten von lockeren Erdstoffgemischen

Heft 12   Kawus SCHAYEGAN   1975
Einfluß von Bodenkonsistenz und Reifeninnendruck auf die fahrdynamischen Grundwerte von EM-Reifen

Heft 13   Curt HEUMANN   1975
Dynamische Einflüsse bei der Schnittkraftbestimmung in standfesten Böden

Heft 14   Hans-Josef KRÄMER   1976
Untersuchung der bearbeitungstechnischen Bodenkennwerte mit schwerem Ramm-Druck-Sondiergerät zur Beurteilung des Maschineneinsatzes im Erdbau

Heft 15   Friedrich ULBRICHT   1977
Baggerkraft bei Eimerkettenschwimmbaggern – Untersuchungen zur Einsatzdimensionierung

Heft 16   Bertold KETTERER   1977
Einfluß der Geschwindigkeit auf den Schneidvorgang in rolligen Böden
- vergriffen -

Heft 17   Joachim HORNUNG / Thomas TRÜMPER   1977
Entwicklungstendenzen lärmarmer Tiefbauverfahren für den innerstädtischen Einsatz

Heft 18   Joachim HORNUNG   1978
Geometrisch bedingte Einflüsse auf den Vorgang des maschinellen Reißens von Fels – untersucht an Modellen

Heft 19   Thomas TRÜMPER   1978
Einsatzoptimierung von Tunnelvortriebsmaschinen

Heft 20   Günther GUTH   1978
Optimierung von Bauverfahren – dargestellt an Beispielen aus dem Seehafenbau

Heft 21	Klaus LAUFER	1978
	Gesetzmäßigkeiten in der Mechanik des drehenden Bohrens im Grenzbereich zwischen Locker- und Festgestein	
	- vergriffen -	

Heft 22	Urs BRUNNER	1979
	Submarines Bauen – Entwicklung eines Bausystems für den Einsatz auf dem Meeresboden	
	- vergriffen -	

Heft 23	Volker SCHULER	1979
	Drehendes Bohren in Lockergestein – Gesetzmäßigkeiten und Nutzanwendung	
	- vergriffen -	

| Heft 24 | Christian BENOIT | 1980 |
| | *Die Systemtechnik der Unterwasserbaustelle im Offshore-Bereich* | |

| Heft 25 | Bernhard WÜST | 1980 |
| | *Verbesserung der Umweltfreundlichkeit von Maschinen, insbesondere von Baumaschinen-Antrieben* | |

| Heft 26 | Hans-Josef KRÄMER | 1981 |
| | *Geräteseitige Einflussparameter bei Ramm- und Drucksondierungen und ihre Auswirkungen auf den Eindringwiderstand* | |

| Heft 27 | Bertold KETTERER | 1981 |
| | *Modelluntersuchungen zur Prognose von Schneid- und Planierkräften im Erdbau* | |

| Heft 28 | Harald BEITZEL | 1981 |
| | *Gesetzmäßigkeiten zur Optimierung von Betonmischern* | |

| Heft 29 | Bernhard WÜST | 1982 |
| | *Einfluß der Baustellenarbeit auf die Lebensdauer von Turmdrehkranen* | |

| Heft 30 | Hans PINNOW | 1982 |
| | *Einsatz großer Baumaschinen und bisher nicht erfasster Sonderbauformen in lärmempfindlichen Gebieten* | |

Heft 52   Christian MEYSENBURG   2002
Ermittlung von Grundlagen für das Controlling in öffentlichen Bauverwaltungen

Heft 53   Matthias BURCHARD   2002
Grundlagen der Wettbewerbsvorteile globaler Baumärkte und Entwicklung eines Marketing Decision Support Systems (MDSS) zur Unternehmensplanung

Heft 54   Jarosław JURASZ   2003
Geometric Modelling for Computer Integrated Road Construction

Heft 55   Sascha GENTES   2003
Optimierung von Standardbaumaschinen zur Rettung Verschütteter

Heft 56   Gerhard W. SCHMIDT   2003
Informationsmanagement und Transformationsaufwand im Gebäudemanagement

Heft 57   Karl Ludwig KLEY   2004
Positionierungslösung für Straßenwalzen – Grundlage für eine kontinuierliche Qualitätskontrolle und Dokumentation der Verdichtungsarbeit im Asphaltbau

Heft 58   Jochen WENDEBAUM   2004
Nutzung der Kerntemperaturvorhersage zur Verdichtung von Asphaltmischgut im Straßenbau

Heft 59   Frank FIEDRICH   2004
Ein High-Level-Architecture-basiertes Multiagentensystem zur Ressourcenoptimierung nach Starkbeben

Heft 60   Joachim DEDEKE   2005
Rechnergestützte Simulation von Bauproduktionsprozessen zur Optimierung, Bewertung und Steuerung von Bauplanung und Bauausführung

Heft 61   Michael OTT   2007
Fertigungssystem Baustelle – Ein Kennzahlensystem zur Analyse und Beurteilung der Produktivität von Prozessen

**Die bei KIT Scientific Publishing verlegten Hefte (ab Heft 63) sind unter www.uvka.de
als PDF frei verfügbar oder als Druckausgabe bestellbar.**

Sonderhefte Reihe F – Forschung

Institutsintern verlegt

Heft 1   Vorträge anlässlich der Tagung   1972
Forschung für den Baubetrieb
(15. / 16. Juni 1972)

Heft 2   Vorträge anlässlich der Tagung   1974
Forschung für den Baubetrieb
(11. / 12. Juni 1974)

Heft 3   Vorträge anlässlich der Tagung   1979
Forschung für den Baubetrieb
(12. / 13. Juni 1979)

Heft 4   Vorträge anlässlich der Tagung   1983
Forschung für die Praxis
(15. / 16. Juni 1983)

Heft 5   Vorträge anlässlich der Tagung   1987
Baumaschinen für die Praxis
(04. / 05. Juni 1987)

Heft 6   Vorträge anlässlich der Tagung   1992
Forschung und Entwicklung für die maschinelle Bauausführung
(26. Juni 1992)
- vergriffen -

Bei KIT Scientific Publishing Karlsruhe verlegt

Heft 7   Stefan SENITZ   2007
Abschlusssymposium 2007 Graduiertenkolleg „Naturkatastrophen"
– Verständnis, Vorsorge und Bewältigung von Naturkatastrophen

Heft 7 ist unter www.uvka.de als PDF frei verfügbar oder als Druckausgabe bestellbar (ISBN 978-3-86644-145-3).

Reihe G – Gäste

Institutsintern verlegt

Heft 1   Zbigniew KORZEN   1981
Ähnlichkeitsbetrachtungen der Bodenbearbeitungsvorgänge

Heft 2   Yoshinori TAKADA   1983
Untersuchungen zur Abschätzung der Aufreißleistung von Reißraupen – Wechselwirkung von Maschine und Fels beim Reißen

Heft 3   Geza JANDY   1986
Systemtechnik (Systems Engineering)

Günter KÜHN
Was ist Systemtechnik, und was nützt sie dem Bauingenieur?

Heft 4   Piotr DUDZINSKI   1987
Konstruktionsmerkmale bei Lenksystemen an mobilen Erdbaumaschinen mit Reifenfahrwerken

Heft 5   Yoshitaka OJIRO   1988
Impact-Reißen – Untersuchungen über die Optimierung der Betriebsparameter mit Hilfe der Modellsimulation

Reihe G wird künftig in Reihe F fortgesetzt.

Reihe L – Lehre und Allgemeines

Institutsintern verlegt

| Heft 1 | Günter KÜHN
Baubetrieb in Karlsruhe
- vergriffen - | 1972 |

Heft 1   Günter KÜHN   1972
Baubetrieb in Karlsruhe
- vergriffen -

Heft 2   Dieter KARLE   1971
Afrika-Exkursion Gabun – Kamerun
- vergriffen -

Heft 3   Gabrielle und Uwe GRIESBACH   1975
Studenten berichten: 52.000 km Afrika – Asien

Heft 4   Günter KÜHN   1976
Letzte Fragen und ihre Antworten – auch für das Leben auf der Baustelle
- vergriffen -

Heft 5   Festschrift 1967 – 1977   1977
zum 10jährigen Bestehen des Instituts für Maschinenwesen im Baubetrieb

Heft 6   Günter KÜHN   1978
Baumaschinenforschung in Karlsruhe – Rückblick auf eine zehnjährige Institutstätigkeit

Heft 7   Günter KÜHN   1979
Baubetriebsausbildung in Karlsruhe

Heft 8   Bertold KETTERER/Hans-Josef KRÄMER   1980
Studenten-Exkursionen Saudi-Arabien 1978/79

Heft 9   Hans-Josef KRÄMER   1980
Baubetrieb – Studium und Berufserfahrung – Referate bei Seminaren für Bauingenieursstudenten

Heft 10   Christian BENOIT   1980
Studenten-Exkursion Brasilien 1980

Reihe L wird künftig in Reihe V fortgesetzt.

Reihe U – Untersuchungen

Institutsintern verlegt

| Heft 1 | Günter KÜHN | 1973 |

Günter KÜHN
Monoblock- oder Verstellausleger
- vergriffen -

Heft 2   Roland HERR   1974
Untersuchungen der Ladeleistung von Hydraulikbaggern im Feldeinsatz

Heft 3   Thomas TRÜMPER   1975
Einsatzstudie hydraulischer Schaufelradbagger SH 400

Reihe U wird künftig in Reihe F fortgesetzt.

Reihe V – Vorlesungen und Mitteilungen

Institutsintern verlegt

Heft 1	Heinrich MÜLLER *Management im Baubetrieb*	1974
Heft 2	Erwin RICKEN *Baubetriebswirtschaft B* **- vergriffen -**	1974
Heft 3	Thomas TRÜMPER *Elektrotechnik* **- vergriffen -**	1975
Heft 4	Albrecht GÖHRING *Zusammenfassung des Seminars Anorganische Chemie*	1975
Heft 5	Joachim HORNUNG *Netzplantechnik* **- vergriffen -**	1975
Heft 6	Günter KÜHN *Baubetriebstechnik I* *Teil A: Baubetrieb* *Teil B: Hochbautechnik* **- vergriffen -**	1988
Heft 7	Günter KÜHN *Baubetriebstechnik II* *Teil A: Tiefbau* *Teil B: Erdbau*	1985
Heft 8	Bernhard WÜST *Maschinentechnik I*	1982
Heft 9	Norbert WARDECKI *Maschinentechnik II*	1983
Heft 10	Fritz HEINEMANN *Einführung in die Baubetriebswirtschaftslehre* **- vergriffen -**	1991

| Heft 11 | Fritz GEHBAUER | 1989 |
| | *Wer soll die Zukunft gestalten, wenn nicht wir?* | |

| Heft 12 | Die Studenten | 1989 |
| | *Studenten-Exkursion 1989 Chile – Argentinien - Brasilien* | |

| Heft 13 | Mitgliederverzeichnis – Gesellschaft der Freunde des Instituts | 1996 |

| Heft 14 | *Das Institut* | 1996 |

Heft 15	Die Studenten	1990
	Studenten-Exkursion 1990 Deutschland – Dänemark – Norwegen –	
	Belgien	

Heft 16	Fritz GEHBAUER	1990
	Baubetriebstechnik I	
	Teil A: Baubetrieb	
	Teil B: Hochbau	
	Teil C: Schlüsselfertigbau	

Heft 17	Fritz GEHBAUER	1994
	Baubetriebstechnik II	
	Teil A: Erdbau	
	Teil B: Tiefbau	

| Heft 18 | Die Studenten | 1991 |
| | *Studenten-Exkursion 1991 Deutschland – Polen* | |

Heft 19	Die Studenten	1992
	Studenten-Exkursion 1992 Südostasien – Bangkok – Hongkong –	
	Taipeh	

| Heft 20 | Alfred WELTE | 2001 |
| | *Naßbaggertechnik – Ein Sondergebiet des Baubetriebs* | |

| Heft 21 | Die Studenten | 1993 |
| | *Studenten-Exkursion 1993 Großbritannien* | |

| Heft 22 | Die Studenten | 1994 |
| | *Studenten-Exkursion 1994 Österreich* | |

Veröffentlichungen des Instituts für Technologie und Management im Baubetrieb

Heft 23	Die Studenten *Studenten-Exkursion 1995 Deutschland* **- vergriffen -**	1995
Heft 24	Die Studenten *Studenten-Exkursion 1996 Neue Bundesländer*	1996
Heft 25	Herbert FEGER *Betonbereitung* *Teil 1 der Vorlesung Betonbereitung und -transport*	1997
Heft 26	Herbert FEGER *Betontransport* *Teil 2 der Vorlesung Betonbereitung und -transport*	1997
Heft 27	Die Studenten *Studenten-Exkursion 1997 Deutschland – Tschechien*	1997
Heft 27	Fritz GEHBAUER *Baubetriebsplanung und Grundlagen der Verfahrenstechnik im* *Hoch-, Tief- und Erdbau Band I* *Teil A: Baubetrieb* *Teil B: Hochbau* *Teil C: Schlüsselfertigbau*	2004
	überarbeitete Neuauflage	2009
Heft 28	Die Studenten *Studenten-Exkursion 1998 Deutschland*	1998
Heft 28	Fritz GEHBAUER *Baubetriebsplanung und Grundlagen der Verfahrenstechnik im* *Hoch-, Tief- und Erdbau Band II* *Teil A: Erdbau* *Teil B: Tiefbau*	2004
	überarbeitete Neuauflage	2009
Heft 29	Die Studenten *Studenten-Exkursion 1999 Deutschland – Schweiz – Frankreich*	1999

Heft 30	Fritz GEHBAUER *Baubetriebswirtschaftslehre* **- vergriffen -**	2001
Heft 31	Die Studenten *Studenten-Exkursion 2000 Deutschland – Rhein/Main – Ruhr*	2000
Heft 32	Die Studenten *Studenten-Exkursion 2001 Goldisthal – Berlin – Hannover*	2001
Heft 33	Die Studenten *Studenten-Exkursion 2002 Essen – Hamburg – Hannover*	2002
Heft 34	Die Studenten *Studenten-Exkursion 2003 Zürich – Luzern – München*	2003
Heft 35	Die Studenten *Studenten-Exkursion 2004 Köln – Hamburg – Hannover*	2004
Heft 36	Die Studenten *Studenten-Exkursion 2005 Schweiz – Österreich – Deutschland*	2005
Heft 37	Die Studenten *Studenten-Exkursion 2006 Innsbruck – Wien*	2006
Heft 38	Die Studenten *Studenten-Exkursion 2007 Köln – Amsterdam*	2007
Heft 39	Die Studenten *Studenten-Exkursion 2008 – Dubai*	2008
Heft 40	Die Studenten *Studenten-Exkursion 2009 – Japan*	2009